CHAPMAN & HALL/CRC
Texts in Statistical Science Series

Series Editors
Bradley P. Carlin, *University of Minnesota, USA*
Julian J. Faraway, *University of Bath, UK*
Martin Tanner, *Northwestern University, USA*
Jim Zidek, *University of British Columbia, Canada*

Analysis of Failure and Survival Data
P. J. Smith

**The Analysis of Time Series —
An Introduction, Sixth Edition**
C. Chatfield

**Applied Bayesian Forecasting and Time Series
Analysis**
A. Pole, M. West and J. Harrison

**Applied Nonparametric Statistical Methods,
Fourth Edition**
P. Sprent and N.C. Smeeton

**Applied Statistics — Handbook of GENSTAT
Analysis**
E.J. Snell and H. Simpson

Applied Statistics — Principles and Examples
D.R. Cox and E.J. Snell

Applied Stochastic Modelling, Second Edition
B.J.T. Morgan

Bayesian Data Analysis, Second Edition
A. Gelman, J.B. Carlin, H.S. Stern
and D.B. Rubin

**Bayesian Ideas and Data Analysis: An Introduction
for Scientists and Statisticians**
R. Christensen, W. Johnson, A. Branscum,
and T.E. Hanson

**Bayesian Methods for Data Analysis,
Third Edition**
B.P. Carlin and T.A. Louis

Beyond ANOVA — Basics of Applied Statistics
R.G. Miller, Jr.

**Computer-Aided Multivariate Analysis,
Fourth Edition**
A.A. Afifi and V.A. Clark

A Course in Categorical Data Analysis
T. Leonard

A Course in Large Sample Theory
T.S. Ferguson

Data Driven Statistical Methods
P. Sprent

Decision Analysis — A Bayesian Approach
J.Q. Smith

Design and Analysis of Experiment with SAS
J. Lawson

**Elementary Applications of Probability Theory,
Second Edition**
H.C. Tuckwell

Elements of Simulation
B.J.T. Morgan

**Epidemiology — Study Design and
Data Analysis, Second Edition**
M. Woodward

Essential Statistics, Fourth Edition
D.A.G. Rees

Exercises and Solutions in Biostatistical Theory
L.L. Kupper, B.H. Neelon, and S.M. O'Brien

**Extending the Linear Model with R — Generalized
Linear, Mixed Effects and Nonparametric Regression
Models**
J.J. Faraway

A First Course in Linear Model Theory
N. Ravishanker and D.K. Dey

**Generalized Additive Models:
An Introduction with R**
S. Wood

Graphics for Statistics and Data Analysis with R
K.J. Keen

**Interpreting Data — A First Course
in Statistics**
A.J.B. Anderson

**Introduction to General and Generalized
Linear Models**
H. Madsen and P. Thyregod

**An Introduction to Generalized
Linear Models, Third Edition**
A.J. Dobson and A.G. Barnett

Introduction to Multivariate Analysis
C. Chatfield and A.J. Collins

**Introduction to Optimization Methods and Their
Applications in Statistics**
B.S. Everitt

Introduction to Probability with R
K. Baclawski

**Introduction to Randomized Controlled Clinical
Trials, Second Edition**
J.N.S. Matthews

**Introduction to Statistical Inference and Its
Applications with R**
M.W. Trosset

Introduction to Statistical Methods for Clinical Trials
T.D. Cook and D.L. DeMets

Large Sample Methods in Statistics
P.K. Sen and J. da Motta Singer

Linear Models with R
J.J. Faraway

Logistic Regression Models
J.M. Hilbe

Markov Chain Monte Carlo — Stochastic Simulation for Bayesian Inference, Second Edition
D. Gamerman and H.F. Lopes

Mathematical Statistics
K. Knight

Modeling and Analysis of Stochastic Systems, Second Edition
V.G. Kulkarni

Modelling Binary Data, Second Edition
D. Collett

Modelling Survival Data in Medical Research, Second Edition
D. Collett

Multivariate Analysis of Variance and Repeated Measures — A Practical Approach for Behavioural Scientists
D.J. Hand and C.C. Taylor

Multivariate Statistics — A Practical Approach
B. Flury and H. Riedwyl

Pólya Urn Models
H. Mahmoud

Practical Data Analysis for Designed Experiments
B.S. Yandell

Practical Longitudinal Data Analysis
D.J. Hand and M. Crowder

Practical Statistics for Medical Research
D.G. Altman

A Primer on Linear Models
J.F. Monahan

Probability — Methods and Measurement
A. O'Hagan

Problem Solving — A Statistician's Guide, Second Edition
C. Chatfield

Randomization, Bootstrap and Monte Carlo Methods in Biology, Third Edition
B.F.J. Manly

Readings in Decision Analysis
S. French

Sampling Methodologies with Applications
P.S.R.S. Rao

Statistical Analysis of Reliability Data
M.J. Crowder, A.C. Kimber, T.J. Sweeting, and R.L. Smith

Statistical Methods for Spatial Data Analysis
O. Schabenberger and C.A. Gotway

Statistical Methods for SPC and TQM
D. Bissell

Statistical Methods in Agriculture and Experimental Biology, Second Edition
R. Mead, R.N. Curnow, and A.M. Hasted

Statistical Process Control — Theory and Practice, Third Edition
G.B. Wetherill and D.W. Brown

Statistical Theory, Fourth Edition
B.W. Lindgren

Statistics for Accountants
S. Letchford

Statistics for Epidemiology
N.P. Jewell

Statistics for Technology — A Course in Applied Statistics, Third Edition
C. Chatfield

Statistics in Engineering — A Practical Approach
A.V. Metcalfe

Statistics in Research and Development, Second Edition
R. Caulcutt

Stochastic Processes: An Introduction, Second Edition
P.W. Jones and P. Smith

Survival Analysis Using S — Analysis of Time-to-Event Data
M. Tableman and J.S. Kim

The Theory of Linear Models
B. Jørgensen

Time Series Analysis
H. Madsen

Time Series: Modeling, Computation, and Inference
R. Prado and M. West

Texts in Statistical Science

Introduction to General and Generalized Linear Models

Henrik Madsen

Technical University of Denmark
Lyngby, Denmark

Poul Thyregod

Technical University of Denmark
Lyngby, Denmark

CRC Press
Taylor & Francis Group
Boca Raton London New York

CRC Press is an imprint of the
Taylor & Francis Group an **informa** business

A CHAPMAN & HALL BOOK

CRC Press
Taylor & Francis Group
6000 Broken Sound Parkway NW, Suite 300
Boca Raton, FL 33487-2742

© 2011 by Taylor and Francis Group, LLC
CRC Press is an imprint of Taylor & Francis Group, an Informa business

No claim to original U.S. Government works

Printed in the United States of America on acid-free paper
10 9 8 7 6 5 4 3 2 1

International Standard Book Number: 978-1-4200-9155-7 (Hardback)

Library of Congress Cataloging-in-Publication Data

Madsen, Henrik, 1955-
    Introduction to general and generalized linear models / Henrik Madsen, Poul Thyregod.
       p. cm. -- (Chapman & Hall/CRC texts in statistical science series)
    Includes bibliographical references and index.
    ISBN 978-1-4200-9155-7 (hardcover : alk. paper)
    1. Linear models (Statistics) 2. Generalized estimating equations. I. Thyregod, Poul, 1939-2008. II. Title.

QA276.M315 2010
519.5'35--dc22
                                         2010029753

Visit the Taylor & Francis Web site at
http://www.taylorandfrancis.com

and the CRC Press Web site at
http://www.crcpress.com

Contents

Preface   xi
   Notation   xiii

1 Introduction   1
   1.1 Examples of types of data   2
   1.2 Motivating examples   3
   1.3 A first view on the models   5

2 The likelihood principle   9
   2.1 Introduction   9
   2.2 Point estimation theory   10
   2.3 The likelihood function   14
   2.4 The score function   17
   2.5 The information matrix   18
   2.6 Alternative parameterizations of the likelihood   20
   2.7 The maximum likelihood estimate (MLE)   21
   2.8 Distribution of the ML estimator   22
   2.9 Generalized loss-function and deviance   23
   2.10 Quadratic approximation of the log-likelihood   23
   2.11 Likelihood ratio tests   25
   2.12 Successive testing in hypothesis chains   27
   2.13 Dealing with nuisance parameters   33
   2.14 Problems   38

3 General linear models   41
   3.1 Introduction   41
   3.2 The multivariate normal distribution   42
   3.3 General linear models   44
   3.4 Estimation of parameters   48
   3.5 Likelihood ratio tests   53
   3.6 Tests for model reduction   58
   3.7 Collinearity   64
   3.8 Inference on parameters in parameterized models   70
   3.9 Model diagnostics: residuals and influence   73

3.10 Analysis of residuals 77
3.11 Representation of linear models 78
3.12 General linear models in R 81
3.13 Problems 83

4 Generalized linear models   **87**
4.1 Types of response variables 89
4.2 Exponential families of distributions 90
4.3 Generalized linear models 99
4.4 Maximum likelihood estimation 102
4.5 Likelihood ratio tests 111
4.6 Test for model reduction 115
4.7 Inference on individual parameters 116
4.8 Examples 117
4.9 Generalized linear models in R 152
4.10 Problems 153

5 Mixed effects models   **157**
5.1 Gaussian mixed effects model 159
5.2 One-way random effects model 160
5.3 More examples of hierarchical variation 174
5.4 General linear mixed effects models 179
5.5 Bayesian interpretations 185
5.6 Posterior distributions 191
5.7 Random effects for multivariate measurements 192
5.8 Hierarchical models in metrology 197
5.9 General mixed effects models 199
5.10 Laplace approximation 201
5.11 Mixed effects models in R 218
5.12 Problems 219

6 Hierarchical models   **225**
6.1 Introduction, approaches to modeling of overdispersion . . 225
6.2 Hierarchical Poisson Gamma model 226
6.3 Conjugate prior distributions 233
6.4 Examples of one-way random effects models 237
6.5 Hierarchical generalized linear models 242
6.6 Problems 243

7 Real life inspired problems   **245**
7.1 Dioxin emission 246
7.2 Depreciation of used cars 249
7.3 Young fish in the North Sea 250
7.4 Traffic accidents 251
7.5 Mortality of snails 252

A Supplement on the law of error propagation   **255**
   A.1   Function of one random variable 255
   A.2   Function of several random variables 255

B Some probability distributions   **257**
   B.1   The binomial distribution model 259
   B.2   The Poisson distribution model 262
   B.3   The negative binomial distribution model 264
   B.4   The exponential distribution model 266
   B.5   The gamma distribution model 268
   B.6   The inverse Gaussian distribution model 275
   B.7   Distributions derived from the normal distribution 280
   B.8   The Gamma-function 284

C List of symbols   **285**

Bibliography   **287**

Index   **293**

Preface

This book contains an introduction to general and generalized linear models using the popular and powerful likelihood techniques. The aim is to provide a flexible framework for the analysis and model building using data of almost any type. This implies that the more well-known analyses based on Gaussian data like regression analysis, analysis of variance and analysis of covariance are generalized to a much broader family of problems that are linked to, for instance, binary, positive, integer, ordinal and qualitative data.

By using parallel descriptions in two separate chapters of general and generalized linear models, the book facilitates a unique comparison between these two important classes of models and, furthermore, presents an easily accessible introduction to the more advanced concepts related to generalized linear models.

Likewise, the concept of hierarchical models is illustrated separately – in one chapter a description of Gaussian based hierarchical models, such as the mixed effects linear models is outlined, and in another chapter an introduction is presented to the generalized concept of those hierarchical models that are linked to a much broader class of problems connected to various types of data and related densities. The book also introduces new concepts for mixed effects models thereby enabling more flexibility in the model building and in the allowed data structures.

Throughout the book the statistical software R is used. Examples show how the problems are solved using R, and for each of the chapters individual guidelines are provided in order to facilitate the use of R when solving the relevant type of problems.

Theorems are used to emphasize the most important results. Proofs are provided, only, if they clarify the results. Problems on a smaller scale are dealt with at the end of most of the chapters, and a separate chapter with real life inspired problems is included as the final chapter of the book.

During the sequence of chapters, more advanced models are gradually introduced. With such an approach, the relationship between general and generalized linear models and methods becomes more apparent.

The last chapter of this book is devoted to problems inspired by real life situations. At the home page http://www.imm.dtu.dk/~hm/GLM solutions to the problems are found. The homepage also contains additional exercises – called assignments – and a complete set of data for the examples used in the book. Furthermore, a collection of slides for an introductory course on general, generalized and mixed effects models can be found on the homepage.

The contents of this book are mostly based on a comprehensive set of material developed by Professor Poul Thyregod during his series of lectures at the Section of Mathematical Statistics at the Technical University of Denmark (DTU). Poul was the first person in Denmark who received a PhD in mathematical statistics. Poul was also one of the few highly skilled in mathematical statistics who was fully capable of bridging the gap between theory and practice within statistics and data analysis. He possessed the capability to link statistics to real problems and to focus on the real added value of statistics — in order to help us understand the real world a bit better. The ability to work with engineers and scientists, to be part of the discovery process, and to be able to communicate so distinctly what statistics is all about is clearly a gift. Poul possessed that gift. I am grateful to be one of a long list of students who had the privilege of learning from his unique capabilities. Sadly, Poul passed away in the summer of 2008, which was much too early in his life. I hope, however, that this book will reflect his unique talent to establish an easily accessible introduction to theory and practice of modern statistical modeling.

I am grateful to all who have contributed with useful comments, suggestions and contributions. First I would like to thank my colleagues Gilles Guillot, Martin Wæver Pedersen, Stig Mortensen and Anders Nielsen for their helpful and very useful assistance and comments.

In particular, I am grateful to Anna Helga Jónsdóttir for her assistance with text, proofreading, figures, exercises and examples. Without her insistent support this book would never had been completed. Finally, I would like to thank Helle Welling for proofreading, and Morten Høgholm for both proofreading and for proposing and creating a new layout in LaTeX.

Henrik Madsen
Lyngby, Denmark

Notation

All vectors are column vectors. Vectors and matrices are emphasized using a bold font. Lowercase letters are used for vectors and uppercase letters are used for matrices. Transposing is denoted with the upper index ^{T}.

Random variables are always written using uppercase letters. Thus, it is not possible to distinguish between a multivariate random variable (random vector) and a matrix. However, variables and random variables are assigned to letters from the last part of the alphabet (X, Y, Z, U, V, ...), while constants are assigned to letters from the first part of the alphabet (A, B, C, D, ...). From the context it should be possible to distinguish between a matrix and a random vector.

CHAPTER 1

Introduction

This book provides an introduction to methods for statistical modeling using essentially all kind of data. The principles for modeling are based on likelihood techniques. These techniques facilitate our aim of bridging the gap between theory and practice for modern statistical model building.

Each chapter of the book contains examples and guidelines for solving the problems using the statistical software package R, which can be freely downloaded and installed on almost any computer system. We do, however, refer to other software packages as well.

In general the focus is on establishing models that explain the variation in data in such a way that the obtained models are well suited for predicting the outcome for given values of some explanatory variables. More specifically we will focus on *formulating, estimating, validating and testing models* for predicting the *mean value* of the random variables. However, by the considered approach we will consider the complete stochastic model for the data which includes an appropriate choice of the *density* describing the variation of the data. It will be demonstrated that this approach facilitates adequate methods for describing also the uncertainty of the predictions.

By the approach taken, the theory and practice in relation to widely applied methods for modeling using *regression analysis, analysis of variance* and the *analysis of covariance*, that are all related to Gaussian distributed data, are established in a way which facilitates an easily accessible extension to similar methods applied in the case of, e.g., Poisson, Gamma and Binomial distributed data. This is obtained by using the likelihood approach in both cases, and becomes clear that the *general linear models* are relevant for *Gaussian distributed samples* whereas the *generalized linear models* facilitate a modeling of the variation in a much broader context, namely for all data originating from the so-called *exponential family of densities* including Poisson, Binomial, Exponential, Gaussian, and Gamma distributions.

The presentation of the general and generalized linear models is provided using essentially the same methods related to the likelihood principles, but described in two separate chapters. By a parallel presentation of the methods and models in two chapters, a clear comparison between the two model types is recognized. This parallel presentation is also aiming at providing an easily accessible description of the theory for generalized linear models. This is due to the fact that the book first provides the corresponding or parallel results for the general linear models, which is easier to understand, and in many cases well-known.

The book also contains a first introduction to both mixed effects models (also called mixed models) and hierarchical models. Again, a parallel setup in two separate chapters is provided. The first chapter concentrates on introducing the random effects and, consequently, also the mixed effects in a Gaussian context. The subsequent chapter provides an introduction to non-Gaussian hierarchical models where the considered models again are members of the exponential family of distributions.

To the readers with a theoretical interest it will be obvious that virtually all the results are based on about a handful of results from the likelihood theory, and that the results that are valid for finite samples for the general linear models are valid asymptotically in the case of generalized linear models. The necessary likelihood theory is described in the chapter following the Introduction.

1.1   Examples of types of data

Let us first illustrate the power of the methods considered in this book by listing some of the types of data which can be modelled using the described techniques. In practice several types of response variables are seen as indicated by the examples listed below:

i) *Continuous data* (e.g., $y_1 = 2.3$, $y_2 = -0.2$, $y_3 = 1.8$, ..., $y_n = 0.8$). Normal (Gaussian) distributed. Used, e.g., for air temperatures in degrees Celsius. An example is found in Example 2.18 on page 14.

ii) *Continuous positive data* (e.g., $y_1 = 0.0238$, $y_2 = 1.0322$, $y_3 = 0.0012$, ..., $y_n = 0.8993$). Log-normally distributed. Often used for concentrations.

iii) *Count data* (e.g., $y_1 = 57$, $y_2 = 67$, $y_3 = 54$, ..., $y_n = 59$). Poisson distributed. Used, e.g., for number of accidents — see Example 4.7 on page 123 on page 123.

iv) *Binary (or quantal) data* (e.g., $y_1 = 0$, $y_2 = 0$, $y_3 = 1$, ..., $y_n = 0$), or proportion of counts (e.g. $y_1 = 15/297$, $y_2 = 17/242$, $y_3 = 2/312$, ..., $y_n = 144/285$). Binomial distribution — see Example 4.6 on page 118 or Example 4.14 on page 140.

v) *Nominal data* (e.g., "Very unsatisfied", "Unsatisfied", "Neutral", "Satisfied", "Very satisfied"). Multinomial distribution — see Example 4.12 on page 133.

The reader will also become aware that the data of a given type might look alike, but the (appropriate) statistical treatment is different!

1.2 Motivating examples

The Challenger disaster

On January 28, 1986, Space Shuttle Challenger broke apart 73 seconds into its flight and the seven crew members died. The disaster was due to a disintegration of an O-ring seal in the right rocket booster. The forecast for January 28, 1986 indicated an unusually cold morning with air temperatures around 28 degrees F (-1 degrees C).

During a teleconference on January 27, one of the engineers, Morton Thiokol, responsible for the shuttle's rocket booster, expressed concern due to the low temperature.

The planned launch on January 28, 1986 was launch number 25. During the previous 24 launches problems with the O-ring were observed in 6 cases. Figure 1.1 shows the relationship between observed sealing problems and the air temperature. A model of the probability for O-ring failure as a function of the air temperature would clearly have shown that given the forecasted air temperature, problems with the O-rings were very likely to occur.

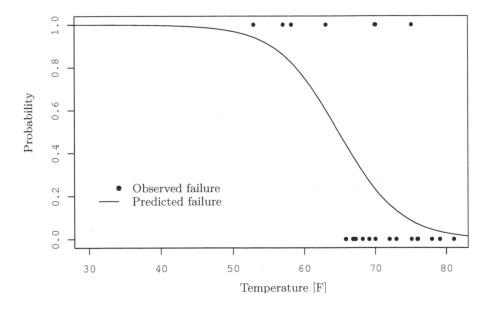

Figure 1.1: *Observed failure of O-rings in 6 out of 24 launches along with predicted probability for O-ring failure.*

Table 1.1: *Incidence of Torsade de Pointes by dose for high risk patients.*

Index	Daily dose [mg]	Number of subjects	Number showing TdP	Fraction showing TdP
i	x_i	n_i	z_i	p_i
1	80	69	0	0
2	160	832	4	0.5
3	320	835	13	1.6
4	480	459	20	4.4
5	640	324	12	3.7
6	800	103	6	5.8

QT prolongation for drugs

In the process of drug development it is required to perform a study of potential prolongation of a particular interval of the electrocardiogram (ECG), the QT interval. The QT interval is defined as the time required for completion of both ventricular depolarization and repolarization. The interval has gained clinical importance since a prolongation has been shown to induce potentially fatal ventricular arrhythmia such as Torsade de Pointes (TdP). The arrhythmia causes the QRS complexes, another part of the ECG, to swing up and down around the baseline of the ECG in a chaotic fashion. This probably caused the name which means "twisting of the points" in French. A number of drugs have been reported to prolong the QT interval, both cardiac and non-cardiac drugs. Recently, both previously approved as well as newly developed drugs have been withdrawn from the market or have had their labeling restricted because of indication of QT prolongation. Table 1.1 shows results from a clinical trial where a QT prolonging drug was given to high risk patients. The patients were given the drug in six different doses and the number of incidents of Torsade de Points counted.

It is reasonable to consider the *fraction*, $Y_i = \frac{z_i}{n_i}$, of incidences of Torsade de Points as the interesting variable. A natural distributional assumption is the binomial distribution, $Y_i \sim B(n_i, p_i)/n_i$, where n_i is the number of subjects given the actual dosage and p_i is the fraction showing Torsade de Pointes.

▶ **Remark 1.1 – A bad model**
Obviously the fraction, p_i is higher for a higher daily dosage of the drug. However, a linear model of the form $Y_i = p_i + \epsilon_i$ where $p_i = \beta_0 + \beta_1 x_i$ does not reflect that, p_i is between zero and one, and the model for the fraction, Y_i (as 'mean plus noise') is clearly not adequate, since the observations are between zero and one.

It is, thus, clear that the distribution of ϵ_i and then the variance of observations must be dependent on p_i. Also, the problem with the homogeneity of the variance indicates that a traditional "mean plus noise" model is not adequate here. ◀

▸ **Remark 1.2 – A correct model**
Instead we will now formulate a model for transformed values of the observed
fractions p_i.

Given that $Y_i \sim B(n_i, p_i)/n_i$ we have that

$$E[Y_i] = p_i \tag{1.1}$$

$$\mathrm{Var}[Y_i] = \frac{p_i(1 - p_i)}{n_i} \tag{1.2}$$

i.e., the variance is now a function of the mean value. Later on the so-called
mean value function $V(E[Y_i])$ will be introduced which relates the variance to
the mean value.

A successful construction is to consider a function, the so-called *link function*
of the mean value $E[Y]$. In this case we will use the *logit*-transformation

$$g(p_i) = \log\left(\frac{p_i}{1 - p_i}\right) \tag{1.3}$$

and we will formulate a *linear model* for the transformed values. A plot of
the observed logits, $g(p_i)$ as a function of the concentration indicates a linear
relation of the form

$$g(p_i) = \beta_0 + \beta_1 x_i \tag{1.4}$$

After having estimated the parameters, i.e., we have obtained $(\widehat{\beta}_0, \widehat{\beta}_1)$, it is now
possible to use the inverse transformation, which gives the predicted fraction
$\widehat{p}$ of subjects showing Torsade de Pointes as a function of a daily dose, x using
the *logistic function*:

$$\widehat{p} = \frac{\exp\left(\widehat{\beta}_0 + \widehat{\beta}_1 x\right)}{1 + \exp(\widehat{\beta}_0 + \widehat{\beta}_1 x)}. \tag{1.5}$$

This approach is called *logistic regression*.

It is easily seen that this model will ensure that the fraction is between
zero and one, and we also see that we have established a reasonable description
of the relation between the mean and the variance of the observations. ◂

1.3 A first view on the models

As mentioned previously, we will focus on statistical methods to formulate
models for predicting the expected value of the *outcome, dependent,* or *response
variable,* Y_i as a function of the known *independent variables,* $x_{i1}, x_{i2}, \ldots, x_{ik}$.
These k variables are also called *explanatory,* or *predictor variables* or *covariates*.
This means that we shall focus on models for the expectation $E[Y_i]$.

Previously we have listed examples of types of response variables. Also
the explanatory variables might be labeled as *continuous, discrete, categorical,
binary, nominal,* or *ordinal*. To predict the response, a typical model often
includes a combination of such types of variables.

Since we are going to use a likelihood approach, a specification of the probability distribution of Y_i is a very important part when specifying the model.

General linear models

In Chapter 3, which considers *general linear models*, the expected value of the response variable Y is linked linearly to the explanatory variables by an equation of the form

$$E[Y_i] = \beta_1 x_{i1} + \cdots + \beta_k x_{ik} . \qquad (1.6)$$

It will be shown that for Gaussian data it is reasonable to build a model directly for the expectation as shown in (1.6), and this relates to the fact that for Gaussian distributed random variables, all conditional expectations are linear (see e.g., Madsen (2008)).

▶ **Remark 1.3**

In model building, models for the mean value are generally considered. However, for some applications, models for, say, the 95% quantile might be of interest. Such models can be established by, e.g., *quantile regression*; see Koenker (2005).   ◄

Generalized linear models

As indicated by the motivating example above it is, however, often more reasonable to build a linear model for a transformation of the expected value of the response. This approach is more formally described in connection with the *generalized linear models* in Chapter 4, where a link between the expected value of response and the explanatory variables is of the form

$$g(E[Y_i]) = \beta_1 x_{i1} + \ldots + \beta_k x_{ik} . \qquad (1.7)$$

The function $g(.)$ is called the *link function* and the right hand side of (1.7) is called the *linear component* of the model.

Thus, a full specification of the model contains a specification of

1. The *probability density* of Y. In Chapter 3 this will be the Gaussian density, i.e., $Y \sim N(\mu, \sigma^2)$, whereas in Chapter 4 the probability density will belong to the *exponential family of densities*, which includes the Gaussian, Poisson, Binomial, Gamma, and other distributions.

2. The smooth monotonic *link function* $g(.)$. Here we have some freedom, but the so-called *canonical link* function is directly linked to the used density. As indicated in the discussion related to (1.6) no link function is needed for Gaussian data – or the link is the identity.

3. The *linear component*. See the discussion above.

In statistical modeling it is very useful to formulate the model for all n observations $\boldsymbol{Y} = (Y_1, Y_2, \ldots, Y_n)^T$.

Let us introduce the known *model vector* $\boldsymbol{x}_i = (x_{i1}, x_{i2}, \ldots, x_{ik})^T$ for the ith observation, and unknown *parameter vector* $\boldsymbol{\beta} = (\beta_1, \beta_2, \ldots, \beta_k)^T$. Then the model for all n observations can be written as

$$
\begin{pmatrix} g(\mathrm{E}[Y_1]) \\ \vdots \\ g(\mathrm{E}[Y_n]) \end{pmatrix} = \begin{pmatrix} \boldsymbol{x}_1^T \\ \vdots \\ \boldsymbol{x}_n^T \end{pmatrix} \boldsymbol{\beta} \tag{1.8}
$$

or

$$
g(\mathrm{E}[\boldsymbol{Y}]) = \boldsymbol{X}\boldsymbol{\beta} \tag{1.9}
$$

where the matrix $\boldsymbol{X}$ of known coefficients is called the *design matrix*.

As indicated in the formulation above, the parameter vector $\boldsymbol{\beta}$ is *fixed*, but *unknown*, and the typical goal is to obtain an *estimate* $\widehat{\boldsymbol{\beta}}$ of *beta*. Models with fixed parameters are called *fixed effects models*.

Suppose that we are not interested in the individual (fixed) parameter estimates, but rather in the variation of the underlying true parameter. This leads to an introduction of the *random effects models* which will be briefly introduced in the following section.

Hierarchical models

In Chapters 5 and 6 the important concept of *hierarchical models* is introduced. The Gaussian case is introduced in Chapter 5, and this includes the so-called linear mixed effects models. This Gaussian and linear case is a natural extension of the general linear models. An extension of the generalized linear models are found in Chapter 6 which briefly introduces the generalized hierarchical models.

Let us first look at the Gaussian case. Consider for instance the test of ready made concrete. The concrete are delivered by large trucks. From a number of randomly picked trucks a small sample is taken, and these samples are analyzed with respect to the strength of concrete. A reasonable model for the variation of the strength is

$$
Y_{ij} = \mu + U_i + \epsilon_{ij} \tag{1.10}
$$

where μ is the overall strength of the concrete and U_i is the deviation of the average for the strength of concrete delivered by the i'th truck, and $\epsilon_{ij} \sim \mathrm{N}(0, \sigma^2)$ the deviation between concrete samples from the same truck.

Here we are typically not interested in the individual values of U_i but rather in the variation of U_i, and we will assume that $U_i \sim \mathrm{N}(0, \sigma_u^2)$.

The model (1.10) is a *one-way random effects model*. The parameters are now μ, σ_u^2 and σ^2.

Putting $\mu_i = \mu + U_i$ we may formulate (1.10) as a *hierarchical model*, where we shall assume that

$$Y_{ij}|\mu_i \sim \mathrm{N}(\mu_i, \sigma^2) \,, \tag{1.11}$$

and in contrast to the *fixed effects model*, the level μ_i is modeled as a realization of a random variable,

$$\mu_i \sim \mathrm{N}(\mu, \sigma_u^2), \tag{1.12}$$

where the μ_i's are assumed to be mutually independent, and Y_{ij} are *conditionally independent*, i.e., Y_{ij} are mutually independent in the conditional distribution of Y_{ij} for given μ_i.

Let us again consider a model for all n observations and let us further extend the discussion to the vector case of the random effects. The discussion above can now be generalized to the *linear mixed effects model* where

$$\mathrm{E}[\boldsymbol{Y}|\boldsymbol{U}] = \boldsymbol{X}\boldsymbol{\beta} + \boldsymbol{Z}\boldsymbol{U} \tag{1.13}$$

with $\boldsymbol{X}$ and $\boldsymbol{Z}$ denoting known matrices. Note how the mixed effect linear model in (1.13) is a linear combination of *fixed effects*, $\boldsymbol{X}\boldsymbol{\beta}$ and *random effects*, $\boldsymbol{Z}\boldsymbol{U}$. These types of models will be described in Chapter 5.

The non-Gaussian case of the hierarchical models, where

$$g(\mathrm{E}[\boldsymbol{Y}|\boldsymbol{U}]) = \boldsymbol{X}\boldsymbol{\beta} + \boldsymbol{Z}\boldsymbol{U} \tag{1.14}$$

and where $g(.)$ is an appropriate link function will be treated in Chapter 6.

CHAPTER 2

The likelihood principle

2.1 Introduction

Fisher (1922) identified the likelihood function as the key inferential quantity conveying all inferential information in statistical modeling including the uncertainty. In particular, Fisher suggested the method of maximum likelihood to provide a *point estimate* for the parameters of interest, the so-called *maximum likelihood estimate (MLE)*.

Example 2.1 – Likelihood function

Suppose we toss a thumbtack 10 times and observe that 3 times it lands point up. Assuming we know nothing prior to the experiment, what is the probability of landing point up, θ? It is clear that θ cannot be zero and the probability is unlikely to be very high. However, the probability for success $\theta = 0.3$ or $\theta = 0.4$ is likely, since in a binomial experiment with $n = 10$ and $Y = 3$, the number of successes, we get the probabilities $P(Y = 3) = 0.27$ or 0.21 for $\theta = 0.3$ or $\theta = 0.4$, respectively. We have thus found a non-subjective way to compare different values of θ. By considering $P_\theta(Y = 3)$ to be a function of the unknown parameter we have the *likelihood function*:

$$L(\theta) = P_\theta(Y = 3).$$

In a general case with n trials and y successes, the likelihood function is:

$$L(\theta) = P_\theta(Y = y) = \binom{n}{y} \theta^y (1 - \theta)^{n-y}.$$

A sketch of the likelihood function for $n = 10$ and $y = 3$ is shown in Figure 2.1 on the following page. As will be discussed later in the chapter, it is often more convenient to consider the log-likelihood function. The log-likelihood function is:

$$\log L(\theta) = y \log \theta + (n - y) \log(1 - \theta) + \text{const}$$

where const indicates a term that does not depend on θ. By solving $\frac{\partial \log L(\theta)}{\partial \theta} = 0$, it is readily seen that the maximum likelihood *estimate* (MLE) for θ is $\widehat{\theta}(y) = \frac{y}{n}$. In the thumbtack case where we observed $Y = y = 3$ we obtain $\widehat{\theta}(y) = 0.3$. The random variable $\widehat{\theta}(Y) = \frac{Y}{n}$ is called a maximum likelihood *estimator* for θ. Notice the difference between $\widehat{\theta}(y)$ and $\widehat{\theta}(Y)$.

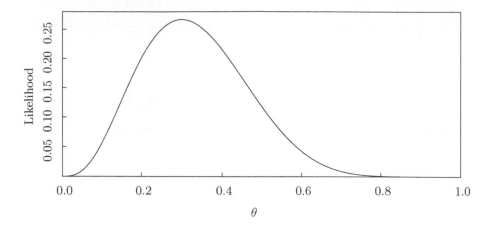

Figure 2.1: *Likelihood function of the success probability θ in a binomial experiment with n = 10 and y = 3.*

The likelihood principle is not just a method for obtaining a point estimate of parameters; it is a method for an objective reasoning with data. It is the entire likelihood function that captures all the information in the data about a certain parameter, not just its maximizer. The likelihood principle also provides the basis for a rich family of methods for selecting the most appropriate model.

Today the likelihood principles play a central role in statistical modeling and inference. Likelihood based methods are inherently computational. In general, numerical methods are needed to find the MLE.

We could view the MLE as a single number representing the likelihood function; but generally, a single number is not enough for summarising the variations of a function. If the (log-)likelihood function is well approximated by a quadratic function it is said to be *regular* and then we need at least two quantities: the location of its maximum and the curvature at the maximum. When our sample becomes large the likelihood function generally becomes regular. The curvature delivers important information about the uncertainty of the parameter estimate.

Before considering the likelihood principles in detail we shall briefly consider some theory related to point estimation.

2.2   Point estimation theory

Assume that the statistical model for the multivariate random variable, $\boldsymbol{Y} = (Y_1, Y_2, \ldots, Y_n)^T$ is given by the parametric family of joint densities

$$\{f_Y(y_1, y_2, \ldots, y_n; \boldsymbol{\theta})\}_{\boldsymbol{\theta} \in \Theta^k} \tag{2.1}$$

with respect to some measure ν_n on $\mathcal{Y}^n$. In the following, the random variable $\boldsymbol{Y}$ will sometimes denote the observations. Assume also that we are given a realization of $\boldsymbol{Y}$ which we shall call the *observation set*, $\boldsymbol{y} = (y_1, y_2, \ldots, y_n)^T$.

We define an *estimator* as a function $\widehat{\boldsymbol{\theta}}(\boldsymbol{Y})$ of the random variable $\boldsymbol{Y}$. For given observations, $\widehat{\boldsymbol{\theta}}(\boldsymbol{y})$ is called an *estimate*. Note that an estimator is a random variable whereas an estimate is a specific number.

Example 2.2 – Estimate and Estimator
In Example 2.1 on page 9 $\widehat{\theta}(y) = y \, / \, n$ is an estimate whereas the random variable $\widehat{\theta}(Y) = Y \, / \, n$ is an estimator. In both cases they are of the maximum likelihood type.

Let us now briefly introduce some properties that are often used to describe point estimators.

DEFINITION 2.1 – UNBIASED ESTIMATOR
Any estimator $\widehat{\boldsymbol{\theta}} = \widehat{\boldsymbol{\theta}}(\boldsymbol{Y})$ is said to be *unbiased* if $\mathrm{E}[\widehat{\boldsymbol{\theta}}] = \boldsymbol{\theta}$ for all $\boldsymbol{\theta} \in \Theta^k$.

Example 2.3 – Unbiased estimator
Consider again the binomial experiment from Example 2.1 where we derived the maximum likelihood estimator

$$\widehat{\theta}(Y) = \frac{Y}{n}. \tag{2.2}$$

Since

$$\mathrm{E}\left[\widehat{\theta}(Y)\right] = \frac{\mathrm{E}\,[Y]}{n} = \frac{n \cdot \theta}{n} = \theta \tag{2.3}$$

it is seen that the estimator is unbiased cf. Definition 2.1.

Another important property is *consistency*.

DEFINITION 2.2 – CONSISTENT ESTIMATOR
An estimator is *consistent* if the sequence $\boldsymbol{\theta}_n(\boldsymbol{Y})$ of estimators for the parameter $\boldsymbol{\theta}$ converges in probability to the true value $\boldsymbol{\theta}$. Otherwise the estimator is said to be inconsistent.

For more details and more precise definitions see, e.g., Lehmann and Casella (1998) p. 332.

DEFINITION 2.3 – MINIMUM MEAN SQUARE ERROR
An estimator $\widehat{\boldsymbol{\theta}} = \widehat{\boldsymbol{\theta}}(\boldsymbol{Y})$ is said to be *uniformly minimum mean square error* if[1]

$$\mathrm{E}\left[(\widehat{\boldsymbol{\theta}}(\boldsymbol{Y}) - \boldsymbol{\theta})(\widehat{\boldsymbol{\theta}}(\boldsymbol{Y}) - \boldsymbol{\theta})^T\right] \leq \mathrm{E}\left[(\tilde{\boldsymbol{\theta}}(\boldsymbol{Y}) - \boldsymbol{\theta})(\tilde{\boldsymbol{\theta}}(\boldsymbol{Y}) - \boldsymbol{\theta})^T\right] \qquad (2.4)$$

for all $\boldsymbol{\theta} \in \Theta^k$ and all other estimators $\tilde{\boldsymbol{\theta}}(\boldsymbol{Y})$.

▸ **Remark 2.1**
In the class of unbiased estimators the minimum mean square estimator is said to be a *minimum variance unbiased estimator* (MVUE) and, furthermore, if the estimators considered are linear functions of the data, the estimator is a *best linear unbiased estimator* (BLUE).   ◂

By considering the class of unbiased estimators it is most often not possible to establish a suitable estimator; we need to add a criterion on the variance of the estimator. A low variance is desired, and in order to evaluate the variance a suitable lower bound is given by the Cramer-Rao inequality.

THEOREM 2.1 – CRAMER-RAO INEQUALITY
Given the parametric density $f_{\boldsymbol{Y}}(\boldsymbol{y}; \boldsymbol{\theta}), \boldsymbol{\theta} \in \Theta^k$, for the observations $\boldsymbol{Y}$. Subject to certain regularity conditions, the variance covariance of any unbiased estimator $\widehat{\boldsymbol{\theta}}(\boldsymbol{Y})$ of $\boldsymbol{\theta}$ satisfies the inequality

$$\mathrm{Var}\left[\widehat{\boldsymbol{\theta}}(\boldsymbol{Y})\right] \geq \boldsymbol{i}^{-1}(\boldsymbol{\theta}) \qquad (2.5)$$

where $\boldsymbol{i}(\boldsymbol{\theta})$ is the Fisher information matrix defined by

$$\boldsymbol{i}(\boldsymbol{\theta}) = \mathrm{E}\left[\left(\frac{\partial \log f_{\boldsymbol{Y}}(\boldsymbol{Y}; \boldsymbol{\theta})}{\partial \boldsymbol{\theta}}\right)\left(\frac{\partial \log f_{\boldsymbol{Y}}(\boldsymbol{Y}; \boldsymbol{\theta})}{\partial \boldsymbol{\theta}}\right)^T\right] \qquad (2.6)$$

and where $\mathrm{Var}\left[\widehat{\boldsymbol{\theta}}(\boldsymbol{Y})\right] = \mathrm{E}[(\widehat{\boldsymbol{\theta}}(\boldsymbol{Y}) - \boldsymbol{\theta})(\widehat{\boldsymbol{\theta}}(\boldsymbol{Y}) - \boldsymbol{\theta})^T]$. The Fisher information matrix is discussed in more detail in Section 2.5 on page 18.

Proof Since $\widehat{\boldsymbol{\theta}}(\boldsymbol{Y})$ is unbiased we have that

$$\mathrm{E}\left[\widehat{\boldsymbol{\theta}}(\boldsymbol{Y})\right] = \boldsymbol{\theta} \qquad (2.7)$$

i.e.,

$$\int \widehat{\boldsymbol{\theta}}(\boldsymbol{y}) f_{Y}(\boldsymbol{y}; \boldsymbol{\theta})\{dy\} = \boldsymbol{\theta} \qquad (2.8)$$

---

[1]Note that the inequality should be understood in the way that the left hand side ÷ right hand side is non-negative definite.

which implies that

$$\frac{\partial}{\partial \theta} \int \widehat{\theta}(y) f_Y(y; \theta)\{dy\} = I. \tag{2.9}$$

Assuming sufficient regularity to allow for differentiation under the integral we obtain

$$\int \widehat{\theta}(y) \frac{\partial}{\partial \theta} f_Y(y; \theta)\{dy\} = I \tag{2.10}$$

or

$$\int \widehat{\theta}(y) \frac{\partial \log f_Y(y; \theta)}{\partial \theta} f_Y(y; \theta)\{dy\} = I \tag{2.11}$$

or

$$E\left[\widehat{\theta}(Y) \frac{\partial \log f_Y(Y; \theta)}{\partial \theta}\right] = I. \tag{2.12}$$

Furthermore, we see that

$$\begin{aligned}
E\left[\frac{\partial \log f_Y(Y; \theta)}{\partial \theta}\right] &= \int \frac{\partial \log f_Y(y; \theta)}{\partial \theta} f_Y(y; \theta)\{dy\} \\
&= \int \frac{\partial f_Y(y; \theta)}{\partial \theta}\{dy\} = \frac{\partial}{\partial \theta} \int f_Y(y; \theta)\{dy\} = 0^T.
\end{aligned} \tag{2.13}$$

Using (2.12) and (2.13) we are able to find the variance (or variance covariance matrix) for $\begin{bmatrix} \widehat{\theta}(Y) \\ \partial \log f_Y(Y; \theta)/\partial \theta \end{bmatrix}$.

$$\begin{aligned}
E\left[\begin{pmatrix} \widehat{\theta}(Y) - \theta \\ (\partial \log f_Y(Y; \theta)/\partial \theta)^T \end{pmatrix} \left((\widehat{\theta}(Y) - \theta)^T \quad \partial \log f_Y(Y; \theta)/\partial \theta\right)\right] \\
= \mathrm{Var}\begin{bmatrix} \widehat{\theta}(Y) & I \\ I & i(\theta) \end{bmatrix}. \tag{2.14}
\end{aligned}$$

This variance matrix is clearly non-negative definite, and we have

$$[I \quad i^{-1}(\theta)] \begin{bmatrix} \mathrm{Var}\left[\widehat{\theta}(Y)\right] & I \\ I & i(\theta) \end{bmatrix} \begin{bmatrix} I \\ i^{-1}(\theta) \end{bmatrix} \geq 0 \tag{2.15}$$

i.e.,

$$\mathrm{Var}\left[\widehat{\theta}(Y)\right] - i^{-1}(\theta) \geq 0 \tag{2.16}$$

which establishes the Cramer-Rao inequality.   ∎

DEFINITION 2.4 – EFFICIENT ESTIMATOR
An unbiased estimator is said to be *efficient* if its covariance is equal to the Cramer-Rao lower bound.

▸ **Remark 2.2 – Dispersion matrix**
The matrix $\mathrm{Var}[\widehat{\theta}(Y)]$ is often called a variance covariance matrix since it contains variances in the diagonal and covariances outside the diagonal. This important matrix will often be termed the *Dispersion matrix* in this book. ◂

2.3   The likelihood function

The likelihood function is built on an assumed parameterized statistical model as specified by a parametric family of joint densities for the observations $Y = (Y_1, Y_2, \ldots, Y_n)^T$. The *likelihood* of any specific value θ of the parameters in a model is (proportional to) the probability of the actual outcome, $Y_1 = y_1, Y_2 = y_2, \ldots, Y_n = y_n$, calculated for the specific value θ. The likelihood function is simply obtained by considering the likelihood as a function of $\theta \in \Theta^k$.

DEFINITION 2.5 – LIKELIHOOD FUNCTION
Given the parametric density $f_Y(\boldsymbol{y}, \boldsymbol{\theta})$, $\boldsymbol{\theta} \in \Theta^k$, for the observations $\boldsymbol{y} = (y_1, y_2, \ldots, y_n)$ the *likelihood function for* $\boldsymbol{\theta}$ is the function

$$L(\boldsymbol{\theta}; \boldsymbol{y}) = c(y_1, y_2, \ldots, y_n) f_Y(y_1, y_2, \ldots, y_n; \boldsymbol{\theta}) \qquad (2.17)$$

where $c(y_1, y_2, \ldots, y_n)$ is a constant.

▸ **Remark 2.3**
The likelihood function is thus the joint probability density for the actual observations considered as a function of $\boldsymbol{\theta}$.   ◂

▸ **Remark 2.4**
The likelihood function is only meaningful up to a multiplicative constant, meaning that we can ignore terms not involving the parameter.   ◂

As illustrated in Example 2.1 on page 9, the likelihood function contains a measure of relative preference for various parameter values. The measure is closely linked to the assumed statistical model, but given the model the likelihood is an objective quantity that provides non-subjective measures of belief about the values of the parameter.

Very often it is more convenient to consider the *log-likelihood* function defined as

$$\ell(\boldsymbol{\theta}; \boldsymbol{y}) = \log(L(\boldsymbol{\theta}; \boldsymbol{y})) \qquad (2.18)$$

where $L(\boldsymbol{\theta}; \boldsymbol{y})$ is given by (2.17). Sometimes the likelihood and the log-likelihood function will be written as $L(\boldsymbol{\theta})$ and $\ell(\boldsymbol{\theta})$, respectively, i.e., the dependency on $\boldsymbol{y}$ is not explicitly mentioned.

▸ **Remark 2.5**
It is common practice, especially when plotting, to normalize the likelihood function to have unit maximum and the log-likelihood to have zero maximum. ◂

Example 2.4 – Likelihood function for mean of normal distribution
An automatic production of a bottled liquid is considered to be stable. A
sample of three bottles was selected randomly from the production and the
volume of the content was measured. The deviation from the nominal volume
of 700.0 ml was recorded. The deviations (in ml) were 4.6, 6.3, and 5.0.
  At first a *model* is formulated

i) Model: C+E (center plus error) model, $Y = \mu + \epsilon$

ii) Data: $Y_i = \mu + \epsilon_i$

iii) Assumptions:

   - Y_1, Y_2, Y_3 are independent
   - $Y_i \sim N(\mu, \sigma^2)$
   - σ^2 is known, $\sigma^2 = 1$.

Thus, there is only one unknown model parameter, $\mu_Y = \mu$.
  The joint probability density function for Y_1, Y_2, Y_3 is

$$
\begin{aligned}
f_{Y_1,Y_2,Y_3}(y_1, y_2, y_3; \mu) &= \frac{1}{\sqrt{2\pi}} \exp\left[-\frac{(y_1 - \mu)^2}{2}\right] \\
&\times \frac{1}{\sqrt{2\pi}} \exp\left[-\frac{(y_2 - \mu)^2}{2}\right] \qquad (2.19) \\
&\times \frac{1}{\sqrt{2\pi}} \exp\left[-\frac{(y_3 - \mu)^2}{2}\right]
\end{aligned}
$$

which for every value of μ is a function of the three variables y_1, y_2, y_3.
  Now, we have the *observations*, $y_1 = 4.6$; $y_2 = 6.3$ and $y_3 = 5.0$, and
establish the likelihood function

$$
\begin{aligned}
L_{4.6,6.3,5.0}(\mu) &= f_{Y_1,Y_2,Y_3}(4.6, 6.3, 5.0; \mu) \\
&= \frac{1}{\sqrt{2\pi}} \exp\left[-\frac{(4.6 - \mu)^2}{2}\right] \\
&\times \frac{1}{\sqrt{2\pi}} \exp\left[-\frac{(6.3 - \mu)^2}{2}\right] \\
&\times \frac{1}{\sqrt{2\pi}} \exp\left[-\frac{(5.0 - \mu)^2}{2}\right].
\end{aligned}
$$

The function depends only on μ. Note that the likelihood function expresses
the infinitesimal probability of obtaining the sample result $(4.6, 6.3, 5.0)$ as a
function of the unknown parameter μ.
  Reducing the expression we find

$$
L_{4.6,6.3,5.0}(\mu) = \frac{1}{(\sqrt{2\pi})^3} \exp\left[-\frac{1.58}{2}\right] \exp\left[-\frac{3(5.3 - \mu)^2}{2}\right]. \qquad (2.20)
$$

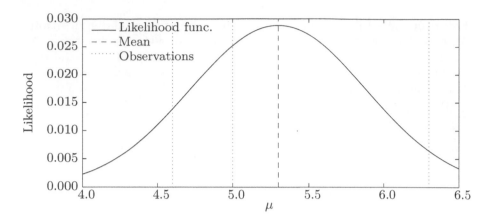

Figure 2.2: *The likelihood function for μ given the observations $y_1 = 4.6$, $y_2 = 6.3$, and $y_3 = 5.0$, as in Example 2.4.*

A sketch of the likelihood function is shown in Figure 2.2. Note that, while the probability density function (2.19) is a function of (y_1, y_2, y_3) which describes the prospective *variation* in data, the likelihood function (2.20) is a function of the unknown parameter μ, describing the relative *plausibility* or likelihood of various values of μ in light of the given data. The likelihood function indicates to which degree the various values of μ are in agreement with the given observations. Note, that the maximum value of the likelihood function (2.20) is obtained for $\hat{\mu} = 5.3$ which equals the sample mean $\overline{y} = \sum_{i=1}^{n} y_i / n$.

Sufficient statistic

The primary goal in analyzing observations is to characterize the information in the observations by a few numbers. A *statistic* $t(Y_1, Y_2, \ldots, Y_n)$ is the result of applying a function (algorithm) to the set of observations. In estimation a sufficient statistic is a statistic that contains all the information in the observations.

DEFINITION 2.6 – SUFFICIENT STATISTIC
A (possibly vector-valued) function $t(Y_1, Y_2, \ldots, Y_n)$ is said to be a *sufficient statistic* for a (possibly vector-valued) parameter, $\boldsymbol{\theta}$, if the probability density function for $t(Y_1, Y_2, \ldots, Y_n)$ can be factorized into a product

$$f_{Y_1, \ldots, Y_n}(y_1, \ldots, y_n; \boldsymbol{\theta}) = h(y_1, \ldots, y_n) g(t(y_1, y_2, \ldots, y_n); \boldsymbol{\theta})$$

with the factor $h(y_1, \ldots, y_n)$ not depending on the parameter $\boldsymbol{\theta}$, and the factor $g(t(y_1, y_2, \ldots, y_n); \boldsymbol{\theta})$ only depending on $y_1, \ldots, y_n$ through the function

$t(\cdot, \cdot, \ldots, \cdot)$. Thus, if we know the value of $t(y_1, y_2, \ldots, y_n)$, the individual values $y_1, \ldots, y_n$ do not contain further information about the value of $\boldsymbol{\theta}$.

Roughly speaking, a statistic is sufficient if we are able to calculate the likelihood function (apart from a factor) only knowing $t(Y_1, Y_2, \ldots, Y_n)$.

Example 2.5 – Sufficiency of the sample mean
Consider again the the situation from Example 2.4. We obtain more general insight if we just use the symbols (y_1, y_2, y_3) for the data values. Using this notation, the likelihood function is

$$
\begin{aligned}
L_{y_1,y_2,y_3}(\mu) &= \frac{1}{\sqrt{2\pi}} \exp\left[-\frac{(y_1-\mu)^2}{2}\right] \\
&\times \frac{1}{\sqrt{2\pi}} \exp\left[-\frac{(y_2-\mu)^2}{2}\right] \\
&\times \frac{1}{\sqrt{2\pi}} \exp\left[-\frac{(y_3-\mu)^2}{2}\right] \\
&= \frac{1}{(\sqrt{2\pi})^3} \exp\left[-\frac{\sum(y_i-\bar{y})^2}{2}\right] \exp\left[-\frac{3(\bar{y}-\mu)^2}{2}\right]
\end{aligned}
$$
(2.21)

which shows that (except for a factor not depending on μ), the likelihood function does only depend on the observations (y_1, y_2, y_3) through the average $\bar{y} = \sum y_i/3$. Thus, whatever be the individual values (y_1, y_2, y_3) underlying a given value of $\bar{y}$, they give rise to, e.g., the same supremum of the likelihood function. Therefore, as long as we are concerned with inference about μ under this normal distribution model, it is sufficient to consider the sample mean $\bar{y} = \sum_{i=1}^{3} y_i/3$, and disregard the individual values (y_1, y_2, y_3). We say that the sample mean, $\overline{Y} = \sum_{i=1}^{3} Y_i/3$ is a sufficient statistic for the parameter μ.

2.4 The score function

DEFINITION 2.7 – SCORE FUNCTION
Consider $\boldsymbol{\theta} = (\theta_1, \cdots, \theta_k)^T \in \Theta^k$, and assume that Θ^k is an open subspace of $\mathbb{R}^k$, and that the log-likelihood is continuously differentiable. Then consider the following first order partial derivative of the log-likelihood function:

$$
\ell'_{\boldsymbol{\theta}}(\boldsymbol{\theta}; \boldsymbol{y}) = \frac{\partial}{\partial\boldsymbol{\theta}} \ell(\boldsymbol{\theta}; \boldsymbol{y}) = \begin{pmatrix} \frac{\partial}{\partial\theta_1}\ell(\boldsymbol{\theta}; \boldsymbol{y}) \\ \vdots \\ \frac{\partial}{\partial\theta_k}\ell(\boldsymbol{\theta}; \boldsymbol{y}) \end{pmatrix}.
$$
(2.22)

The function $\ell'_{\boldsymbol{\theta}}(\boldsymbol{\theta}; \boldsymbol{y})$ is called the *score function*, often written as $S(\boldsymbol{\theta}; \boldsymbol{y})$.

Note, that the score function depends on y. Hence, the score function $\ell'_\theta(\boldsymbol{\theta}; \boldsymbol{Y})$ is a random variable.

THEOREM 2.2
Under normal regularity conditions

$$E_\theta \left[\frac{\partial}{\partial \boldsymbol{\theta}} \ell(\boldsymbol{\theta}; \boldsymbol{Y}) \right] = \boldsymbol{0}. \tag{2.23}$$

Proof Follows by differentiation of

$$\int f_Y(\boldsymbol{y}; \boldsymbol{\theta}) \mu\{dy\} = 1. \tag{2.24}$$

∎

2.5   The information matrix

DEFINITION 2.8 – OBSERVED INFORMATION
The matrix

$$\boldsymbol{j}(\boldsymbol{\theta}; \boldsymbol{y}) = -\frac{\partial^2}{\partial \boldsymbol{\theta} \partial \boldsymbol{\theta}^T} \ell(\boldsymbol{\theta}; \boldsymbol{y}) \tag{2.25}$$

with the elements

$$\boldsymbol{j}(\boldsymbol{\theta}; \boldsymbol{y})_{ij} = -\frac{\partial^2}{\partial \theta_i \partial \theta_j} \ell(\boldsymbol{\theta}; \boldsymbol{y})$$

is called the *observed information* corresponding to the observation $\boldsymbol{y}$ and the parameter $\boldsymbol{\theta}$.

The observed information is thus equal to the Hessian (with opposite sign) of the log-likelihood function evaluated at $\boldsymbol{\theta}$. The Hessian matrix is simply (with opposite sign) the *curvature* of the log-likelihood function.

DEFINITION 2.9 – EXPECTED INFORMATION
The expectation of the observed information

$$\boldsymbol{i}(\boldsymbol{\theta}) = E[\boldsymbol{j}(\boldsymbol{\theta}; \boldsymbol{Y})], \tag{2.26}$$

where $\boldsymbol{j}(\boldsymbol{\theta}; \boldsymbol{Y})$ is given by (2.25), and where the expectation is determined under the distribution corresponding to $\boldsymbol{\theta}$, is called the *expected information*, or the *information matrix* corresponding to the parameter $\boldsymbol{\theta}$. The expected information (2.26) is also known as the *Fisher information matrix*.

By differentiating (2.24) twice, we find

$$\mathrm{E}_\theta\left[\frac{\partial^2}{\partial\boldsymbol{\theta}\partial\boldsymbol{\theta}^T}\ell(\boldsymbol{\theta};\boldsymbol{Y})\right] + \mathrm{E}_\theta\left[\frac{\partial}{\partial\boldsymbol{\theta}}\ell(\boldsymbol{\theta};\boldsymbol{Y})\left(\frac{\partial}{\partial\boldsymbol{\theta}}\ell(\boldsymbol{\theta};\boldsymbol{Y})\right)^T\right] = \mathbf{0}, \qquad (2.27)$$

which gives us the following important result:

Lemma 2.1 (Fisher Information Matrix) *The expected information or Fisher Information Matrix is equal to the dispersion matrix for the score function, i.e.,*

$$\begin{aligned} i(\boldsymbol{\theta}) &= \mathrm{E}_\theta\left[-\frac{\partial^2}{\partial\boldsymbol{\theta}\partial\boldsymbol{\theta}^T}\ell(\boldsymbol{\theta};\boldsymbol{Y})\right] \\ &= \mathrm{E}_\theta\left[\frac{\partial}{\partial\boldsymbol{\theta}}\ell(\boldsymbol{\theta};\boldsymbol{Y})\left(\frac{\partial}{\partial\boldsymbol{\theta}}\ell(\boldsymbol{\theta};\boldsymbol{Y})\right)^T\right] \\ &= \mathrm{D}_\theta[\ell'_\theta(\boldsymbol{\theta};\boldsymbol{Y})] \end{aligned} \qquad (2.28)$$

where $\mathrm{D}[\cdot]$ *denotes the dispersion matrix.*

This fundamental result shows that *the expected information is equal to the dispersion matrix of the score function.* In estimation the information matrix provides a measure for the accuracy obtained in determining the parameters.

Example 2.6 – Score function, observed and expected information
Consider again the situation from Example 2.4 on page 15.
    From (2.20) one obtains the log-likelihood function

$$\ell(\mu; 4.6, 6.3, 5.0) = -\frac{3(5.3 - \mu)^2}{2} + C(4.6, 6.3, 5.0)$$

and, hence, the score function is

$$\ell'_\mu(\mu; 4.6, 6.3, 5.0) = 3 \cdot (5.3 - \mu),$$

with the observed information

$$j(\mu; 4.6, 6.3, 5.0) = 3.$$

In order to determine the expected information it is necessary to perform analogous calculations substituting the data by the corresponding random variables Y_1, Y_2, Y_3.
    According to Equation (2.21), the likelihood function can be written as

$$L_{y_1,y_2,y_3}(\mu) = \frac{1}{(\sqrt{2\pi})^3}\exp\left[-\frac{\sum(y_i - \bar{y})^2}{2}\right]\exp\left[-\frac{3(\bar{y} - \mu)^2}{2}\right]. \qquad (2.29)$$

Introducing the random variables (Y_1, Y_2, Y_3) instead of (y_1, y_2, y_3) in (2.29) and taking logarithms one finds

$$\ell(\mu; Y_1, Y_2, Y_3) = -\frac{3(\overline{Y} - \mu)^2}{2} - 3\log(\sqrt{2\pi}) - \frac{\sum(Y_i - \overline{Y})^2}{2},$$

and, hence, the score function is

$$\ell'_\mu(\mu; Y_1, Y_2, Y_3) = 3(\overline{Y} - \mu),$$

and the observed information

$$j(\mu; Y_1, Y_2, Y_3) = 3.$$

It is seen in this (Gaussian) case that the observed information (curvature of log-likelihood function) does not depend on the observations y_1, y_2, y_3, and hence the expected information is

$$i(\mu) = \mathrm{E}[j(\mu; Y_1, Y_2, Y_3)] = 3.$$

2.6 Alternative parameterizations of the likelihood

DEFINITION 2.10 – THE LIKELIHOOD FUNCTION FOR ALTERNATIVE PARAMETERIZATIONS

The likelihood function does not depend on the actual parameterization. Let $\psi = \psi(\theta)$ denote a one-to-one mapping of $\Omega \subset \mathbb{R}^k$ onto $\Psi \subset \mathbb{R}^k$. The parameterization given by ψ is just an alternative parameterization of the model.

The likelihood and log-likelihood function for the parameterization given by ψ is

$$L_\Psi(\psi; y) = L_\Omega(\theta(\psi); y) \tag{2.30}$$
$$\ell_\Psi(\psi; y) = \ell_\Omega(\theta(\psi); y). \tag{2.31}$$

▸ **Remark 2.6 – Invariance property**
Equation (2.30) gives rise to the very useful invariance property – see Theorem 2.3.   ◂

This means that the likelihood is *not* a joint probability density on Ω, since the Jacobian should have been used in (2.30). The score function and the information matrix depends in general on the parameterization.

Consider another parameterization given as $\theta = \theta(\beta) \in \mathbb{R}^k$ for $\beta \in \mathrm{B} \subset \mathbb{R}^m$ where $m \leq k$.

Then the score function for the parameter set $\boldsymbol{\beta}$ is

$$\ell'_\beta(\boldsymbol{\beta}; \boldsymbol{y}) = \boldsymbol{J}^T \ell'_\theta(\boldsymbol{\theta}(\boldsymbol{\beta}); \boldsymbol{y}) \qquad (2.32)$$

where the Jacobian $\boldsymbol{J}$ is

$$\boldsymbol{J} = \frac{\partial \boldsymbol{\theta}}{\partial \boldsymbol{\beta}} \qquad (2.33)$$

with the elements

$$J_{ij} = \frac{\partial \theta_i}{\partial \beta_j} i = 1, 2, \ldots, k; j = 1, 2, \ldots, m.$$

By using the fact that the expected information is equal to the dispersion matrix of the score function (cf. Lemma 2.1 on page 19) we obtain the expected information with the parameterization given by β:

$$\boldsymbol{i}_\beta(\boldsymbol{\beta}) = -\mathrm{E}\left[\frac{\partial^2}{\partial \boldsymbol{\beta} \partial \boldsymbol{\beta}^T} l(\boldsymbol{\theta}(\boldsymbol{\beta}); \boldsymbol{Y})\right] = \boldsymbol{J}^T \boldsymbol{i}_\theta(\boldsymbol{\theta}(\boldsymbol{\beta})) \boldsymbol{J} \qquad (2.34)$$

where the Jacobian $\boldsymbol{J}$ is given by (2.33).

2.7   The maximum likelihood estimate (MLE)

DEFINITION 2.11 – MAXIMUM LIKELIHOOD ESTIMATE (MLE)
Given the observations $\boldsymbol{y} = (y_1, y_2, \ldots, y_n)$ the *Maximum Likelihood Estimate (MLE)* is a function $\widehat{\boldsymbol{\theta}}(\boldsymbol{y})$ such that

$$L(\widehat{\boldsymbol{\theta}}; \boldsymbol{y}) = \sup_{\boldsymbol{\theta} \in \Theta} L(\boldsymbol{\theta}; \boldsymbol{y}). \qquad (2.35)$$

The function $\widehat{\boldsymbol{\theta}}(\boldsymbol{Y})$ over the sample space of observations $\boldsymbol{Y}$ is called an *ML estimator*.

In practice it is convenient to work with the log-likelihood function $\ell(\boldsymbol{\theta}; \boldsymbol{y})$.

When the supremum is attained at an interior point, then the *score function* can be used to obtain the estimate, since the MLE can be found as the solution to

$$\ell'_\theta(\boldsymbol{\theta}; \boldsymbol{y}) = \boldsymbol{0} \qquad (2.36)$$

which are called the *estimation equations for the ML-estimator*, or, just the ML equations.

THEOREM 2.3 – INVARIANCE PROPERTY
Assume that $\widehat{\boldsymbol{\theta}}$ is a maximum likelihood estimator for $\boldsymbol{\theta}$, and let $\boldsymbol{\psi} = \boldsymbol{\psi}(\boldsymbol{\theta})$ denote a one-to-one mapping of $\Omega \subset \mathbb{R}^k$ onto $\Psi \subset \mathbb{R}^k$. Then the estimator $\boldsymbol{\psi}(\widehat{\boldsymbol{\theta}})$ is a maximum likelihood estimator for the parameter $\boldsymbol{\psi}(\boldsymbol{\theta})$.

Proof Follows directly from the definition.   ■

The principle can be generalized to the case where the mapping is not one-to-one.

2.8   Distribution of the ML estimator

Now we will provide results which can be used for inference under very general conditions. As the price for the generality, the results are only asymptotically valid.

THEOREM 2.4 – DISTRIBUTION OF THE ML ESTIMATOR
We assume that $\widehat{\boldsymbol{\theta}}$ *is consistent. Then, under some regularity conditions,*

$$\widehat{\boldsymbol{\theta}} - \boldsymbol{\theta} \to \mathrm{N}(0, \boldsymbol{i}(\boldsymbol{\theta})^{-1}) \tag{2.37}$$

where $\boldsymbol{i}(\boldsymbol{\theta})$ *is the expected information or the information matrix – see* (2.26).

Proof Omitted.   ■

Asymptotically the variance of the estimator is seen to be equal to the Cramer-Rao lower bound (Theorem 2.1) for any unbiased estimator. The practical significance of this result is that the MLE makes efficient use of the available data for large datasets.

In practice, we would use

$$\widehat{\boldsymbol{\theta}} \sim \mathrm{N}(\theta, \boldsymbol{j}^{-1}(\widehat{\boldsymbol{\theta}})) \tag{2.38}$$

where $\boldsymbol{j}(\widehat{\boldsymbol{\theta}})$ is the observed (Fisher) information – see (2.25). This means that asymptotically

i) $\mathrm{E}[\widehat{\boldsymbol{\theta}}] = \boldsymbol{\theta}$

ii) $D[\widehat{\boldsymbol{\theta}}] = \boldsymbol{j}^{-1}(\widehat{\boldsymbol{\theta}})$

The standard error of $\widehat{\theta}_i$ is given by the standard deviation

$$\widehat{\sigma}_{\widehat{\theta}_i} = \sqrt{\mathrm{Var}_{ii}[\widehat{\boldsymbol{\theta}}]} \tag{2.39}$$

where $\mathrm{Var}_{ii}[\widehat{\boldsymbol{\theta}}]$ is the i'th diagonal term of $\boldsymbol{j}^{-1}(\widehat{\boldsymbol{\theta}})$. Hence, an estimate of the dispersion (variance-covariance matrix) of the estimator is

$$D[\widehat{\boldsymbol{\theta}}] = \boldsymbol{j}^{-1}(\widehat{\boldsymbol{\theta}}). \tag{2.40}$$

An estimate of the uncertainty of the individual parameter estimates is obtained by decomposing the dispersion matrix as follows:

$$D[\widehat{\boldsymbol{\theta}}] = \widehat{\boldsymbol{\sigma}}_{\widehat{\boldsymbol{\theta}}} \boldsymbol{R} \widehat{\boldsymbol{\sigma}}_{\widehat{\boldsymbol{\theta}}} \tag{2.41}$$

into $\widehat{\boldsymbol{\sigma}}_{\widehat{\boldsymbol{\theta}}}$, which is a diagonal matrix of the standard deviations of the individual parameter estimates, and $\boldsymbol{R}$, which is the corresponding correlation matrix. The value R_{ij} is thus the estimated correlation between $\widehat{\theta}_i$ and $\widehat{\theta}_j$.

A test of an individual parameter $\mathcal{H}_0 : \theta_i = \theta_{i,0}$ is given by the *Wald statistic*:

$$Z_i = \frac{\widehat{\theta}_i - \theta_{i,0}}{\widehat{\sigma}_{\widehat{\theta}_i}} \tag{2.42}$$

which under $\mathcal{H}_0$ is approximately $N(0, 1)$-distributed.

2.9   Generalized loss-function and deviance

The log-likelihood function is used as a generalized loss-function in the case of observations described by the generalized densities like the Gamma, Poisson, log-normal, and others.

DEFINITION 2.12 – DEVIANCE
Assume the density $f_{(Y,\mu(X))}(\boldsymbol{y})$ of $\boldsymbol{Y}$ is indexed by a parameter, $\mu(\boldsymbol{X})$ that depends on the covariate(s), $\boldsymbol{X}$. Then we have the *generalized loss function*, *cross entropy loss*, or *deviance*.

$$D(\boldsymbol{Y}; \mu(\boldsymbol{X})) = -2 \log f_{(Y,\mu(X))}(\boldsymbol{y}). \tag{2.43}$$

This is an important aspect of generalization to a large class of densities which we will consider later on in this book. The factor 2 implies that for Gaussian observations the generalized loss function equals the classical sum of squared error loss function.

2.10   Quadratic approximation of the log-likelihood

A second-order Taylor expansion around $\widehat{\theta}$ provides us with a quadratic approximation of the normalized log-likelihood around the MLE.

Consider, for a moment, the scalar case. Taking a second-order Taylor expansion around $\widehat{\theta}$ we get

$$\ell(\theta) \approx \ell(\widehat{\theta}) + \ell'(\widehat{\theta})(\theta - \widehat{\theta}) - \frac{1}{2}j(\widehat{\theta})(\theta - \widehat{\theta})^2 \tag{2.44}$$

and then

$$\log \frac{L(\theta)}{L(\widehat{\theta})} \approx -\frac{1}{2}j(\widehat{\theta})(\theta - \widehat{\theta})^2. \tag{2.45}$$

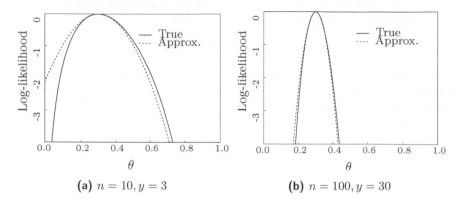

(a) $n = 10, y = 3$ **(b)** $n = 100, y = 30$

Figure 2.3: *Quadratic approximation of the log-likelihood function as in Example 2.7.*

In the case of normality the approximation is exact which means that a quadratic approximation of the log-likelihood corresponds to normal approximation of the $\widehat{\theta}(Y)$ estimator. If the log-likelihood function is well approximated by a quadratic function it is said to be *regular*.

Example 2.7 – Quadratic approximation of the log-likelihood
Consider again the situation from Example 2.1 where the log-likelihood function is
$$\ell(\theta) = y \log \theta + (n - y) \log(1 - \theta) + const.$$
The score function is
$$\ell'(\theta) = \frac{y}{\theta} - \frac{n - y}{1 - \theta},$$
and the observed information
$$j(\theta) = \frac{y}{\theta^2} + \frac{n - y}{(1 - \theta)^2}.$$
For $n = 10$, $y = 3$ and $\widehat{\theta} = 0.3$ we obtain
$$j(\widehat{\theta}) = 47.6.$$
The log-likelihood function and the corresponding quadratic approximation are shown in Figure 2.3a. The approximation is poor as can be seen in the figure. By increasing the sample size to $n = 100$, but still with $\widehat{\theta} = 0.3$, the approximation is much better as seen in Figure 2.3b.

At the point $\widehat{\theta} = \frac{y}{n}$ we have
$$j(\widehat{\theta}) = \frac{n}{\widehat{\theta}(1 - \widehat{\theta})}.$$

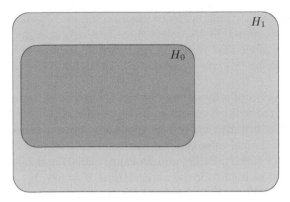

Figure 2.4: *Classical situation: Testing a null hypothesis against an alternative.*

and we find the variance of the estimate

$$\text{Var}[\widehat{\theta}] = j^{-1}(\widehat{\theta}) = \frac{\widehat{\theta}(1 - \widehat{\theta})}{n}.$$

Until now we have seen how the likelihood function can be used to find both point estimates of unknown parameters (the maximum of the (log-)likelihood function) and estimates of the precision/variance of the parameter estimates (from the Hessian of the log-likelihood function). The power of the likelihood principles will be further demonstrated in the following section.

2.11 Likelihood ratio tests

In this section we describe methods for testing hypotheses using the likelihood function. The basic idea is to determine the maximum likelihood estimates under both a null and alternative hypothesis. However, we shall first briefly describe hypothesis testing and the notation used in this book.

It is assumed that a *sufficient model* with $\theta \in \Omega$ exists. In classical *hypothesis testing* we consider some theory or assumption about the parameters $\mathcal{H}_0 : \theta \in \Omega_0$ where $\Omega_0 \subset \Omega$, and the purpose of the testing is to analyze whether the observations provide sufficient evidence to reject this theory or assumption. If not, we accept the null hypothesis.

More formally consider the *null hypothesis* $\mathcal{H}_0 : \theta \in \Omega_0$ against the *alternative hypothesis* $\mathcal{H}_1 : \theta \in \Omega \setminus \Omega_0$ $(\Omega_0 \subseteq \Omega)$, where $\dim(\Omega_0) = m$ and $\dim(\Omega) = k$. Notice that Ω_0 and $\Omega \setminus \Omega_0$ are distinct sets. The setup is illustrated in Figure 2.4.

The evidence against $\mathcal{H}_0$ is measured by the so-called *p-value*. The *p*-value is the probability under $\mathcal{H}_0$ of observing a value of the test statistic equal to or more extreme than the actually observed test statistic. Hence, a small *p*-value (say ≤ 0.05) leads to a strong evidence against $\mathcal{H}_0$, and $\mathcal{H}_0$ is then said to be *rejected*.[2] Likewise, we retain $\mathcal{H}_0$ unless there is a strong evidence against this hypothesis. Remember that rejecting $\mathcal{H}_0$ when $\mathcal{H}_0$ is true is called a *Type I error*, while retaining $\mathcal{H}_0$ when the truth is actually $\mathcal{H}_1$ is called a *Type II error*.

The *critical region* of a hypothesis test is the set of all outcomes which, if they occur, cause the null hypothesis to be rejected and the alternative hypothesis accepted. It is usually denoted by C. For a more detailed introduction to testing and statistical inference we can refer to Cox and Hinkley (2000), Casella and Berger (2002), Bickel and Doksum (2000) and Lehmann (1986).

Again a statistical model for the observations $\boldsymbol{Y} = (Y_1, Y_2, \ldots, Y_n)$ is given by a parametric family of joint densities $\{f_Y(y_1, y_2, \ldots, y_n; \boldsymbol{\theta})\}_{\boldsymbol{\theta} \in \Theta^k}$.

DEFINITION 2.13 – LIKELIHOOD RATIO

Consider the hypothesis $\mathcal{H}_0 \colon \boldsymbol{\theta} \in \Omega_0$ against the alternative $\mathcal{H}_1 \colon \boldsymbol{\theta} \in \Omega \setminus \Omega_0$ ($\Omega_0 \subseteq \Omega$), where $\dim(\Omega_0) = m$ and $\dim(\Omega) = k$.

For given observations $y_1, y_2, \ldots, y_n$ the *likelihood ratio* is defined as

$$\lambda(\boldsymbol{y}) = \frac{\sup_{\boldsymbol{\theta} \in \Omega_0} L(\boldsymbol{\theta}; \boldsymbol{y})}{\sup_{\boldsymbol{\theta} \in \Omega} L(\boldsymbol{\theta}; \boldsymbol{y})}. \tag{2.46}$$

Clearly, if λ is small, then the data are seen to be more plausible under the alternative hypothesis than under the null hypothesis. Hence, the hypothesis ($\mathcal{H}_0$) is rejected for small values of λ. It is sometimes possible to transform the likelihood ratio into a statistic, the exact distribution of which is known under $\mathcal{H}_0$. This is for instance the case for the General Linear Model for Gaussian data which is considered in Chapter 3.

In most cases, however, we must use the following important result regarding the asymptotic behavior.

THEOREM 2.5 – WILK'S LIKELIHOOD RATIO TEST

For $\lambda(\boldsymbol{y})$ defined by (2.46), then under the null hypothesis $\mathcal{H}_0$ (as above), the random variable $-2 \log \lambda(\boldsymbol{Y})$ converges in law to a χ^2 random variable with $(k - m)$ degrees of freedom, i.e.,

$$-2 \log \lambda(\boldsymbol{Y}) \to \chi^2(k - m) \tag{2.47}$$

under $\mathcal{H}_0$.

Proof Omitted. Briefly, however, it follows from the quadratic approximations discussed in Section 2.10, and the fact that for any (now multivariate) statistic

---

[2] Please remember that a large p-value is not an evidence in favor of $\mathcal{H}_0$ since this value can be due to a low power of the test.

T of interest (with dim $T = p$) it follows under appropriate conditions that asymptotically

$$(\widehat{T} - \mathrm{E}[T])^T V^{-1}(\widehat{T} - \mathrm{E}[T]) \sim \chi^2(p) \tag{2.48}$$

where V is the asymptotic variance-covariance matrix for the (p dimensional) statistic.   ∎

Note, that in both cases the model must be estimated under both $\mathcal{H}_0$ and the alternative hypothesis. An alternative is the *Lagrange Multiplier test* (LM test) which only requires evaluations under the restricted model, so, if the restricted model is much simpler, then LM might be attractive.

▶ **Remark 2.7**
Let us introduce

$$L_0 = \sup_{\boldsymbol{\theta} \in \Omega_0} L(\boldsymbol{\theta}; \boldsymbol{y}) \qquad \text{and} \qquad L = \sup_{\boldsymbol{\theta} \in \Omega_{\text{full}}} L(\boldsymbol{\theta}; \boldsymbol{y})$$

We then notice that

$$\begin{aligned} -2\log \lambda(\boldsymbol{Y}) &= -2(\log L_0 - \log L) \\ &= 2(\log L - \log L_0). \end{aligned} \tag{2.49}$$

The statistic $\mathrm{D} = -2\log \lambda(\boldsymbol{Y}) = 2(\log L - \log L_0)$ is called the *deviance* by Nelder and Wedderburn (1972).   ◀

2.12 Successive testing in hypothesis chains

Hypothesis chains

Consider a *chain* of hypotheses specified by a sequence of parameter spaces

$$\mathbb{R} \subseteq \Omega_M \cdots \subset \Omega_2 \subset \Omega_1 \subseteq \mathbb{R}^n. \tag{2.50}$$

For each parameter space Ω_i we define a hypothesis $\mathcal{H}_i : \boldsymbol{\theta} \in \Omega_i$ with $\dim(\Omega_i) < \dim(\Omega_{i-1})$.
    Clearly,

$$\mathcal{H}_M \Rightarrow \cdots \Rightarrow \mathcal{H}_2 \Rightarrow \mathcal{H}_1 \tag{2.51}$$

as indicated in Figure 2.5 on the following page.

▶ **Remark 2.8 – Notation used for hypotheses**
The notation used above in (2.50) and (2.51) will only be used for hypothesis chains. In the standard case with a model and an alternative we will use the traditional notation $\mathcal{H}_o$ for the null hypothesis, related to a model with the parameter set Ω_0, and $\mathcal{H}_1$ for the alternative hypothesis with parameter set Ω_1 (see Section 2.11).   ◀

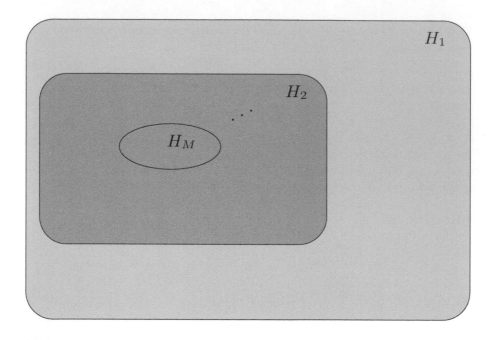

Figure 2.5: *Illustration of the chain (2.51) of hypotheses.*

DEFINITION 2.14 – PARTIAL LIKELIHOOD RATIO TEST
Assume that the hypothesis $\mathcal{H}_i$ allows the sub-hypothesis $\mathcal{H}_{i+1} \subset \mathcal{H}_i$.

The *partial likelihood ratio test* for $\mathcal{H}_{i+1}$ under $\mathcal{H}_i$ is the likelihood ratio test for the hypothesis $\mathcal{H}_{i+1}$ under the assumption that the hypothesis $\mathcal{H}_i$ holds. The likelihood ratio test statistic for this partial test is

$$\lambda_{\mathcal{H}_{i+1}|\mathcal{H}_i}(\boldsymbol{y}) = \frac{\sup_{\boldsymbol{\theta} \in \Omega_{i+1}} L(\boldsymbol{\theta}; \boldsymbol{y})}{\sup_{\boldsymbol{\theta} \in \Omega_i} L(\boldsymbol{\theta}; \boldsymbol{y})}.$$

When $\mathcal{H}_{i+1}$ holds, it follows from Theorem 2.5 that the distribution of $-2 \log \lambda_{\mathcal{H}_{i+1}|\mathcal{H}_i}(\boldsymbol{Y})$ approaches a $\chi^2(f)$ distribution with $f = \dim(\Omega_i) - \dim(\Omega_{i+1})$.

▶ **Remark 2.9 – Null model and Full model**
In the following the concepts of a *null model* and a *full model* becomes useful. For the null model, we are using the equality membership in the left side of Equation (2.50), i.e., $\Omega_M = \Omega_{null} = \mathbb{R}$, i.e., a model with only one parameter is used for fitting the observations. For the full model, we are using the equality membership in the right side of Equation (2.50), i.e., $\Omega_1 = \Omega_{full} = \mathbb{R}^n$, i.e., the dimension is equal to the number of observations and, hence, the model fits each observation perfectly.   ◀

THEOREM 2.6 – PARTITIONING INTO A SEQUENCE OF PARTIAL TESTS
Consider a chain of hypotheses as specified by (2.50) and (2.51).

Now, assume that $\mathcal{H}_1$ holds, and consider the minimal hypothesis $\mathcal{H}_M$: $\boldsymbol{\theta} \in \Omega_M$ with the alternative $\mathcal{H}_1 : \boldsymbol{\theta} \in \Omega_1 \setminus \Omega_M$. The likelihood ratio test statistic $\lambda_{\mathcal{H}_M | \mathcal{H}_1}(\boldsymbol{y})$ for this hypothesis may be factorized into a chain of partial likelihood ratio test statistics $\lambda_{\mathcal{H}_{i+1} | \mathcal{H}_i}(\boldsymbol{y})$ for $\mathcal{H}_{i+1}$ given $\mathcal{H}_i$, $i = 1, \ldots, M$.

Proof The likelihood ratio test statistic (2.46) for $\mathcal{H}_M$ is

$$\lambda_{\mathcal{H}_M | \mathcal{H}_1}(\boldsymbol{y}) = \frac{\sup_{\boldsymbol{\theta} \in \Omega_M} L(\boldsymbol{\theta}; \boldsymbol{y})}{\sup_{\boldsymbol{\theta} \in \Omega_1} L(\boldsymbol{\theta}; \boldsymbol{y})}.$$

By successive multiplication by

$$\frac{\sup_{\boldsymbol{\theta} \in \Omega_i} L(\boldsymbol{\theta}; \boldsymbol{y})}{\sup_{\boldsymbol{\theta} \in \Omega_i} L(\boldsymbol{\theta}; \boldsymbol{y})} = 1$$

we obtain

$$
\begin{aligned}
\lambda_{\mathcal{H}_M | \mathcal{H}_1}(\boldsymbol{y}) &= \frac{\sup_{\boldsymbol{\theta} \in \Omega_M} L(\boldsymbol{\theta}; \boldsymbol{y})}{\sup_{\boldsymbol{\theta} \in \Omega_1} L(\boldsymbol{\theta}; \boldsymbol{y})} \\
&= \frac{\sup_{\boldsymbol{\theta} \in \Omega_M} L(\boldsymbol{\theta}; \boldsymbol{y})}{\sup_{\boldsymbol{\theta} \in \Omega_{M-1}} L(\boldsymbol{\theta}; \boldsymbol{y})} \cdots \frac{\sup_{\boldsymbol{\theta} \in \Omega_3} L(\boldsymbol{\theta}; \boldsymbol{y})}{\sup_{\boldsymbol{\theta} \in \Omega_2} L(\boldsymbol{\theta}; \boldsymbol{y})} \frac{\sup_{\boldsymbol{\theta} \in \Omega_2} L(\boldsymbol{\theta}; \boldsymbol{y})}{\sup_{\boldsymbol{\theta} \in \Omega_1} L(\boldsymbol{\theta}; \boldsymbol{y})} \\
&= \lambda_{\mathcal{H}_M | \mathcal{H}_{M-1}}(\boldsymbol{y}) \cdots \lambda_{\mathcal{H}_3 | \mathcal{H}_2}(\boldsymbol{y}) \lambda_{\mathcal{H}_2 | \mathcal{H}_1}(\boldsymbol{y})
\end{aligned}
$$

which shows that the test of $\mathcal{H}_M$ under $\mathcal{H}_1$ may be decomposed into a chain of *partial likelihood ratio tests* for $\mathcal{H}_{i+1}$ assuming that $\mathcal{H}_i$ holds.   ∎

▶ **Remark 2.10 – The partial tests "corrects" for the effect of the parameters that are in the model at that particular stage**
When interpreting the test statistic corresponding to a particular stage in the hierarchy of models, one often says that the effect of the parameters that are in the model at that stage is "controlled for", or "corrected for".   ◀

Example 2.8 – Sub-hypotheses in a two-factor model
Consider a two-factor experiment with r levels of factor *Length* and c levels of factor *Thick*. The usual two-factor model for the mean value $\mu_{ij} = \mathrm{E}[Y_{ijk}]$ is

$$\mu_{ij} = \mu_0 + \alpha_i + \beta_j + \gamma_{ij} \quad i = 1, 2, \ldots, r; j = 1, 2, \ldots, c$$

with suitable restrictions on the parameters.

Figure 2.6 on the next page illustrates the two hypothesis chains that may be formulated for such an experiment. First a test for vanishing interaction terms ($\gamma_{ij} = 0$). Following "acceptance" in this test one may either test for vanishing effect of *Thick*, viz. a model $\mu_{ij} = \mu + \alpha_i$, and – if accepted –

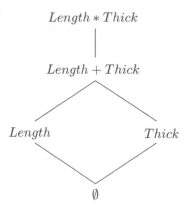

Figure 2.6: *Inclusion diagram corresponding to a two-factor experiment. In the upper level we have used the same notation as used by the software R, i.e., Length∗Thick denotes a two-factor model with interaction between the two factors.*

subsequently test for vanishing effect of *Length*, ending with a null model (left hand side of the diagram), or one may start by testing a hypothesis of vanishing effect of *Length* (model $\mu_{ij} = \mu + \beta_j$) and finally testing whether the model may be reduced to the null model. The experiment will be discussed in more detail in Example 4.14 on page 140.

Example 2.9 – Hypothesis chains for a three-factor experiment
Figure 2.7 illustrates analogously possible hypothesis chains for a three-factor experiment. Notice the difference in complexity between the possible hypothesis chains for a two-factor model (Figure 2.6) and a three-factor model (Figure 2.7).

In practice, it may well happen that one works down a path until the model can not be further reduced following that path, but going back a few steps and choosing another branch might result in a greater simplification of the model.

Now it is clear that for models with a large number of explanatory variables (factors), well-defined strategies for testing hypothesis are useful.

Strategies for variable selection in hypothesis chains

Typically, one of the following principles for model selection is used:

a) Forward selection: Starting with a null model, the procedure adds, at each step, the variable that would give the lowest p-value of the variable not yet included in the model. The procedure stops when all variables have been added, or when no variables meet the pre-specified limit for the *p*-value.

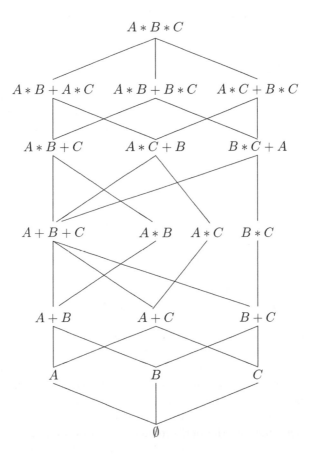

Figure 2.7: *Inclusion diagram corresponding to three-factor models. The notation is the same as used by the software R, i.e., A∗B denotes a two-factor model with interaction between the two factors.*

b) Backward selection: Starting with a full model, variables are step by step deleted from the model until all remaining variables meet a specified limit for their p-values. At each step, the variable with the largest p-value is deleted.

c) Stepwise selection: This is a modification of the forward selection principle. Variables are added to the model step by step. In each step, the procedure also examines whether variables already in the model can be deleted.

d) Best subset selection: For $k = 1, 2, \ldots$ up to a user-specified limit, the procedure identifies a specified number of best models containing k variables.

Other strategies for model selection

All of the methods described in the previous section are so-called *in-sample* methods for model selection and for evaluating the model performance. They are characterized by the fact that the model complexity is evaluated using the same observations as those used for estimating the parameters of the model. It is said that the *training data* is used for evaluating the performance of the model. In this case it is clear that any extra parameter will lead to a reduction of the loss function.

In the in-sample case statistical tests are used to access the significance of extra parameters, and when the improvement is small in some sense (as it will be described in later chapters), the parameters are considered to be non-significant. This is the classical statistical approach which will also be taken in this book.

Let us, however, for a moment briefly describe an alternative and very popular method for model selection often called *statistical learning*, or *data mining*, where it is assumed that we have a data-rich situation so that only a part of the data is needed for model estimation.

Ultimately we are seeking the so-called *generalized performance* of a model which is defined as the expected performance of an independent set of observations. The expected performance can be evaluated as the expected value of the generalized loss function. The expected prediction error on a set of independent observations is called the *test error* or *generalization error*.

In a *data-rich situation*, the performance can be evaluated by splitting the total set of observations in three parts: A *training set* (used for estimating the parameters), a *validation set* (used for *out-of-sample* model selection), and a *test set* (used for assessing the generalized performance, i.e., the performance on new data). A typical split of data is 50% for training, and, say, 25% for both validation and testing. A typical behavior of the (possibly generalized) training and test error as a function of the model complexity is shown in Figure 2.8.

Since data is often scarce, another procedure called *cross-validation (CV)*, sometimes called *rotation estimation*, is often used to assess how the results of a statistical analysis will generalize to an independent dataset. In K-fold cross-validation we first split the data in K parts of approximately the same

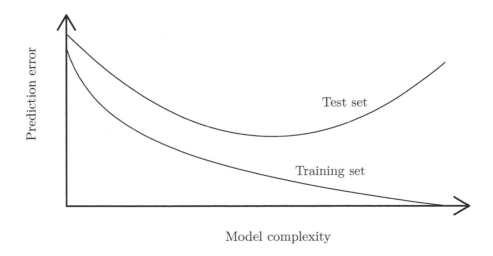

Figure 2.8: *A typical behavior of the (possibly generalized) training and test prediction error as a function of the model complexity.*

size. Then the analysis is performed on one subset (the *training set*), and the complementary $K - 1$ subsets (the *validation* or *test set*) are used for validating the analysis. In order to reduce the variability, multiple rounds of cross-validation are performed using each of the K subsets as the training set and the rest for validation. The final validation result is obtained by averaging the results over all the rounds. For the value of K typical values chosen are 3, 5 or n, where the latter corresponds to *leave-one-out* cross-validation.

In this book the methods for testing and model selection are related to *nested models*. For non-nested (or completely different) models – or even different modeling procedures – cross validation can be used to compare the performance characteristics.

For further reading about statistical learning, data mining, and cross-validation see, e.g., Hastie, Tibshirani, and Friedman (2001) and Geisser (1993).

2.13   Dealing with nuisance parameters

Nuisance parameters

In many cases, the likelihood function is a function of many parameters but our interest focuses on the estimation on one or a subset of the parameters, with the others being considered as *nuisance parameters*. For the Gaussian model for instance, we are often interested in μ but not in σ^2, and then σ^2 is considered as a nuisance parameter used for scaling only, i.e., used to account for the variability in the data.

Methods are needed to summarize the likelihood on a subset of the parameters by eliminating the nuisance parameters. Accounting for the extra uncertainty due to unknown nuisance parameters is an essential consideration, especially in small-sample cases.

In the literature several methods have been proposed to eliminate such nuisance parameters so that the likelihood can be written as a function of the parameters of interest only; the most important methods being profile, marginal and conditional likelihoods. However, there is no single technique that is acceptable in all situations, see e.g., Bayarri, DeGroot, and Kadane (1987) and Berger, Liseo, and Wolpert (1999).

Profile likelihood

It is often possible to write some parameters as functions of other parameters, thereby reducing the number of independent parameters. The function is the parameter value which maximises the likelihood given the value of the other parameters. This procedure is called concentration of the parameters and results in the concentrated likelihood function, also occasionally known as the maximized likelihood function, but most often called the *profile likelihood*.

For example, consider a regression analysis model with normally distributed errors. The most likely value of the error variance is the variance of the residuals. The residuals depend on all other parameters. Hence the variance parameter can be written as a function of the other parameters.

Unlike the marginal and conditional likelihoods, the profile likelihood is not a true likelihood as it is not based directly on a probability distribution and this leads to some less satisfactory properties. Some attempts have been made to improve this, resulting in modified profile likelihood.

The idea of profile likelihood can also be used to compute *confidence intervals* that often have better small-sample properties than those based on asymptotic standard errors calculated from the full likelihood; see also Section 2.13.

Hence, a likelihood approach to eliminating a nuisance parameter is to replace it by its maximum likelihood estimate at each fixed value of the parameter of interest, and the resulting likelihood is called the *profile likelihood*.

DEFINITION 2.15 – PROFILE LIKELIHOOD

Assume that the statistical model for the observations $Y_1, Y_2, \ldots, Y_n$ is given by the family of joint densities (2.1) where $\boldsymbol{\theta} = (\boldsymbol{\tau}, \boldsymbol{\zeta})$ and $\boldsymbol{\tau}$ denoting the parameter of interest. Then the *profile likelihood function* for $\boldsymbol{\tau}$ is the function

$$L_P(\boldsymbol{\tau}; \boldsymbol{y}) = \sup_{\boldsymbol{\zeta}} L((\boldsymbol{\tau}, \boldsymbol{\zeta}); \boldsymbol{y}) \qquad (2.52)$$

with the joint likelihood $L((\boldsymbol{\tau}, \boldsymbol{\zeta}); \boldsymbol{y})$ given by (2.17) and where the maximization is performed at a fixed value of $\boldsymbol{\tau}$.

It should be emphasized that at fixed $\boldsymbol{\tau}$, the maximum likelihood estimate of $\boldsymbol{\zeta}$ is generally a function of $\boldsymbol{\tau}$. Therefore, we can also write

$$L_P(\boldsymbol{\tau}; \boldsymbol{y}) = L((\boldsymbol{\tau}, \widehat{\boldsymbol{\zeta}}_\tau); \boldsymbol{y}). \tag{2.53}$$

We shall use a similar notation for profile likelihoods as for ordinary likelihoods, except that we shall indicate the profiling by the subscript P for profile. In particular we shall let $\ell_P(\boldsymbol{\tau}; \boldsymbol{y})$ denote the profile log-likelihood function.

Consider the partitioning of the score vector

$$\ell' = (\ell'_\tau, \ell'_\zeta)$$

where $\ell'_\tau = \partial \ell((\boldsymbol{\tau}, \boldsymbol{\zeta}))/\partial \boldsymbol{\tau}$ is a p-dimensional vector and $\ell'_\zeta = \partial \ell((\boldsymbol{\tau}, \boldsymbol{\zeta}))/\partial \boldsymbol{\zeta}$ is a $(k-p)$-dimensional vector. Under regularity conditions, $\widehat{\boldsymbol{\zeta}}_\tau$ is the solution in $\boldsymbol{\zeta}$ of $\ell'_\zeta(\boldsymbol{\tau}, \boldsymbol{\zeta}) = 0$.

The *profile score* function is

$$\frac{\partial}{\partial \boldsymbol{\tau}} \ell_P(\boldsymbol{\tau}; \boldsymbol{y}) = \ell'_\tau(\boldsymbol{\tau}, \widehat{\boldsymbol{\zeta}}_\tau) + \ell'_\zeta(\boldsymbol{\tau}, \widehat{\boldsymbol{\zeta}}_\tau) \frac{\partial}{\partial \boldsymbol{\tau}} \widehat{\boldsymbol{\zeta}}_\tau.$$

The second summand on the right-hand side vanishes, because $\ell'_\zeta(\boldsymbol{\tau}, \widehat{\boldsymbol{\zeta}}_\tau) = 0$. Therefore, the profile score is

$$\frac{\partial}{\partial \boldsymbol{\tau}} \ell_P(\boldsymbol{\tau}; \boldsymbol{y}) = \ell'_\tau(\boldsymbol{\tau}, \widehat{\boldsymbol{\zeta}}_\tau) \tag{2.54}$$

Although the profile likelihood function, $L_P(\boldsymbol{\tau})$, is an interesting tool for inference, it is not a genuine likelihood and, therefore, the universal properties of maximum-likelihood estimators are not necessarily carried over to maximum profile likelihood estimators.

In particular, the profile score function (2.54) does not have zero expectation, see Example 2.10.

Example 2.10 – Profile likelihood for the variance of a normal distribution

Assume that $Y_1, Y_2, \ldots, Y_n$ are independent, identically distributed random variables with $Y_i \sim \mathrm{N}(\mu, \sigma^2)$ with $\mu \in \mathbb{R}$ and $\sigma^2 \in \mathbb{R}_+$ unknown parameters.

The maximum likelihood estimate of the variance σ^2 is

$$\widehat{\sigma}^2 = \frac{1}{n} \sum_{i=1}^{n} (y_i - \overline{y})^2.$$

Since $\sum_{i=1}^{n} (y_i - \overline{y})^2 / \sigma^2 \sim \chi^2_{n-1}$ it follows that

$$\mathrm{E}[\widehat{\sigma}^2] = \frac{n-1}{n} \sigma^2.$$

For small values of n this is a severe underestimate. Furthermore, the profile log-likelihood of the variance

$$\ell_P(\sigma^2; \boldsymbol{y}) = -\frac{n}{2}\log(\sigma^2) - \frac{1}{2\sigma^2}\sum_{i=1}^{n}(y_i - \overline{y})^2$$

is the same as the log-likelihood of σ^2 if the mean μ is known to be $\overline{y}$. This means that we do not pay a price for not knowing μ. Hence, the bias of $\widehat{\sigma}^2$ is only a symptom of a potentially more serious problem. The bias itself can be traced from the score function

$$\ell'_{\sigma^2}(\sigma^2; \boldsymbol{y}) = \frac{\partial}{\partial\sigma^2}\ell_P(\sigma^2; \boldsymbol{y}) = -\frac{n}{2\sigma^2} + \frac{\sum_{i=1}^{n}(y_i - \overline{y})^2}{2(\sigma^2)^2}.$$

This yields

$$\mathrm{E}[\ell'_{\sigma^2}(\sigma^2; \boldsymbol{Y})] = -1/(2\sigma^2) \neq 0.$$

Thus, the profile score function does not satisfy the zero mean property (2.23).

The most important theoretical methods to eliminate nuisance parameters are via marginalizing or conditioning, and unlike the profile likelihood, the resulting likelihoods are based on the probability of the observed quantities and, hence, the resulting marginal and conditional likelihoods are true likelihoods. As will be demonstrated, these methods do not lead to the overly optimistic precision level of the profile likelihood. The marginal likelihood is introduced below.

Profile likelihood based confidence intervals

Consider again $\boldsymbol{\theta} = (\boldsymbol{\tau}, \boldsymbol{\zeta})$. From the discussion in the previous section and the result given in (2.46) on page 26 it is seen that

$$\left\{\boldsymbol{\tau}; \frac{L_P(\boldsymbol{\tau}; \boldsymbol{y})}{L(\widehat{\boldsymbol{\theta}}, \boldsymbol{y})} > \exp(-\frac{1}{2}\chi_{1-\alpha}^2(p))\right\} \tag{2.55}$$

defines a set of values of $\boldsymbol{\tau}$ (p-dimensional) that constitutes a $100(1 - \alpha)\%$ confidence region for $\boldsymbol{\tau}$. For normal distributions the confidence is exact and else it is an approximation. Likelihood base confidence intervals have the advantage of possibly being asymmetric when this is relevant, where other methods are using a quadratic (and thus symmetric) approximation – see also Figure 2.7 on page 24.

Marginal likelihood

DEFINITION 2.16 – MARGINAL LIKELIHOOD

As above, assume that the statistical model for the observations $Y_1, Y_2, \ldots, Y_n$ is given by the family of joint densities (2.1) where $\boldsymbol{\theta} = (\boldsymbol{\tau}, \boldsymbol{\zeta})$ and $\boldsymbol{\tau}$ denoting the parameter of interest. Let (U, V) be a sufficient statistic for (τ, ζ) for which the factorization

$$f_{U,V}(u, v; (\tau, \zeta)) = f_U(u; \tau) f_{V|U=u}(v; u, \tau, \zeta)$$

holds. Provided that the likehood factor which corresponds to $f_{V|U=u}(\cdot)$ can be neglected, inference about τ can be based on the marginal model for U with density $f_U(u; \tau)$. The corresponding likelihood function

$$L_M(\tau; u) = f_U(u; \tau) \tag{2.56}$$

is called the *marginal likelihood function* based on U.

Example 2.11 – Marginal likelihood for the variance of a normal distribution

Let us again consider the situation from Example 2.10 on page 35 and the sample variance

$$s^2 = \frac{1}{n-1} \sum_{i=1}^{n} (y_i - \overline{y})^2.$$

It is then evident that $\overline{Y}$ and S^2 are independent. However, $\overline{Y} \sim \mathrm{N}(\mu, \sigma^2/n)$ and $(n-1)S^2 \sim \sigma^2 \chi^2_{n-1}$ so the parameters do not separate clearly.

$$f_{\overline{Y}, S^2}(\overline{y}, s^2; \mu, \sigma^2) = f_{\overline{Y}}(\overline{y}; \mu, \sigma^2) f_{S^2}(s^2; \sigma^2)$$

If we are interested in σ^2, and μ is unknown, it is intuitive that the observed $\overline{y}$ does not carry any information about σ^2. Therefore, we can ignore $\overline{y}$, and concentrate on the marginal likelihood

$$\ell_M(\sigma^2; s^2) = -\frac{n-1}{2} \log(\sigma^2) - \frac{(n-1)s^2}{2\sigma^2} \tag{2.57}$$

that is obtained from the $\sigma^2 \chi^2_{n-1}/(n-1)$ distribution of S^2.

Differentiating $\ell_M(\sigma^2; s^2)$ above with respect to σ^2 we obtain the marginal score function

$$\frac{\partial}{\partial \sigma^2} \ell_M(\sigma^2; s^2) = -\frac{n-1}{2\sigma^2} + \frac{(n-1)s^2}{2(\sigma^2)^2} \tag{2.58}$$

leading to the maximum marginal likelihood estimate

$$\widehat{\sigma}^2_M = s^2.$$

Since $\mathrm{E}[S^2] = \sigma^2$ it follows that the marginal score function (2.58) satisfies

$$\mathrm{E}\left[\frac{\partial}{\partial \sigma^2} \ell_M(\sigma^2; S^2)\right] = 0.$$

▶ **Remark 2.11**

As $\mathrm{E}[S^2] = \sigma^2$ it follows that the maximum marginal likelihood estimate provides an unbiased estimator of σ^2.   ◀

An alternative method to eliminate nuisance parameters is to consider a conditional likelihood:

Conditional likelihood

Assume that it is possible to find a sufficient statistic for the nuisance parameters, and then by setting up a likelihood conditioning on this statistic will result in a likelihood that does not depend on the nuisance parameters.

DEFINITION 2.17 – CONDITIONAL LIKELIHOOD

As above, assume that the statistical model for the observations $Y_1, Y_2, \ldots, Y_n$ is given by the family of joint densities (2.1) where $\theta = (\tau, \zeta)$ and τ denoting the parameter of interest. Let $U = U(\boldsymbol{Y})$ be a statistic so that the factorization

$$f_Y(\boldsymbol{y}; (\tau, \zeta)) = f_U(u; \tau, \zeta) f_{Y|U=u}(\boldsymbol{y}; u, \tau)$$

holds. Provided that the likehood factor which corresponds to $f_U(\cdot)$ can be neglected, inference about τ can be based on the conditional model for $Y|U = u$ with density $f_{Y|U=u}(\cdot)$. The corresponding likelihood function

$$L_C(\tau) = L_C(\tau; \boldsymbol{y}|u) = f_{Y|U=u}(\boldsymbol{y}; u, \tau) \tag{2.59}$$

is called the *conditional likelihood function* based on conditioning on u.

2.14 Problems

Please note that some of the problems are designed in such a way that they can be solved by hand calculations while others require access to a computer.

Exercise 2.1

For assessing the bacterial density in water the following method is sometimes used: Two samples of 1 ml are prepared and one of them is diluted with sterile water to a concentration of $1/10$ of the original source. Assume that the bacterial density in the source is λ bact/ml and assume further that bacterias are randomly distributed in the source. It is natural to use the Poisson distribution to model the bacterial count in the two samples, then $Y_1 \sim \mathrm{P}(\lambda)$ and $Y_2 \sim \mathrm{P}(\lambda/10)$ with Y_1 and Y_2 denoting the bacterial count in the undiluted and the diluted sample, respectively, and Y_1 and Y_2 are independent.

Question 1 Consider the estimator

$$T_1 = \frac{Y_1 + 10Y_2}{2}.$$

Verify that this estimator is *unbiased*, i.e., that $E[T_1] = \lambda$. Determine the variance of this estimator, and compare it to using simply Y_1 as an estimator for λ. One might think that the estimator Y_1 would be improved by including also the information, Y_2, from the diluted sample. Try to explain why this is not the case.

Question 2 Now consider an unbiased estimator

$$T_w = wY_1 + (1 - w)Y_2 \quad \text{with } 0 \le w \le 1.$$

Derive an expression for the variance of this estimator, and determine the value of the weight, w, that minimizes the variance for any specific value of λ. The estimator corresponding to this value of w is termed the MVU-estimator (Minimum Variance Unbiased) estimator. What is the variance of this MVU estimator?

Question 3 Derive the likelihood function corresponding to specific values, y_1 and y_2 of the bacterial counts. Derive the maximum likelihood estimator, and compare it to the MVU-estimator above.

Observe that the estimator has a simple interpretation: The two samples contain in total 1.1 ml of the original source, and $Y_1 + Y_2$ bacteria were found in these 1.1 ml; hence, the estimator for the bacterial density is $(Y_1 + Y_2)/1.1$ bact/ml.

Exercise 2.2
Assume that $Y_1, \ldots, Y_n$ are i.i.d. log-normally distributed random variables; i.e., $\log Y_i \sim N(\mu, \sigma^2)$; $i = 1, \ldots, n$.

Question 1 Show that

$$\mu_Y = E[Y_i] = \exp\left(\mu + \frac{\sigma^2}{2}\right)$$

$$\sigma_Y^2 = E[Y_i - \mu_Y] = \mu_Y^2(\exp(\sigma^2) - 1)$$

Question 2 Show that the MLE for (μ_Y, σ_Y^2) is

$$\hat{\mu}_Y = \exp\left(\frac{1}{n}\sum_{i=1}^{n} X_t + \frac{1}{2}s\right)$$

$$\hat{\sigma}_Y^2 = \hat{\mu}_Y^2(\exp(s) - 1)$$

where

$$s = \frac{1}{n}\sum_{i=1}^{n}\left(X_i - \frac{1}{n}\sum_{k=1}^{n} X_k\right)^2$$

$$X_k = \log Y_k, \quad k = 1, \ldots, n$$

Exercise 2.3
Show that the following is valid:

$$
\begin{aligned}
i(\boldsymbol{\theta}) &= \mathrm{E}\left[\left(\frac{\partial \log f_{\boldsymbol{Y}}(\boldsymbol{y};\boldsymbol{\theta})}{\partial \boldsymbol{\theta}}\right)\left(\frac{\partial \log f_{\boldsymbol{Y}}(\boldsymbol{y};\boldsymbol{\theta})}{\partial \boldsymbol{\theta}}\right)^{T}\right] \\
&= -\mathrm{E}\left[\frac{\partial^2 \log f_{\boldsymbol{Y}}(\boldsymbol{y};\boldsymbol{\theta})}{\partial \boldsymbol{\theta}^2}\right]
\end{aligned}
\tag{2.60}
$$

Exercise 2.4
In a customer survey, a company has randomly chosen 10 customers and asked them their opinion about some service the company provides. Among the 10 customers, 3 were unsatisfied with the service. Write down expressions for the likelihood function, the log-likelihood function, the score function, the observed and the expected information.

Exercise 2.5
The following are measurements of tensile strength on laboratory prepared soil samples:

75   78   86   82   76   82   76   75   73   83   92   77

Assume that the data are an i.i.d. sample from $N(\theta, \sigma^2)$, where σ^2 is assumed to be known at the observed sample variance. Write down and sketch the likelihood function for θ if

Question 1 all of the data is given

Question 2 only the sample mean is given

Exercise 2.6
Sometimes it is convenient to use a maximum likelihood method in order to find a transformation that may linearize models, correct for heteroscedasticity and normalize the distribution of the residuals. This will be used in this exercise, where the so-called Box and Cox transformation, Box and Tiao (1975), is considered. The model considered in Box and Tiao (1975) is the following:

$$
\boldsymbol{y}^\lambda = \boldsymbol{X}\boldsymbol{\beta} + \boldsymbol{\epsilon}
$$

where the transformed value is

$$
Y_i^{(\lambda)} = \begin{cases} \dfrac{Y_i^\lambda - 1}{\lambda} & \text{for } \lambda \neq 0 \\[2mm] \log(Y_i) & \text{for } \lambda = 0 \end{cases}
$$

and $\boldsymbol{\epsilon} \in N_n(0, \sigma^2 \boldsymbol{I}_n)$. Derive the MLE for λ.

CHAPTER 3

General linear models

3.1 Introduction

Several of the methods considered during elementary courses on statistics can be collected under the heading *general linear models* or *classical* GLM. The classical GLM leads to a unique way of describing the variations of experiments with a *continuous* variable. The classical GLMs include

i) *regression analysis*
   Explanatory variables: *Continuous* or *categorical.*

ii) *analysis of variance* ANOVA
   Explanatory variables: *Ordinal* or *nominal* (i.e., *discrete*).

iii) *analysis of covariance* ANCOVA
   Explanatory variables: *Nominal* or *continuous.*

Classical GLMs have been investigated since the days of Carl Friedrich Gauss (1777–1855). The models have an intuitive appeal because effects are modelled as additive, and also because ordinary least squares (OLS) estimation methods lead to explicit solutions for the estimates. Ordinary least squares estimation method is sometimes invoked because use of the method does not require distributional assumptions.

In mathematical statistics, classical GLMs are naturally studied in the framework of the multivariate normal distribution. The multivariate normal distribution is well suited for modeling linear mean-value structure for observations in a vector space that is equipped with an inner product. This is the approach we will take in this chapter.

In this chapter we shall consider the set of n observations as a sample from a n-dimensional normal distribution. In this framework of *linear models* we shall see that classical GLMs may be understood as models restricting the n-dimensional mean-value of the observation to lie in a linear (affine) subspace of $\mathbb{R}^n$.

In the normal distribution model, maximum-likelihood estimation of mean value parameters may be interpreted geometrically as *projection* on an appropriate subspace, and the likelihood-ratio test statistics for model reduction may be expressed in terms of *norms* of these projections.

In particular we shall demonstrate how a test for model reduction may be formulated as tests in a chain of hypotheses, corresponding to partitioning the projection on the innermost hypothesis into a sequence of successive projections, and a comparison of the corresponding deviances.

We shall introduce various techniques for model-checking, and, finally, we shall discuss different equivalent representations of classical GLMs.

3.2   The multivariate normal distribution

Let $Y = (Y_1, Y_2, \ldots, Y_n)^T$ be a random vector with $Y_1, Y_2, \ldots, Y_n$ independent identically distributed (i.i.d.) $N(0, 1)$ random variables. Note that $E[Y] = 0$ and the variance-covariance matrix $\text{Var}[Y] = I$.

Using the introduced random vector Y with i.i.d. $N(0, 1)$ random variables, we are able to define the multivariate normal distribution.

DEFINITION 3.1 – MULTIVARIATE NORMAL DISTRIBUTION
Z has a k-dimensional multivariate normal distribution if Z has the same distribution as $AY + b$ for some n, some $k \times n$ matrix A, and some k dimensional vector b. We indicate the multivariate normal distribution by writing $Z \sim N(b, AA^T)$.

Since A and b are fixed, we have $E[Z] = b$ and $\text{Var}[Z] = AA^T$. Let us assume that the variance-covariance matrix is known apart from a constant factor, σ^2, i.e., $\text{Var}[Z] = \sigma^2 \Sigma$. The density for the k-dimensional random vector Z with mean μ and covariance $\sigma^2 \Sigma$ is

$$f_Z(z) = \frac{1}{(2\pi)^{k/2} \sigma^k \sqrt{\det \Sigma}} \exp\left[-\frac{1}{2\sigma^2} (z - \mu)^T \Sigma^{-1} (z - \mu) \right] \tag{3.1}$$

where Σ is seen to be (a) symmetric and (b) positive semi-definite. We write $Z \sim N_k(\mu, \sigma^2 \Sigma)$.

▸ **Remark 3.1 – Σ known**
Note that it is assumed that Σ is known.   ◂

▸ **Remark 3.2 – Σ is a weight matrix**
As seen from (3.1), Σ is used for scaling of the deviations from the *mean μ* and, hence, the matrix Σ will often be denoted as *weight matrix*. The case $\Sigma = I$ is considered as an unweighted problem.   ◂

The normal density as a statistical model

Consider now the n observations $Y = (Y_1, Y_2, \ldots, Y_n)^T$, and assume that a statistical model is

$$Y \sim N_n(\boldsymbol{\mu}, \sigma^2 \boldsymbol{\Sigma}). \tag{3.2}$$

For reasons which will be clear later on, the variance-covariance matrix for the observations is called the *dispersion matrix* and denoted $D[Y]$, i.e., the dispersion matrix for Y is

$$D[Y] = \sigma^2 \boldsymbol{\Sigma}.$$

DEFINITION 3.2 – INNER PRODUCT AND NORM
The bilinear form

$$\delta_{\Sigma}(\boldsymbol{y}_1, \boldsymbol{y}_2) = \boldsymbol{y}_1^T \boldsymbol{\Sigma}^{-1} \boldsymbol{y}_2 \tag{3.3}$$

defines an *inner product* in $\mathbb{R}^n$. Corresponding to this inner product we can define *orthogonality*, which is obtained when the inner product is zero, and a *norm* defined by $\|\boldsymbol{y}\|_{\Sigma} = \sqrt{\delta_{\Sigma}(\boldsymbol{y}, \boldsymbol{y})}$.

DEFINITION 3.3 – DEVIANCE FOR NORMAL DISTRIBUTED VARIABLES
Let us introduce the notation

$$D(\boldsymbol{y}; \boldsymbol{\mu}) = \delta_{\Sigma}(\boldsymbol{y} - \boldsymbol{\mu}, \boldsymbol{y} - \boldsymbol{\mu}) = (\boldsymbol{y} - \boldsymbol{\mu})^T \boldsymbol{\Sigma}^{-1}(\boldsymbol{y} - \boldsymbol{\mu}) \tag{3.4}$$

to denote the quadratic norm of the vector $(\boldsymbol{y} - \boldsymbol{\mu})$ corresponding to the inner product defined by $\boldsymbol{\Sigma}^{-1}$.

For a normal distribution with $\boldsymbol{\Sigma} = \boldsymbol{I}$, the deviance is just the Residual Sum of Squares (RSS).

Using this notation the normal density is expressed as a density defined on any finite dimensional vector space equipped with the inner product, δ_{Σ}, and we write

$$f(\boldsymbol{y}; \boldsymbol{\mu}, \sigma^2) = \frac{1}{(\sqrt{2\pi})^n \sigma^n \sqrt{\det(\boldsymbol{\Sigma})}} \exp\left[-\frac{1}{2\sigma^2} D(\boldsymbol{y}; \boldsymbol{\mu})\right]. \tag{3.5}$$

Likelihood function

From (3.5) we obtain the likelihood function

$$L(\boldsymbol{\mu}, \sigma^2; \boldsymbol{y}) = \frac{1}{(\sqrt{2\pi})^n \sigma^n \sqrt{\det(\boldsymbol{\Sigma})}} \exp\left[-\frac{1}{2\sigma^2} D(\boldsymbol{y}; \boldsymbol{\mu})\right]. \tag{3.6}$$

Since the likelihood function expresses a measure of "rational belief" about the parameters, and not the probability density for the parameters, the constants are often disregarded.

Score function and information matrix for μ

The log-likelihood function is (apart from an additive constant)

$$
\begin{aligned}
\ell_{\mu,\sigma^2}(\mu,\sigma^2;y) &= -(n/2)\log(\sigma^2) - \frac{1}{2\sigma^2}(y-\mu)^T\Sigma^{-1}(y-\mu) \\
&= -(n/2)\log(\sigma^2) - \frac{1}{2\sigma^2}\mathrm{D}(y;\mu).
\end{aligned}
\tag{3.7}
$$

The score function wrt. μ is thus

$$
\frac{\partial}{\partial\mu}\,\ell_{\mu,\sigma^2}(\mu,\sigma^2;y) = \frac{1}{\sigma^2}\left[\Sigma^{-1}y - \Sigma^{-1}\mu\right] = \frac{1}{\sigma^2}\Sigma^{-1}(y-\mu)
\tag{3.8}
$$

and the observed information (wrt. μ) is

$$
j(\mu;y) = \frac{1}{\sigma^2}\Sigma^{-1}\,.
\tag{3.9}
$$

It is seen that the observed information does not depend on the observations y. Hence, the expected information is

$$
i(\mu) = \frac{1}{\sigma^2}\Sigma^{-1}\,.
$$

The maximum likelihood estimate of μ is found by minimizing the deviance $\mathrm{D}(y;\mu)$ wrt. μ. Since the vector space is equipped with the inner product δ_Σ the estimate is found as the projection onto the (sub-)space spanning μ.

However, for the actual model (3.2) the MLE of μ is seen to be the vector of observations (see (3.8)).

DEFINITION 3.4 – THE FULL (OR SATURATED) MODEL
In the full (or saturated) model μ varies freely in $\mathbb{R}^n$, and, hence, each observation fits the model perfectly, i.e., $\widehat{\mu} = y$.

The full model is used as a benchmark for assessing the fit of any model to the data. This is done by calculating the *deviance* between the model and the data. In the case of normal distributed observations with $\Sigma = I$ we have already noted that the deviance becomes the Residual Sum of Squares (RSS).

3.3 General linear models

Statistical models

Models play an important role in statistical inference. A model is a mathematical way of describing the variation of the observations. In the case of a normal density the observation Y_i is most often written as

$$
Y_i = \mu_i + \epsilon_i
\tag{3.10}
$$

which for all n observations $(Y_1, Y_2, \ldots, Y_n)^T$ can be written on the matrix form

$$Y = \mu + \epsilon. \tag{3.11}$$

By assuming that

$$E[\epsilon] = 0 \tag{3.12}$$

$$D[\epsilon] = \sigma^2 \Sigma \tag{3.13}$$

i.e., a variance-covariance structure for ϵ, as described in Section 3.2 on page 42, we shall write

$$Y \sim N_n(\mu, \sigma^2 \Sigma) \text{ for } y \in \mathbb{R}^n. \tag{3.14}$$

If, for all the n observations, the value of μ_i varies freely, then the model in (3.10) is the full model.

There is a strong link between normally distributed variables and linearity, where the conditional mean $E[Y|X]$ is linear in X. We have the following theorem.

THEOREM 3.1 – PROJECTIONS FOR NORMALLY DISTRIBUTED VARIABLES
Let $(Y^T, X^T)^T$ be a normally distributed random variable with mean

$$\begin{pmatrix} \mu_Y \\ \mu_X \end{pmatrix} \quad \text{and covariance} \quad \begin{pmatrix} \Sigma_{YY} & \Sigma_{YX} \\ \Sigma_{XY} & \Sigma_{XX} \end{pmatrix}$$

Then $Y|X$ is normally distributed with mean

$$E[Y|X] = \mu_Y + \Sigma_{YX}\Sigma_{XX}^{-1}(X - \mu_X) \tag{3.15}$$

and variance

$$\text{Var}[Y|X] = \Sigma_{YY} - \Sigma_{YX}\Sigma_{XX}^{-1}\Sigma_{YX}^T \tag{3.16}$$

Furthermore, the error, $(Y - E[Y|X])$, and X are independent.

Proof Omitted. See, for instance, Jazwinski (1970). ∎

▶ **Remark 3.3**
Without the assumption about normality, the error $(Y - E[Y|X])$ and X were uncorrelated; but since we here consider normally distribution variables, they are also independent. In the geometric illustrations later on we say that $(Y - E[Y|X])$ and X are *orthogonal*. ◀

▶ **Remark 3.4**
A more mathematical treatment of the projection theorem involves inner product spaces; see Madsen (1992) or Brockwell and Davis (1987). ◀

In this chapter we shall study so-called *linear models*, i.e., models where it is assumed that $\boldsymbol{\mu}$ belongs to a linear (or affine) subspace Ω_0 of $\mathbb{R}^n$. In Section 3.2 we introduced the *full model* with $\Omega_{full} = \mathbb{R}^n$, and hence each observation fits the model perfectly, i.e., $\widehat{\boldsymbol{\mu}} = \boldsymbol{y}$.

In general a more restricted model is useful. An example is the most restricted model, namely the *null model* which only describes the variations of the observations by a common mean value for all observations.

DEFINITION 3.5 – THE NULL MODEL
In the null model $\boldsymbol{\mu}$ varies in $\mathbb{R}$, and, hence, a model with one parameter representing a common mean value for all observations. In this case the model-subspace is $\Omega_{null} = \mathbb{R}$, i.e., the dimension is one.

In most cases the variation is described by a model somewhere between the null model and the full model, as considered in the next section.

The general linear model

In the classical GLM the variation is (partly) described by a mathematical model describing a relationship between a *response variable* (or dependent variable) and a set of *explanatory variables* (or independent variables). In this case the dimension of the subspace Ω_0 spanning the variations of $\boldsymbol{\mu}$ depends on the complexity of the model, which is somewhere between the null and the full model.

DEFINITION 3.6 – THE GENERAL LINEAR MODEL
Assume that $Y_1, Y_2, \ldots, Y_n$ is normally distributed as described by (3.2). A *general linear model* for $Y_1, Y_2, \ldots, Y_n$ is a model of the form (3.2), where an affine hypothesis is formulated for $\boldsymbol{\mu}$. The hypothesis is of the form

$$\mathcal{H}_0 \; : \; \boldsymbol{\mu} - \boldsymbol{\mu}_0 \in \Omega_0 \,, \tag{3.17}$$

where Ω_0 is a linear subspace of $\mathbb{R}^n$ of dimension k, and where $\boldsymbol{\mu}_0$ denotes a vector of *a priori known offset values*.

DEFINITION 3.7 – DIMENSION OF GENERAL LINEAR MODEL
The dimension of the subspace Ω_0 for the linear model (3.17) is the *dimension of the model*.

DEFINITION 3.8 – DESIGN MATRIX FOR CLASSICAL GLM
Assume that the linear subspace $\Omega_0 = \text{span}\{x_1, \ldots, x_k\}$, i.e., the subspace is spanned by k vectors $(k < n)$.

Consider a general linear model (3.17) where the hypothesis can be written

$$\mathcal{H}_0 \colon \boldsymbol{\mu} - \boldsymbol{\mu}_0 = \boldsymbol{X}\boldsymbol{\beta} \quad \text{with } \boldsymbol{\beta} \in \mathbb{R}^k, \tag{3.18}$$

where X has full rank. The $n \times k$ matrix X of known deterministic coefficients is called the *design matrix*.

The ith row of the design matrix is given by the *model vector*

$$x_i^T = \begin{pmatrix} x_{i1} \\ x_{i2} \\ \vdots \\ x_{ik} \end{pmatrix}^T , \tag{3.19}$$

for the ith observation.

The *offset* specifies terms in the model for which the coefficients are known. In statistical software the offset may be specified in the model formula. In this chapter we shall, however, assume that the *offset* $\mu_0 = 0$. In the chapter dealing with generalized linear models, the offset term becomes useful, for instance, in Poisson regression.

▸ **Remark 3.5 – Full model**
For the full model $X = I_n$.   ◂

▸ **Remark 3.6 – Null model**
For the null model $X = (1, 1, \ldots, 1)^T$.   ◂

Example 3.1 – Multiple linear regression
The multiple linear regression model

$$y_i = \beta_0 + \beta_1 x_{i1} + \beta_2 x_{i2} + \epsilon_i \tag{3.20}$$

defines a general linear model with

$$x_i = (1, x_{i1}, x_{i2})^T \tag{3.21}$$

$$\beta = (\beta_0, \beta_1, \beta_2)^T. \tag{3.22}$$

▸ **Remark 3.7 – Design matrix not full rank**
The design matrix X need not have full rank. In this case several possible solutions exist. However, all solutions provide the same prediction $\hat{\mu}$. A unique solution can be obtained by defining restrictions on the relations between the parameters.   ◂

DEFINITION 3.9 – ESTIMABLE FUNCTION
Consider a general linear model parameterized as in (3.18). A linear combination $\psi = c^T \beta$ is said to be *estimable* if there exists a linear combination $a^T y$ of the observations such that $E[a^T Y] = c^T \beta$ for all values of β.

If $\boldsymbol{X}$ is of full rank, then all linear combinations are estimable.

Example 3.2 – One-way ANOVA
The models related to the ANalysis Of VAriance (one-way comparison of means) are also a classical GLM.

For r groups the classical GLM is specified by

$$\boldsymbol{x}_i = (1, x_{i1}, x_{i2}, \ldots, x_{ir})^T \tag{3.23}$$

$$\boldsymbol{\beta} = (\beta_0, \beta_1, \beta_2, \ldots, \beta_r)^T \tag{3.24}$$

with $x_{ij} = 1$ if the i'th observation belongs to group j. This design matrix is not of full rank since $\sum_j x_{ij} = 1$. Thus, the model is over-parameterized, and it is necessary to impose some restrictions on the parameters in order to obtain identifiability of parameters. In Section 3.11 we shall consider other parameterizations of this model.

▸ **Remark 3.8 – Fixed effects model**
The model in Example 3.2 is a so-called *fixed effects model* since the effect is specific or linked to the actual group, and a parameter estimate is found (and reasonable) for each group. ◂

Later on in Chapter 5 also *random effects models* will be introduced. This is a *variance components model* which is a kind of *hierarchical model* where the grouping being analyzed consists of different populations and the actual groups are considered as random outcomes of analyzing a (typical small) number of groups within these populations, and the variations between groups are described by parameters characterizing the variation between groups. For Example 3.2 this will lead to estimation of a variance parameter representing the variation between groups rather than estimating r individual mean values.

3.4   Estimation of parameters

Estimation of mean value parameters

Under the hypothesis

$$\mathcal{H}_0 \colon \boldsymbol{\mu} \in \Omega_0,$$

the maximum likelihood estimate for the set $\boldsymbol{\mu}$ is found as the orthogonal projection (with respect to δ_Σ), $p_0(\boldsymbol{y})$ of $\boldsymbol{y}$ onto the linear subspace Ω_0.

THEOREM 3.2 – ML ESTIMATES OF MEAN VALUE PARAMETERS
For a hypothesis of the form

$$\mathcal{H}_0 \colon \boldsymbol{\mu}(\boldsymbol{\beta}) = \boldsymbol{X}\boldsymbol{\beta} \tag{3.25}$$

the maximum likelihood estimated for β is found as a solution to the normal equation

$$X^T \Sigma^{-1} y = X^T \Sigma^{-1} X \widehat{\beta}. \tag{3.26}$$

If X has full rank, the solution is uniquely given by

$$\widehat{\beta} = (X^T \Sigma^{-1} X)^{-1} X^T \Sigma^{-1} y. \tag{3.27}$$

Proof By considering the expression (3.6) for the likelihood function, it is readily seen that the maximum wrt. μ is found by minimizing the deviance $D(y; \mu)$ wrt. $\mu \in \Omega_0$.

Since $D(y; \mu)$ is a norm with the inner product, δ_Σ, the minimum is obtained as the projection onto Ω_0.

In the parameterized case (3.18) we find, using (2.32), since

$$\frac{\partial}{\partial \beta} \mu(\beta) = \frac{\partial}{\partial \beta} X\beta = X$$

that

$$\begin{aligned}
\frac{\partial}{\partial \beta} \ell_\beta(\beta, \sigma^2; y) &= \left[\frac{\partial \mu}{\partial \beta}\right]^T \frac{\partial}{\partial \mu} \ell_\mu(\mu(\beta), \sigma^2; y) \\
&= \frac{1}{\sigma^2} X^T \left[\Sigma^{-1} y - \Sigma^{-1} X\beta\right]
\end{aligned} \tag{3.28}$$

since the score function with respect to μ is given by (3.8). It is seen, that (3.28) equals zero for β satisfying the normal equation (3.26). ∎

We shall consider estimation of the variance, σ^2 in Section 3.4 on page 53.

Properties of the ML estimator

THEOREM 3.3 – PROPERTIES OF THE ML ESTIMATOR
For the ML estimator we have

$$\widehat{\beta} \sim N_k(\beta, \sigma^2 (X^T \Sigma^{-1} X)^{-1}). \tag{3.29}$$

Proof The normality follows simply by the fact that the estimator is a linear combination of the normally distributed observations. The mean and dispersion follows by direct calculation. ∎

It follows from (3.29) that

$$E[\widehat{\beta}] = \beta \tag{3.30}$$

$$D[\widehat{\beta}] = \sigma^2 (X^T \Sigma^{-1} X)^{-1}. \tag{3.31}$$

▸ **Remark 3.9 – Orthogonality**

The maximum likelihood estimate for β is found as the value of β which minimizes the *distance* $\|y - X\beta\|$.

The normal equations (3.26) show that

$$X^T \Sigma^{-1}(y - X\widehat{\beta}) = 0 \tag{3.32}$$

i.e., the *residuals* (3.36) are orthogonal (with respect to Σ^{-1}) to the subspace Ω_0.   ◂

The residuals are thus orthogonal to the fitted – or predicted – values (see also Figure 3.1 on page 52).

▸ **Remark 3.10 – Unknown Σ**

Notice that it has been assumed that Σ is known. If Σ is unknown, one possibility is to use the relaxation algorithm described in Madsen (2008).   ◂

▸ **Remark 3.11 – Weighted to unweighted**

A weighted problem can be expressed as an unweighted problem where $\Sigma = I$ by a suitable transformation of the variables. Since Σ is symmetric and positive semi-definite we are able to find a linear transformation T, determined by $\Sigma = TT^T$, such that ϵ in (3.13) on page 45 can be written as $\epsilon = Te$. That is,

$$Y = X\beta + Te$$

where $\text{Var}[e] = \sigma^2 I$. By multiplying both sides of the above equation by T^{-1} (assuming it exists), the weighted (least squares) problem is brought into an unweighted (least squares) problem.   ◂

Fitted values

Notice, that in this subsection we will assume that $\Sigma = I$.

▸ **Remark 3.12 – Fitted – or predicted – values**

The *fitted* values $\widehat{\mu} = X\widehat{\beta}$ are found as the projection of y (denoted $p_0(y)$) onto the subspace Ω_0 spanned by X, and $\widehat{\beta}$ denotes the local coordinates for the projection.   ◂

DEFINITION 3.10 – PROJECTION MATRIX

A matrix H is a *projection matrix* if and only if (a) $H^T = H$ and (b) $H^2 = H$, i.e., the matrix is *idempotent*.

It is readily seen that the matrix

$$H = X[X^T X]^{-1} X^T \tag{3.33}$$

is a projection matrix. Note, that the projection matrix provides the predicted values $\widehat{\mu}$, since

$$\widehat{\mu} = p_0(y) = X\widehat{\beta} = Hy. \tag{3.34}$$

It follows that the predicted values are normally distributed with

$$D[X\widehat{\beta}] = \sigma^2 X[X^T X]^{-1} X^T = \sigma^2 H. \tag{3.35}$$

The matrix H is often termed the *hat-matrix* since it transforms the observations y to their predicted values symbolized by a "hat" on the μ's.

Residuals

In this subsection we will assume that $\Sigma = I$.
The observed *residuals* are

$$r = y - X\widehat{\beta} = (I - H)y. \tag{3.36}$$

Furthermore, the residuals $r = (I - H)Y$ are normally distributed with

$$D[r] = \sigma^2 (I - H). \tag{3.37}$$

Note, that the individual residuals do not have the same variance.

The residuals belong to a subspace of dimension $n - k$, which is orthogonal to Ω_0. It may be shown (see Theorem 3.4) that the distribution of the residuals r is independent of the fitted values $X\widehat{\beta}$.

THEOREM 3.4 – COCHRAN'S THEOREM
Suppose that $Y \sim N_n(0, I_n)$ *(i.e., standard multivariate Gaussian random variable)*

$$Y^T Y = Y^T H_1 Y + Y^T H_2 Y + \cdots + Y^T H_k Y$$

where H_i *is a symmetric* $n \times n$ *matrix with rank* n_i, $i = 1, 2, \ldots, k$.
Then any one of the following conditions implies the other two:

i) The ranks of the H_i *adds to* n, *i.e.,* $\sum_{i=1}^{k} n_i = n$

ii) Each quadratic form $Y^T H_i Y \sim \chi_{n_i}^2$ *(thus the* H_i *are positive semidefinite)*

iii) All the quadratic forms $Y^T H_i Y$ *are independent (necessary and sufficient condition).*

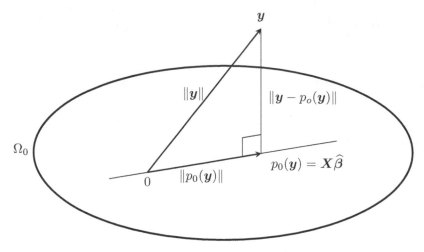

Figure 3.1: *Orthogonality between the residual $(y - X\widehat{\beta})$ and the vector $X\widehat{\beta}$.*

Proof See, e.g., Shao (1999) p. 29.   ■

Example 3.3 – Partitioning using the projection matrix
Assume that a model for zero mean normally i.i.d. observations Y is leading to a projection matrix H. Then Cochran's theorem shows that

$$Y^T Y = Y^T H Y + Y^T (I - H) Y \tag{3.38}$$

i.e., the total variation $Y^T Y$ is separated into the variation described by the model $Y^T H Y$ and the residual variation $Y^T (I-H)Y$. Furthermore the two quadratic forms on the right hand side are independent and χ^2-distributed.

Partitioning of variation

▶ **Remark 3.13 – Partitioning of the variation**
From the discussion above it follows that (Pythagoras)

$$
\begin{aligned}
D(y; X\beta) &= D(y; X\widehat{\beta}) + D(X\widehat{\beta}; X\beta) \\
&= (y - X\widehat{\beta})^T \Sigma^{-1} (y - X\widehat{\beta}) \\
&\quad + (\widehat{\beta} - \beta)^T X^T \Sigma^{-1} X (\widehat{\beta} - \beta) \\
&\geq (y - X\widehat{\beta})^T \Sigma^{-1} (y - X\widehat{\beta}).
\end{aligned}
\tag{3.39}
$$

      ◀

The partitioning and orthogonality are illustrated on Figure 3.1 for $\beta = 0$.

▶ **Remark 3.14 – χ^2-distribution of individual contributions in** (3.39)
Under $\mathcal{H}_0$, as in (3.25), it follows from the normal distribution of $\boldsymbol{Y}$ that

$$\mathrm{D}(\boldsymbol{y}; \boldsymbol{X\beta}) = (\boldsymbol{y} - \boldsymbol{X\beta})^T \boldsymbol{\Sigma}^{-1} (\boldsymbol{y} - \boldsymbol{X\beta}) \sim \sigma^2 \chi_n^2.$$

Furthermore, it follows from the normal distribution of $\boldsymbol{r}$ and of $\widehat{\boldsymbol{\beta}}$ that

$$\mathrm{D}(\boldsymbol{y}; \boldsymbol{X\widehat{\beta}}) = (\boldsymbol{y} - \boldsymbol{X\widehat{\beta}})^T \boldsymbol{\Sigma}^{-1} (\boldsymbol{y} - \boldsymbol{X\widehat{\beta}}) \sim \sigma^2 \chi_{n-k}^2$$

$$\mathrm{D}(\boldsymbol{X\widehat{\beta}}; \boldsymbol{X\beta}) = (\widehat{\boldsymbol{\beta}} - \boldsymbol{\beta})^T \boldsymbol{X}^T \boldsymbol{\Sigma}^{-1} \boldsymbol{X} (\widehat{\boldsymbol{\beta}} - \boldsymbol{\beta}) \sim \sigma^2 \chi_k^2.$$

Moreover, the independence of $\boldsymbol{r}$ and $\boldsymbol{X\widehat{\beta}}$ implies that $\mathrm{D}(\boldsymbol{X\widehat{\beta}}; \boldsymbol{X\beta})$ and $\mathrm{D}(\boldsymbol{y}; \boldsymbol{X\widehat{\beta}})$ are independent.

Thus, (3.39) represents a partition of the $\sigma^2 \chi_n^2$-distribution on the left side into two independent χ^2-distributed variables with $n - k$ and k degrees of freedom, respectively.   ◀

Estimation of the residual variance σ^2

THEOREM 3.5 – ESTIMATION OF THE VARIANCE
Under the hypothesis (3.25) *the maximum marginal likelihood estimator for the variance σ^2 is*

$$\widehat{\sigma}^2 = \frac{\mathrm{D}(\boldsymbol{y}; \boldsymbol{X\widehat{\beta}})}{n - k} = \frac{(\boldsymbol{y} - \boldsymbol{X\widehat{\beta}})^T \boldsymbol{\Sigma}^{-1} (\boldsymbol{y} - \boldsymbol{X\widehat{\beta}})}{n - k}. \tag{3.40}$$

Under the hypothesis $\widehat{\sigma}^2 \sim \sigma^2 \chi_f^2 / f$ with $f = n - k$.

Proof It follows from considerations analogous to Example 2.11 on page 37 and Remark 3.9 on page 50 that the marginal likelihood corresponds to the $\sigma^2 \chi_{n-k}^2$ distribution of $\mathrm{D}(\boldsymbol{y}; \boldsymbol{X\widehat{\beta}})$.   ■

3.5 Likelihood ratio tests

In the classical GLM case the exact distribution of the likelihood ratio test statistic (2.46) may be derived.

Consider the following model for the data $\boldsymbol{Y} \sim \mathrm{N}_n(\boldsymbol{\mu}, \sigma^2 \boldsymbol{\Sigma})$. Let us assume that we have the model

$$\mathcal{H}_1 : \boldsymbol{\mu} \in \Omega_1 \subset \mathbb{R}^n$$

with $\dim(\Omega_1) = m_1$. Now we want to test whether the model may be reduced to a model where $\boldsymbol{\mu}$ is restricted to some subspace of Ω_1, and hence we introduce $\Omega_0 \subset \Omega_1$ as a linear (affine) subspace with $\dim(\Omega_0) = m_0$. The situation is shown in Figure 3.2 on the following page. Then we have:

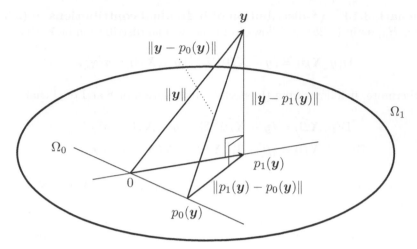

Figure 3.2: *Model reduction. The partitioning of the deviance corresponding to a test of the hypothesis $\mathcal{H}_0 : \mu \in \Omega_0$ under the assumption of $\mathcal{H}_1 : \mu \in \Omega_1$.*

Theorem 3.6 – A test for model reduction
The likelihood ratio test statistic for testing

$$\mathcal{H}_0 : \boldsymbol{\mu} \in \Omega_0 \quad \text{against the alternative} \quad \mathcal{H}_1 : \boldsymbol{\mu} \in \Omega_1 \setminus \Omega_0$$

is a monotone function of

$$F(\boldsymbol{y}) = \frac{\mathrm{D}(p_1(\boldsymbol{y}); p_0(\boldsymbol{y}))/(m_1 - m_0)}{\mathrm{D}(\boldsymbol{y}; p_1(\boldsymbol{y}))/(n - m_1)} \tag{3.41}$$

where $p_1(\boldsymbol{y})$ and $p_0(\boldsymbol{y})$ denote the projections of $\boldsymbol{y}$ on Ω_1 and Ω_0, respectively. Under $\mathcal{H}_0$ we have

$$F \sim \mathrm{F}(m_1 - m_0, n - m_1) \tag{3.42}$$

i.e., large values of F reflect a conflict between the data and $\mathcal{H}_0$, and hence lead to rejection of $\mathcal{H}_0$. The p-value of the test is found as $p = \mathrm{P}[\mathrm{F}(m_1 - m_0, n - m_1) \geq F_{obs}]$, where F_{obs} is the observed value of F given the data.

Proof Under $\mathcal{H}_1$ the maximum of the likelihood-function is found for $\widehat{\boldsymbol{\mu}} = p_1(\boldsymbol{y})$, with corresponding maximum-likelihood estimate $\widehat{\sigma}^2 = \mathrm{D}(\boldsymbol{y}; p_1(\boldsymbol{y}))/n$ such that the maximum of the log-likelihood function is, apart from an additive constant

$$\ell(\widehat{\boldsymbol{\mu}}; \widehat{\sigma}^2) = -(n/2) \log(\widehat{\sigma}^2) - \frac{n}{2}.$$

Analogously it is seen that the maximum of the log-likelihood function under $\mathcal{H}_0$ is

$$\ell(\widehat{\widehat{\boldsymbol{\mu}}}; \widehat{\widehat{\sigma}}^2) = -(n/2) \log(\widehat{\widehat{\sigma}}^2) - \frac{n}{2}$$

with $\widehat{\widehat{\sigma}}^2 = D(\boldsymbol{y}; p_0(\boldsymbol{y}))/n$.

Hence, the value of the likelihood ratio is

$$\lambda(\boldsymbol{y}) = \left[\frac{D(\boldsymbol{y}; p_1(\boldsymbol{y}))}{D(\boldsymbol{y}; p_0(\boldsymbol{y}))}\right]^{n/2},$$

but as $p_1(\boldsymbol{y}) \in \Omega_1$, and $p_0(\boldsymbol{y}) \in \Omega_0 \subset \Omega_1$, then also $p_1(\boldsymbol{y}) - p_0(\boldsymbol{y}) \in \Omega_1$. As moreover the vector of residuals $\boldsymbol{y} - p_1(\boldsymbol{y})$ is orthogonal to Ω_1, it follows that $\boldsymbol{y} - p_1(\boldsymbol{y})$ and $p_1(\boldsymbol{y}) - p_0(\boldsymbol{y})$ are orthogonal, and there we have the Pythagorean relation

$$D(\boldsymbol{y}; p_0(\boldsymbol{y})) = D(\boldsymbol{y}; p_1(\boldsymbol{y})) + D(p_1(\boldsymbol{y}); p_0(\boldsymbol{y})) \tag{3.43}$$

such that we have

$$\lambda(\boldsymbol{y}) = \left[\frac{D(\boldsymbol{y}; p_1(\boldsymbol{y}))}{D(\boldsymbol{y}; p_1(\boldsymbol{y})) + D(p_1(\boldsymbol{y}); p_0(\boldsymbol{y}))}\right]^{n/2}$$

$$= \left[\frac{1}{1 + \dfrac{D(p_1(\boldsymbol{y}); p_0(\boldsymbol{y}))}{D(\boldsymbol{y}; p_1(\boldsymbol{y}))}}\right]^{n/2}$$

$$= \left[\frac{1}{1 + \dfrac{m_1 - m_0}{n - m_1}F}\right]^{n/2}$$

with F given by (3.41). Thus, the larger the value of F, the smaller the value of the likelihood ratio test statistic $\lambda(\boldsymbol{y})$. The fact that F is F-distributed follows from the fact that F is a ratio between two independent χ^2-distributed random quantities. ∎

Traditionally, the partitioning of the variation is presented in a Deviance table (or an *ANalysis Of VAriance table*, ANOVA) as illustrated in Table 3.1. The table reflects the partitioning above in the test for a *model reduction*, and what happens is, loosely spoken, that the deviance between the variation of the model from the hypothesis is measured using the deviance of the observations from the model as a reference. Under $\mathcal{H}_0$ they both are χ^2-distributed, and it is readily seen that they are orthogonal and thus independent, which means that the ratio is F-distributed. If the test quantity is large this shows evidence against the model reduction tested using $\mathcal{H}_0$.

▸ **Remark 3.15 – The test is a conditional test**
It should be noted that the test has been derived as a *conditional test*. It is a test for the hypothesis $\mathcal{H}_0 : \boldsymbol{\mu} \in \Omega_0$ under the assumption that $\mathcal{H}_1 : \boldsymbol{\mu} \in \Omega_1$ is true. The test does in no way assess whether $\mathcal{H}_1$ is in agreement with the data. On the contrary in the test the residual variation under $\mathcal{H}_1$ is used to estimate σ^2, i.e., to assess $D(\boldsymbol{y}; p_1(\boldsymbol{y}))$. ◂

Table 3.1: *Deviance table corresponding to a test for model reduction as specified by* $\mathcal{H}_0$. *For* $\Sigma = \mathbf{I}$ *this corresponds to an analysis of variance table, and then "Deviance" is equal to the "Sum of Squared deviations (SS)."*

Source	f	Deviance	Test statistic, F
Model versus hypothesis	$m_1 - m_0$	$\|p_1(\mathbf{y}) - p_0(\mathbf{y})\|^2$	$\dfrac{\|p_1(\mathbf{y}) - p_0(\mathbf{y})\|^2 / (m_1 - m_0)}{\|\mathbf{y} - p_1(\mathbf{y})\|^2 / (n - m_1)}$
Residual under model	$n - m_1$	$\|\mathbf{y} - p_1(\mathbf{y})\|^2$	
Residual under hypothesis	$n - m_0$	$\|\mathbf{y} - p_0(\mathbf{y})\|^2$	

▸ **Remark 3.16 – The test does not depend on the particular parameterization of the hypotheses**

Note that the test does only depend on the two subspaces Ω_1 and Ω_0, but not on how the subspaces have been parametrized (the particular choice of basis, i.e., the design matrix). Therefore it is sometimes said that the test is *coordinate free.*   ◂

Initial test for model "sufficiency"

▸ **Remark 3.17 – Test for model "sufficiency"**

In practice, one often starts with formulating a rather comprehensive model (also termed *sufficient model*), and then uses Theorem 3.6 to test whether the model may be reduced to the *null model* with $\Omega_{null} = \mathbb{R}$, i.e., $\dim(\Omega_{null}) = 1$. Thus, one formulates the hypotheses

$$\mathcal{H}_{null} : \boldsymbol{\mu} \in \mathbb{R}$$

with the alternative

$$\mathcal{H}_1 : \boldsymbol{\mu} \in \Omega_1 \setminus \mathbb{R}$$

where $\dim \Omega_1 = k$.

The hypothesis is a hypothesis of "Total homogeneity," viz. that all observations are satisfactorily represented by their common mean.

The test for $\mathcal{H}_{null}$ is a special case of Theorem 3.6 on page 54 with $\Omega_0 = \mathbb{R}$. The partitioning (3.43) is

$$\begin{aligned}
(\mathbf{y} - \overline{\mathbf{y}})^T \boldsymbol{\Sigma}^{-1} (\mathbf{y} - \overline{\mathbf{y}}) &= (\mathbf{y} - \mathbf{X}\widehat{\boldsymbol{\beta}})^T \boldsymbol{\Sigma}^{-1} (\mathbf{y} - \mathbf{X}\widehat{\boldsymbol{\beta}}) \\
&\quad + (\mathbf{X}\widehat{\boldsymbol{\beta}} - \overline{\mathbf{y}})^T \boldsymbol{\Sigma}^{-1} (\mathbf{X}\widehat{\boldsymbol{\beta}} - \overline{\mathbf{y}})
\end{aligned} \tag{3.44}$$

Table 3.2: *Deviance table corresponding to the test for model reduction to the null model.*

Source	f	Deviance	Test statistic, F
Model $\mathcal{H}_{null}$	$k-1$	$\|p_1(\boldsymbol{y}) - p_{null}(\boldsymbol{y})\|^2$	$\dfrac{\|p_1(\boldsymbol{y}) - p_{null}(\boldsymbol{y})\|^2/(k-1)}{\|\boldsymbol{y} - p_1(\boldsymbol{y})\|^2/(n-k)}$
Residual under $\mathcal{H}_1$	$n-k$	$\|\boldsymbol{y} - p_1(\boldsymbol{y})\|^2$	
Total	$n-1$	$\|\boldsymbol{y} - p_{null}(\boldsymbol{y})\|^2$	

with

$$\overline{\boldsymbol{y}} = \mathbf{1}_n [\mathbf{1}_n^T \Sigma^{-1} \mathbf{1}_n]^{-1} \mathbf{1}_n \Sigma^{-1} \boldsymbol{y}$$

denoting the *precision weighted* average of the observations (we have put $\mathbf{1}_n = (1, 1, \ldots, 1)^T$).

When $\Sigma = \boldsymbol{I}$, the expression for the deviations simplifies to *sums of squared deviations*, and $\overline{\boldsymbol{y}}$ reduces to the simple average of the y's. ◀

Again the information is presented in the format of a Deviance table (or an *ANalysis Of VAriance table*, ANOVA) as illustrated in Table 3.2. The table reflects the partitioning above. Under $\mathcal{H}_{null}$ we have $F \sim F(k-1, n-k)$, and hence large values of F would indicate rejection of the hypothesis $\mathcal{H}_{null}$. The p-value of the test is $p = P[F(k-1, n-k) \geq F_{obs}]$.

Coefficient of determination, R^2

Standard software for analyzing data according to the general linear model often outputs the *coefficient of determination* R^2 and the adjusted coefficient of determination R^2_{adj}, where the former is defined as

$$R^2 = \frac{D(p_1(\boldsymbol{y}); p_{null}(\boldsymbol{y}))}{D(\boldsymbol{y}; p_{null}(\boldsymbol{y}))} = 1 - \frac{D(\boldsymbol{y}; p_1(\boldsymbol{y}))}{D(\boldsymbol{y}; p_{null}(\boldsymbol{y}))}. \tag{3.45}$$

Clearly, $0 \leq R^2 \leq 1$. Many researchers use R^2 as an indicator of goodness of fit of the linear model. The intuition behind R^2 is as follows: Suppose you want to predict Y. If you do not know the x's, then the best prediction is $\overline{y}$. The variability corresponding to this prediction is expressed by the *total variation*. However, if the model is utilized for the prediction, then the prediction error is reduced to the *residual variation*. In other words, R^2 expresses the fraction of the total variation that is explained by the model.

Clearly, as more variables are added to the model, $D(\boldsymbol{y}; p_1(\boldsymbol{y}))$ will decrease, and R^2 will increase.

The *adjusted coefficient of determination* aims to correct this deficiency,

$$R_{adj}^2 = 1 - \frac{D(\boldsymbol{y}; p_1(\boldsymbol{y}))/(n-k)}{D(\boldsymbol{y}; p_{null}(\boldsymbol{y}))/(n-1)} . \tag{3.46}$$

It charges a penalty for the number of variables in the model. As more variables are added to the model, $D(\boldsymbol{y}; p_1(\boldsymbol{y}))$ decreases, but the corresponding degrees of freedom also decreases. Thus, the numerator in (3.46) may increase if the reduction in the residual deviance caused by the additional variables does not compensate for the loss in the degrees of freedom.

The following relation between R^2 and R_{adj}^2 should be noted:

$$R_{adj}^2 = 1 - \left[\frac{n-1}{n-k}\right] \frac{D(\boldsymbol{y}; p_1(\boldsymbol{y}))}{D(\boldsymbol{y}; p_{null}(\boldsymbol{y}))} = 1 - \left[\frac{n-1}{n-k}\right] (1 - R^2).$$

Many authors warn against using R^2 to indicate the quality of the model as R^2 does only provide a crude measure of the in-sample goodness of the fit of the model, but does not tell how well the model will predict future observations – see also Section 2.12 on page 32 for issues related to *cross validation*, *out-of-sample testing* and *data mining*, as well as Section 3.10 on page 77 for the important step in model building called *residual analysis*.

A long list of alternatives to R^2 for performance evaluation of statistical models for prediction is given in Madsen et al. (2005).

3.6   Tests for model reduction

When a rather comprehensive model (a *sufficient model*) $\mathcal{H}_1$ has been formulated, and an initial investigation as in Section 3.5 on page 56 has demonstrated that at least some of the terms in the model are needed to explain the variation in the response, then the next step is to investigate whether the model may be reduced to a simpler model (corresponding to a smaller subspace), viz. to test whether all the terms are *necessary*.

Successive testing, Type I partition

Sometimes the practical problem to be solved by itself suggests a *chain* of hypotheses, one being a sub-hypothesis of the other as illustrated in Figure 2.5 on page 28. In other cases, the statistician will establish the chain using the general rule that more complicated terms (e.g., interactions) should be removed before simpler terms as illustrated in Figure 2.7 on page 31.

In the case of a classical GLM, such a chain of hypotheses corresponds to a sequence of linear parameter-spaces, $\Omega_i \subset \mathbb{R}^n$, one being a subspace of the other.

$$\mathbb{R} \subseteq \Omega_M \subset \ldots \subset \Omega_2 \subset \Omega_1 \subset \mathbb{R}^n , \tag{3.47}$$

where

$$\mathcal{H}_i : \boldsymbol{\mu} \in \Omega_i, \quad i = 2, \ldots, M$$

with the alternative

$$\mathcal{H}_{i-1}: \ \boldsymbol{\mu} \in \Omega_{i-1} \setminus \Omega_i \qquad i = 2, \dots, M.$$

Given a chain of hypotheses the total model deviance from a sufficient model, i.e., $D(p_M(\boldsymbol{y}); p_1(\boldsymbol{y}))$, can be partitioned into a sum of incremental deviances when the model is reduced:

THEOREM 3.7 – PARTITIONING OF TOTAL MODEL DEVIANCE
Given a chain of hypotheses that have been organised in a hierarchical manner like in (3.47) then the model deviance $D(p_1(\boldsymbol{y}); p_M(\boldsymbol{y}))$ corresponding to the initial model $\mathcal{H}_1$ may be partitioned as a sum of contributions with each term

$$D(p_{i+1}(\boldsymbol{y}); p_i(\boldsymbol{y})) = D(\boldsymbol{y}; p_{i+1}(\boldsymbol{y})) - D(\boldsymbol{y}; p_i(\boldsymbol{y})) \tag{3.48}$$

representing the increase in residual deviance $D(\boldsymbol{y}; p_i(\boldsymbol{y}))$ when the model is reduced from $\mathcal{H}_i$ to the next lower model $\mathcal{H}_{i+1}$.

Proof Follows from Theorem 2.6 on page 29 (partitioning of tests) and from repeated applications of Theorem 3.6 on page 54. ∎

Let us assume that an initial (*sufficient*) model with the projection $p_1(\boldsymbol{y})$ has been found as shown in Table 3.2 on page 57. Then by using Theorem 3.7, and hence by partitioning corresponding to a chain of models we obtain:

$$\|p_1(\boldsymbol{y}) - p_M(\boldsymbol{y})\|^2 = \|p_1(\boldsymbol{y}) - p_2(\boldsymbol{y})\|^2 + \|p_2(\boldsymbol{y}) - p_3(\boldsymbol{y})\|^2$$
$$+ \cdots + \|p_{M-1}(\boldsymbol{y}) - p_M(\boldsymbol{y})\|^2. \tag{3.49}$$

It is common practice for statistical software to print a table showing this partitioning of the model deviance $D(p_1(\boldsymbol{y}); p_M(\boldsymbol{y}))$ corresponding to the chain of successive projections. The partitioning of the model deviance is from this sufficient or total model to lower order models, and very often the most simple model is the null model with dim $= 1$ – see Table 3.4. This is called *Type I partitioning*.

Example 3.4 – Type I partition
The efficacy of detergent powder is often assessed by washing pieces of cloth that have been stained with specified amounts of various types of fat. In order to assess the staining process an experiment was performed where three technicians (β) applied the same amount of each of two types of oil (α) to pieces of cloth, and subsequently measured the area of the stained spot. The results are shown in Table 3.3. The data are analyzed by a two-way model $Y_{ij} \sim N(\mu_{ij}, \sigma^2)$ with the additive mean value structure:

$$\mathcal{H}_1: \ \mu_{ij} = \mu + \alpha_i + \beta_j \ ; \ i = 1, 2; \quad j = 1, 2, 3 \tag{3.50}$$

Table 3.3: *The area of the stained spot after application of the same amount of two types of oil to pieces of cloth.*

Type of fat	Technician		
	A	D	E
Lard	46.5	43.9	53.7
Olive	55.4		61.7

with suitable restrictions on α_i and β_j to assure identifiability.

Now, we formulate the chain

$$\mathcal{H}_3 \subset \mathcal{H}_2 \subset \mathcal{H}_1$$

with $\mathcal{H}_2: \ \beta_j = 0, j = 1, 2, 3$

$$\mathcal{H}_2: \ \mu_{ij} = \mu + \alpha_i \ ; \ i = 1, 2; \quad j = 1, 2, 3 \qquad (3.51)$$

and

$$\mathcal{H}_3: \ \mu_{ij} = \mu \ ; \ i = 1, 2; \quad j = 1, 2, 3 \qquad (3.52)$$

It is seen that $\mathcal{H}_2$ assumes no difference between the technician, and $\mathcal{H}_3$ assumes further no difference between the type of fat.

The partitioning of the model deviance under this chain is:

```
Call:
lm(formula = Area ~ Fat + Techn, data = fat)

Residual standard error: 0.45 on 1 degrees of freedom
  (1 observation deleted due to missingness)
Multiple R-squared: 0.999,      Adjusted R-squared: 0.996
F-statistic: 335.7 on 3 and 1 DF,  p-value: 0.0401


Analysis of Variance Table

Response: Area
          Df  Sum Sq Mean Sq F value  Pr(>F)
Fat        1 132.720 132.720  655.41 0.02485
Techn      2  71.189  35.595  175.78 0.05326
Residuals  1   0.202   0.202
```

We could have formulated the chain as

$$\mathcal{H}_3 \subset \mathcal{H}_2 \subset \mathcal{H}_1$$

with $\mathcal{H}_2: \ \alpha_i = 0, i = 1, 2$

$$\mathcal{H}_2: \ \mu_{ij} = \mu + \beta_j \ ; \ i = 1, 2; \quad j = 1, 2, 3 \qquad (3.53)$$

and

$$\mathcal{H}_3 : \quad \mu_{ij} = \mu \ ; \quad i = 1, 2; \quad j = 1, 2, 3 \tag{3.54}$$

It is seen that $\mathcal{H}_2$ assumes no difference between the type of fat, and $\mathcal{H}_3$ assumes further no difference between the technicians.

The partitioning of the model deviance under this chain is:

```
Call:
lm(formula = Area ~ Techn + Fat, data = fat)

Residual standard error: 0.45 on 1 degrees of freedom
  (1 observation deleted due to missingness)
Multiple R-squared: 0.999,        Adjusted R-squared: 0.996
F-statistic: 335.7 on 3 and 1 DF,  p-value: 0.0401


Analysis of Variance Table

Response: Area
          Df  Sum Sq Mean Sq F value  Pr(>F)
Techn      2 132.507  66.254  327.18 0.03906
Fat        1  71.402  71.402  352.60 0.03387
Residuals  1   0.203   0.203
```

It is noticed that the p-values are different dependent on which chain is chosen.

We can conclude this section on Type I partitioning by summing up: The usual partitioning, which is the Type I test, corresponds to a successive projection corresponding to a chain of hypotheses reflecting a chain of linear parameter subspaces, such that spaces of lower dimensions are embedded in the higher dimensional spaces.

This implies that the effect at any stage (typically the effect of a new variable) in the chain is evaluated after all previous variables in the model have been accounted for. Hence, the Type I deviance table depends on the order of which the variables enters the model.

Reduction of model using partial tests

In the previous section we considered a fixed layout of a chain of models. However, a particular model can be formulated along a high number of different chains as previously illustrated in Figure 2.7 on page 31 for a three-way model.

Let us now consider some other types of test which by construction do not depend on the order in which the variables enter the model. Consider a given model $\mathcal{H}_i$. This model can be reduced along different chains. More particularly we will consider the *partial likelihood ratio test*:

Table 3.4: *Illustration of Type I partitioning of the total model deviance* $\|p_1(\boldsymbol{y}) - p_M(\boldsymbol{y})\|^2$. *In the table it is assumed that* $\mathcal{H}_M$ *corresponds to the null model where the dimension is* dim $= 1$.

Source	f	Deviance	Test
$\mathcal{H}_M$	$m_{M-1} - m_M$	$\|p_{M-1}(\boldsymbol{y}) - p_M(\boldsymbol{y})\|^2$	$\dfrac{\|p_{M-1}(\boldsymbol{y}) - p_M(\boldsymbol{y})\|^2/(m_{M-1} - m_M)}{\|\boldsymbol{y} - p_1(\boldsymbol{y})\|^2/(n - m_1)}$
$\ldots$	$\ldots$	$\ldots$	$\ldots$
$\ldots$	$\ldots$	$\ldots$	$\ldots$
$\mathcal{H}_3$	$m_2 - m_3$	$\|p_2(\boldsymbol{y}) - p_3(\boldsymbol{y})\|^2$	$\dfrac{\|p_2(\boldsymbol{y}) - p_3(\boldsymbol{y})\|^2/(m_2 - m_3)}{\|\boldsymbol{y} - p_1(\boldsymbol{y})\|^2/(n - m_1)}$
$\mathcal{H}_2$	$m_1 - m_2$	$\|p_1(\boldsymbol{y}) - p_2(\boldsymbol{y})\|^2$	$\dfrac{\|p_1(\boldsymbol{y}) - p_2(\boldsymbol{y})\|^2/(m_1 - m_2)}{\|\boldsymbol{y} - p_1(\boldsymbol{y})\|^2/(n - m_1)}$
Residual under $\mathcal{H}_1$	$n - m_1$	$\|\boldsymbol{y} - p_1(\boldsymbol{y})\|^2$	

DEFINITION 3.11 – PARTIAL LIKELIHOOD RATIO TEST
Consider a sufficient model as represented by $\mathcal{H}_i$. Assume now that the hypothesis $\mathcal{H}_i$ allows the different sub-hypotheses $\mathcal{H}_{i+1}^A \subset \mathcal{H}_i$; $\mathcal{H}_{i+1}^B \subset \mathcal{H}_i$; $\ldots$; $\mathcal{H}_{i+1}^S \subset \mathcal{H}_i$.

A *partial likelihood ratio test* for $\mathcal{H}_{i+1}^J$ under $\mathcal{H}_i$ is the (conditional) test for the hypotheses $\mathcal{H}_{i+1}^J$ given $\mathcal{H}_i$.

It follows from Theorem 3.6 that the numerator in the F-test quantity for the partial test is found as the normalized deviance between the two models, i.e., $\widehat{\boldsymbol{\mu}}$ under $\mathcal{H}_i$ and $\widehat{\widehat{\boldsymbol{\mu}}}$ under $\mathcal{H}_{i+1}^J$

$$F(\boldsymbol{y}) = \frac{\mathrm{D}(\widehat{\boldsymbol{\mu}}; \widehat{\widehat{\boldsymbol{\mu}}})/(m_i - m_{i+1})}{\|\boldsymbol{y} - p_i(\boldsymbol{y})\|^2/(n - m_i)} , \tag{3.55}$$

where $\|\boldsymbol{y} - p_i(\boldsymbol{y})\|^2/(n - m_i)$, which for $\boldsymbol{\Sigma} = \boldsymbol{I}$ is the variance of the residuals under $\mathcal{H}_i$.

Simultaneous testing, Type III partition

▸ **Remark 3.18 – Type III partition**
The Type III partition is obtained as the partial test for all factors. ◂

Thus it is clear that the Type III partitioning gives the deviance that would be obtained for each variable if it was entered last into the model. That is, the effect of each variable is evaluated after all other factors have been accounted for. Therefore the result for each term is equivalent to what is obtained with Type I analysis when the term enters the model as the last one in the ordering.

a) **Pros:** Not sample size dependent: Effect estimates are not a function of the frequency of observations in any group (i.e., for unbalanced data, where we have unequal numbers of observations in each group). When there are no missing cells in the design, these subpopulation means are least squares means, which are the best linear-unbiased estimates of the marginal means for the design.

b) **Cons:** Testing main effects in the presence of interactions but not appropriate for designs with missing cells: for ANOVA designs with missing cells, Type III sums of squares generally do not test hypotheses about least squares means, but instead test hypotheses that are complex functions of the patterns of missing cells in higher-order containing interactions and that are ordinary not Type III: marginal or orthogonal.

There is NO consensus on which type should be used for unbalanced designs, but most statisticians generally recommend Type III, which is the default in most software packages such as SAS, SPSS, JMP, Minitab, Stata, Statista, Systat, and Unistat while R, S-Plus, Genstat, and Mathematica use Type I.

Type I SS also called *sequential sum of squares*, whereas Type III is called *marginal sum of squares*. Unlike the Type I SS, the Type III SS will NOT sum to the Sum of Squares for the model corrected only for the mean (Corrected Total SS).

▶ **Remark 3.19 – Correction for effects**
The partial test corrects for all other variables in the model.　　　　　　◀

3.7　Collinearity

When some predictors are linear combinations of others, then $X^T X$ is singular, and there is (exact) *collinearity*. In this case there is no unique estimate of β. When $X^T X$ is close to singular, there is collinearity (some texts call it *multicollinearity*).

There are various ways to detect collinearity:

i) Examination of the correlation matrix for the estimates may reveal strong pairwise collinearities

ii) Considering the change of the variance of the estimate of other parameters when removing a particular parameter.

When collinearity occurs, the variances are large and thus estimates are likely to be far from the true value. *Ridge regression* is an effective counter measure because it allows better interpretation of the regression coefficients by imposing some bias on the regression coefficients and shrinking their variance.

Ridge regression, also called *Tikhonov regularization* is a commonly used method of regularization of ill-posed problems, see Tikhonov and Arsemin (1977) and Hoerl (1962).

▶ **Remark 3.20 – Ridge regression and regularization techniques**
If the design matrix X is ill-conditioned or singular (yielding a large number of solutions) a Tikhonov regularized version of (3.27) is obtained by using some suitably chosen *Tikhonov matrix*, Γ and considering the regularized estimate

$$\widehat{\beta} = (X^T X + \Gamma^T \Gamma)^{-1} X^T y. \tag{3.56}$$

In some cases $\Gamma = I$, or in the Ridge regression case $\Gamma = I\sqrt{\lambda}$, where λ is chosen small enough to ensure stable solutions to (3.56).

The technique is also relevant for problems with a large (compared to the number of observations) number of unknown parameters. In practical statistical applications this is often the case in chemometrics, which is an area of methods for statistical analysis of data produced typically from chemical or optical instrumentation; see, e.g., Brown, Tauler, and Walczak (2009) and Kramer (1998), but includes in general a wide area of disciplines like molecular biology and calibration; see Martens and Naes (1989).

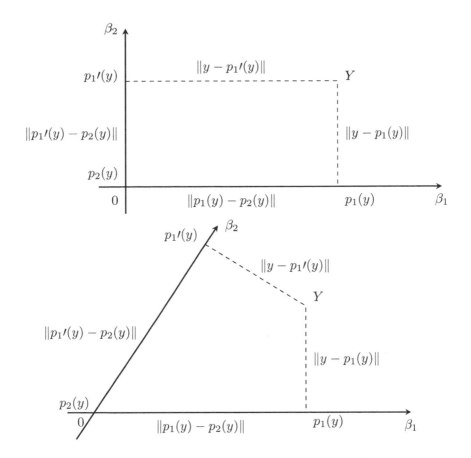

Figure 3.3: *Projections in the case of orthogonal (top) and non-orthogonal (bottom) columns in the design matrix, $\boldsymbol{X}$.*

Some other regularization techniques used in the case of a large number of unknown parameters are the Partial Least Squares (PLS) (see Martens and Naes (1989)), mean subset regularization (see Öjelund et al. (2002)), and calibration with so-called absolute shrinkage (see Öjelund, Madsen, and Thyregod (2001)).   ◀

▶ **Remark 3.21 – Orthogonal parameterization**
Consider now the case $\boldsymbol{X}_2'\boldsymbol{U} = \boldsymbol{0}$, i.e., the space spanned by the columns in $\boldsymbol{U}$ is orthogonal on Ω_2 spanned by $\boldsymbol{X}_2$. Let $\boldsymbol{X}_2\widehat{\boldsymbol{\alpha}}$ denote the projection on Ω_2, and hence we have that $\widehat{\boldsymbol{\alpha}} = \widehat{\widehat{\boldsymbol{\alpha}}}$, is independent of $\widehat{\boldsymbol{\gamma}}$ coming from the projection defined by $\boldsymbol{U}$. See also Figure 3.3.

In this case (3.55) can be simplified, since we have that

$$D(\widehat{\boldsymbol{\mu}}; \widehat{\widehat{\boldsymbol{\mu}}}) = \|U\widehat{\boldsymbol{\gamma}}\|$$

which corresponds to the case of orthogonal columns shown in Figure 3.3.

In this case the F-test (3.41) for model reduction is simplified to a test only based on $\widehat{\boldsymbol{\gamma}}$, i.e.,

$$F(\boldsymbol{y}) = \frac{\|U\widehat{\boldsymbol{\gamma}}\|/(m_1 - m_2)}{s_1^2} = \frac{(U\widehat{\boldsymbol{\gamma}})'\boldsymbol{\Sigma}^{-1}U\widehat{\boldsymbol{\gamma}}/(m_1 - m_2)}{s_1^2}.$$

where $s_1^2 = \|\boldsymbol{y} - p_1(\boldsymbol{y})\|^2/(n - m_1)$. ◄

▶ **Remark 3.22**
If the considered test is related to a subspace which is orthogonal to the space representing the rest of the parameters of the model, then the test quantity for model reduction does not depend on which parameters enter the rest of the model. ◄

We are now ready to conclude:

THEOREM 3.8
The Type I and III partitioning of the deviance are identical if the design matrix $\boldsymbol{X}$ is orthogonal.

In practice it should be noticed that the model used for testing, i.e., the *sufficient model*, should be reevaluated every now and then. For instance if the model can be reduced along some path, then a new sufficient model with more degrees of freedom for the residual deviance can be obtained.

Later on in the phase of model building you might discover that by considering another sub-chain of models some other parameters could enter the model leading to a lower deviance of the residuals.

Inclusion diagrams for a chain of hypotheses

In the following we shall illustrate the use of Type I and Type III deviance tables, and how different paths can be chosen. The various possibilities are illustrated by *inclusion diagrams*.

Comparison of two (or more) regression lines

In Figure 3.4 the inclusion diagram for linear regression is shown. The figure shows that we could either first reduce to parallel lines by testing for the same slope, and then testing for the same intercept, or we could start first by testing for the same intercept and then for identical slope. Let us consider an example.

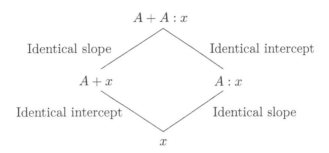

Figure 3.4: *Inclusion diagram for simple linear regression. Possible hypothesis chains for comparison of two regression lines.*

Table 3.5: *Cerebral blood flow (CBF) and age for exposed painters and control group.*

Exposed	CBF	48	41	36	35	43	44	28	28	31	
	Age	23	35	37	38	40	40	45	52	58	
Control	CBF	55	51	44	45	48	50	49	34	31	33	44
	Age	31	32	34	38	38	38	46	48	53	57	39

Example 3.5 – Cerebral blood flow (CBF) for house painters
In the 1970's organic solvents were still used for house painting. Yet, the potential damage to the users was being debated.

For obvious reasons *controlled experiments* on humans were not a feasible way to assess the potential effect of organic solvents, and therefore it was necessary to rely on *observational studies*.

In one study (Arlien-Søborg et al. (1982)) cerebral blood flow (CBF) was studied as an indicator of brain activity. The study comprised 11 unexposed controls, and 9 house painters occupationally exposed to organic solvents for a mean of 22 years. The exposed group had mild to moderate intellectual impairment, and no or only minor cerebral atrophy was seen in a CT-scan of the brain.

Flow was calculated as the initial slope of the wash-out curve after injection of the isotope ^{133}Xe. The flow was measured in ml/100 g/min. The data are shown in Table 3.5. The average flow was 36.8 ml/100 g/min in the group of exposed painters, and 45.4 ml/100 g/min in the controls, thus a slightly lower flow in the group of painters. However, the t-test statistic[1] for

---
[1]The standard small-sample test for comparing two samples under the assumption of equal variances.

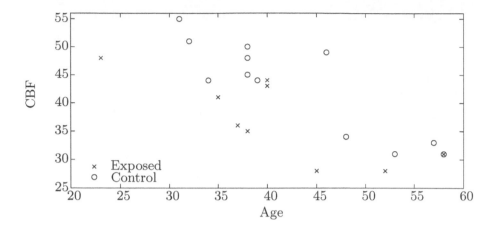

Figure 3.5: *Cerebral blood flow (CBF) vs age for exposed painters and control group.*

testing the hypothesis of identical means in the two groups is $t = 1.99$ with 18 degrees of freedom corresponding to a p-value of $p = \mathrm{P}[|t(18)| \geq 1.99] = 0.0615$. Thus the test fails to reject the hypothesis of identical means.

A closer look at the data reveals that the persons in the control group were slightly older than in the group of exposed painters, the average age in the two groups differs by 0.4 years.

It is well known that brain activity decreases with age, and thus it is conceivable that the effect of the organic solvents has been masked by the older ages in the control group.

Figure 3.5 shows the recorded values of CBF plotted against age. It is evident that CBF decreases with age. A comparison of the two CBF in the groups model should therefore adjust for the age differences.

Initial model

We start by formulating a rather comprehensive model, viz. that the CBF in the two groups decreases linearly with age (in the age-interval under consideration), intercept and slope being specific for each group. Letting (Y_{ij}, x_{ij}) denote the j'th observation in group i, $i = 1, 2$, the model may be formulated in a parametric form as

$$\mathcal{H}_1: \quad \mathrm{E}[Y_{ij}] = \beta_{1i} + \beta_{2i} x_{ij} \quad, i = 1, 2 \quad, j = 1, \ldots, n_i.$$

In terms of a model formula (see Section 3.12) the model is specified as

```
fit1<-lm(formula = CBF ~ Group + Age + Age:Group, data = painters)
```

with a categorical explanatory variable `Group`, a real-valued explanatory variable `Age`, and an interaction term `Age:Group` indicating that there is a different coefficient to `Age` for each value of `Group`.

```
Residual standard error: 4.593 on 16 degrees of freedom
Multiple R-Squared: 0.74
F-statistic: 15.18 on 3 and 16 degrees of freedom,
the p-value is 6.121e-05
```

The initial F-test indicates that the model provides a significant explanation of the variation in CBF.

In order to assess whether all terms are necessary, we formulate the Type I partition of the model deviance:

```
Analysis of Variance Table

Response: CBF

Terms added sequentially (first to last)
          Df Sum of Sq  Mean Sq  F Value      Pr(F)
   Group   1  234.9111 234.9111 11.13682  0.0041787
     Age   1  705.7162 705.7162 33.45705  0.0000279
Age:Group   1   19.6815  19.6815  0.93308  0.3484397
Residuals  16  337.4912  21.0932
```

It is seen that the interaction term Age:Group may be deleted from the model.

Formally: A test of the hypothesis

$$\mathcal{H}_2: \quad \mathrm{E}[Y_{ij}] = \beta_{1i} + \beta_2 x_{ij} \ , i = 1, 2, \ , j = 1, \ldots, n_i$$

fails to reject $\mathcal{H}_2$. The hypothesis corresponds to equal slope in the two groups, viz. $\mathcal{H}_2$: $\beta_{21} = \beta_{22} = \beta_2$.

Now, the model is reduced to $\mathcal{H}_2$, and reestimated which leads to the following partition of model deviation:

```
fit2<-lm(formula = CBF ~ Group + Age, data = painters)

Analysis of Variance Table

Response: CBF

Terms added sequentially (first to last)
          Df Sum of Sq  Mean Sq  F Value       Pr(F)
   Group   1  234.9111 234.9111 11.18083 0.003848431
     Age   1  705.7162 705.7162 33.58928 0.000021509
Residuals  17  357.1727  21.0102
```

Note that a contribution with 1 degree of freedom has moved from the model deviance to the residual deviance. The possible hypotheses chains for this example have been illustrated in Figure 3.4 on page 67. We have chosen the path on the left hand side of the diagram.

Example 3.6 – Alternative hypotheses chains
Instead we could have chosen to follow the path to the right in Figure 3.4 on page 67, that is start to test the hypothesis

$$\mathcal{H}_2: \quad \mathrm{E}[Y_{ij}] = \beta_1 + \beta_{2j}x_{ij} \ , i = 1, 2, \ , j = 1, \ldots, n_i.$$

The hypothesis corresponds to equal intercepts in the two groups. In order to do this in R we need to fit a model with different intercepts and different slopes and another model with equal intercepts and different slopes and then compare the models using a likelihood ratio test with the **anova** function:

```
fit3<-lm(CBF~Age+Age:Group+Group,data=painters)
fit4<-lm(CBF~Age+Age:Group,data=painters)
```

```
> anova(fit3,fit4)
Analysis of Variance Table


Model 1: CBF ~ Age + Age:Group + Group
Model 2: CBF ~ Age + Age:Group
  Res.Df    RSS Df Sum of Sq      F Pr(>F)
1     16 337.49
2     17 396.44 -1    -58.95 2.7947 0.114
```

It is seen that the hypothesis can not be rejected. Now, the model is reduced to $\mathcal{H}_2$, and reestimated which leads to the following partition of model deviation:

```
Analysis of Variance Table


Response: CBF
          Df Sum Sq Mean Sq F value    Pr(>F)
Age        1 687.82  687.82 29.4949 4.494e-05 ***
Age:Group  1 213.54  213.54  9.1568   0.00762 **
Residuals 17 396.44   23.32
```

Comparing this to the results from Example 3.5 it is noted that following the two different paths results in different conclusions.

3.8   Inference on individual parameters in parameterized models

THEOREM 3.9 – TEST OF INDIVIDUAL PARAMETERS
A hypothesis $\beta_j = \beta_j^0$ related to specific values of the parameters is evaluated using the test quantity

$$t_j = \frac{\widehat{\beta}_j - \beta_j^0}{\widehat{\sigma}\sqrt{(X^T\Sigma^{-1}X)_{jj}^{-1}}} \ , \tag{3.57}$$

where $(\boldsymbol{X}^T\boldsymbol{\Sigma}^{-1}\boldsymbol{X})_{jj}^{-1}$ denotes the j'th diagonal element of $(\boldsymbol{X}^T\boldsymbol{\Sigma}^{-1}\boldsymbol{X})^{-1}$. The test quantity is compared with the quantiles of a $\mathrm{t}(n-m_0)$ distribution.

The hypothesis is rejected for large values t_j, i.e., for

$$|t_j| > \mathrm{t}_{1-\alpha/2}(n-m_0)$$

and the p-value is found as $p = 2\mathrm{P}[\mathrm{t}(n-m_0) \geq |t_j|]$.

In the special case of the hypothesis $\beta_j = 0$ the test quantity is

$$t_j = \frac{\hat{\beta}_j}{\hat{\sigma}\sqrt{(\boldsymbol{X}^T\boldsymbol{\Sigma}^{-1}\boldsymbol{X})_{jj}^{-1}}}. \tag{3.58}$$

Proof Omitted. See Section 2.8.   ∎

The difference between this test and the Wald test, is that the latter approximates the t-distribution with the normal distribution. In R the estimates and the values of the test statistics are produced by the extractor **summary** on a linear model object.

Example 3.7

Consider the situation in Example 3.5 on page 67. Under $\mathcal{H}_1$ we obtain the parameter estimates and their corresponding standard deviations:

```
Coefficients:
                Estimate Std. Error t value Pr(>|t|)
(Intercept)      76.7739     7.1315  10.766 9.75e-09
GroupExposed    -16.4294     9.8277  -1.672 0.114015
Age              -0.7941     0.1695  -4.685 0.000248
GroupExposed:Age  0.2259     0.2338   0.966 0.348440
```

which shows that we can drop the coefficient corresponding to `Age:Group` (the difference between the slopes) from the model.

Re-estimating under $\mathcal{H}_2$ leads to the following estimates:

```
Coefficients:
             Estimate Std. Error t value Pr(>|t|)
(Intercept)   71.8755     5.0044  14.363 6.15e-11
GroupExposed  -7.1481     2.0607  -3.469  0.00294
Age           -0.6754     0.1165  -5.796 2.15e-05
```

Confidence intervals and confidence regions

Confidence intervals for individual parameters

A $100(1-\alpha)\%$ confidence interval for β_j is found as

$$\hat{\beta}_j \pm \mathrm{t}_{1-\alpha/2}(n-m_0)\hat{\sigma}\sqrt{(\boldsymbol{X}^T\boldsymbol{\Sigma}^{-1}\boldsymbol{X})_{jj}^{-1}}. \tag{3.59}$$

Simultaneous confidence regions for model parameters

It follows from the distribution of $\widehat{\boldsymbol{\beta}}$ (3.29) that

$$(\widehat{\boldsymbol{\beta}} - \boldsymbol{\beta})^T (\boldsymbol{X}^T \boldsymbol{\Sigma}^{-1} \boldsymbol{X})(\widehat{\boldsymbol{\beta}} - \boldsymbol{\beta}) \leq m_0 \widehat{\sigma}^2 F_{1-\alpha}(m_0, n - m_0). \qquad (3.60)$$

These regions are ellipsoidally shaped regions in $\mathbb{R}^{m_0}$. They may be visualised in two dimensions at a time.

Prediction

In this subsection we will assume that $\boldsymbol{\Sigma} = \boldsymbol{I}$. For predictions in the case of correlated observations we refer to the literature on time series analysis, see Madsen (2008), Box and Jenkins (1970/1976) and Ljung (1976).

Known parameters

Consider the linear model with *known parameters*.

$$Y_i = \boldsymbol{X}_i^T \boldsymbol{\theta} + \varepsilon_i \qquad (3.61)$$

where $\mathrm{E}[\varepsilon_i] = 0$ and $\mathrm{Var}[\varepsilon_i] = \sigma^2$ (i.e., constant). The prediction for a future value $Y_{n+\ell}$ given the independent variable $\boldsymbol{X}_{n+\ell} = \boldsymbol{x}_{n+\ell}$ is

$$\widehat{Y}_{n+\ell} = \mathrm{E}[Y_{n+\ell} | \boldsymbol{X}_{n+\ell} = \boldsymbol{x}_{n+\ell}] = \boldsymbol{x}_{n+\ell}^T \boldsymbol{\theta} \qquad (3.62)$$

$$\mathrm{Var}[Y_{n+\ell} - \widehat{Y}_{n+\ell}] = \mathrm{Var}[\varepsilon_{n+\ell}] = \sigma^2. \qquad (3.63)$$

Unknown parameters

Most often the parameters are unknown but assume that there exist some estimates of $\boldsymbol{\theta}$. Assume also that the estimates are found by the estimator

$$\widehat{\boldsymbol{\theta}} = (\boldsymbol{X}^T \boldsymbol{X})^{-1} \boldsymbol{X}^T \boldsymbol{Y} \qquad (3.64)$$

then the variance of the prediction error can be stated.

THEOREM 3.10 – PREDICTION IN THE GENERAL LINEAR MODEL
Assume that the unknown parameters $\boldsymbol{\theta}$ in the linear model are estimated by using the least squares method (3.64), then the prediction *is*

$$\widehat{Y}_{n+\ell} = \mathrm{E}[Y_{n+\ell} | \boldsymbol{X}_{n+\ell} = \boldsymbol{x}_{n+\ell}] = \boldsymbol{x}_{n+\ell}^T \widehat{\boldsymbol{\theta}}. \qquad (3.65)$$

The variance of the prediction error $e_{n+\ell} = Y_{t+\ell} - \widehat{Y}_{n+\ell}$ *becomes*

$$\mathrm{Var}[e_{n+\ell}] = \mathrm{Var}[Y_{n+\ell} - \widehat{Y}_{n+\ell}] = \sigma^2 [1 + \boldsymbol{x}_{n+\ell}^T (\boldsymbol{X}^T \boldsymbol{X})^{-1} \boldsymbol{x}_{n+\ell}]. \qquad (3.66)$$

Proof (3.65) follows immediately since the mean of the prediction error is 0.

$$
\begin{aligned}
\mathrm{Var}[Y_{n+\ell} - \widehat{Y}_{n+\ell}] &= \mathrm{Var}[\boldsymbol{x}_{n+\ell}^T\boldsymbol{\theta} + \varepsilon_{n+\ell} - \boldsymbol{x}_{n+\ell}^T\widehat{\boldsymbol{\theta}}] \\
&= \mathrm{Var}[\boldsymbol{x}_{n+\ell}^T(\boldsymbol{\theta} - \widehat{\boldsymbol{\theta}}) + \varepsilon_{n+\ell}] \\
&= \boldsymbol{x}_{n+\ell}^T\,\mathrm{Var}[\widehat{\boldsymbol{\theta}}]\boldsymbol{x}_{n+\ell} + \sigma^2 + 2\,\mathrm{Cov}[\boldsymbol{x}_{n+\ell}^T(\boldsymbol{\theta} - \widehat{\boldsymbol{\theta}}), \varepsilon_{t+\ell}] \\
&= \sigma^2 + \boldsymbol{x}_{n+\ell}^T\,\mathrm{Var}[\widehat{\boldsymbol{\theta}}]\boldsymbol{x}_{n+\ell}.
\end{aligned}
$$

The result follows now from the fact that $\mathrm{Var}[\widehat{\boldsymbol{\theta}}] = \sigma^2(\boldsymbol{X}^T\boldsymbol{X})^{-1}$. ∎

If we use an estimate for σ^2 (applying $\boldsymbol{\Sigma} = \boldsymbol{I}$) then a $100(1-\alpha)\%$ *confidence interval for the future $Y_{n+\ell}$* is given as

$$
\begin{aligned}
\widehat{Y}_{n+\ell} &\pm t_{\alpha/2}(n-k)\sqrt{\mathrm{Var}[e_{n+\ell}]} \\
&= \widehat{Y}_{n+\ell} \pm t_{\alpha/2}(n-k)\widehat{\sigma}\sqrt{1 + \boldsymbol{x}_{n+\ell}^T(\boldsymbol{X}^T\boldsymbol{X})^{-1}\boldsymbol{X}_{n+\ell}}
\end{aligned}
\tag{3.67}
$$

where $t_{\alpha/2}(n-k)$ is the $\alpha/2$ quantile in the t distribution with $(n-k)$ degrees of freedom and n is the number of observations used in estimating the k unknown parameters.

A confidence interval for a future value is also called a *prediction interval*.

3.9 Model diagnostics: residuals and influence

Notice, that in this section we will assume that $\boldsymbol{\Sigma} = \boldsymbol{I}$.

Residuals, standardization and studentization

The residuals denote the difference between the observed value y_i and the value $\widehat{\mu}_i$ fitted by the model. It follows from (3.37) that these raw residuals do not have the same variance. The variance of r_i is $\sigma^2(1 - h_{ii})$ with h_{ii} denoting the i'th diagonal element in the hat-matrix. Therefore, in order to make a meaningful comparison of the residuals, it is usual practice to rescale them by dividing by the estimated standard deviation.

DEFINITION 3.12 – STANDARDIZED RESIDUAL
The standardized residual is

$$
r_i^{rs} = \frac{r_i}{\sqrt{\widehat{\sigma}^2(1 - h_{ii})}}
\tag{3.68}
$$

▸ **Remark 3.23 – Standardization does not imply that the variance is 1**

It is a usual convention in statistics that *standardization* of a random variable means transforming to a variable with mean 0, and variance 1. Often standardization takes place by dividing the variable by its standard deviation. In (3.68) we have only divided by an estimate of the standard deviation, so although we have achieved equal variance for the standardized residuals, the variance is not 1.
◂

Residuals are often used to identify *outliers*, i.e., observations that for some reasons do not fit into the general pattern, and possibly should be excluded. However, if a particular observation y_i is such a contaminating observation giving rise to a large residual r_i, then that observation would also inflate the estimate $\widehat{\sigma}$ of the standard deviation in the denominator of (3.68) thereby masking the effect of the contamination. Therefore, it is advantageous to scale the residual with an estimate of the variance, σ^2, that does not include the i'th observation.

DEFINITION 3.13 – STUDENTIZED RESIDUAL
The studentized residual is

$$r_i^{rt} = \frac{r_i}{\sqrt{\widehat{\sigma}_{(i)}^2 (1 - h_{ii})}} \tag{3.69}$$

where $\widehat{\sigma}_{(i)}^2$ denotes the estimate for σ^2 determined by deleting the i'th observation.

It may be shown, see, e.g., Jørgensen (1993), that $\widehat{\sigma}_{(i)}^2 \sim \sigma^2 \chi^2(f)/f$-distribution with $f = n - m_0 - 1$ and therefore, as r_i is independent of $\widehat{\sigma}_{(i)}^2$ that the studentized residual follows a t(f)-distribution when $\mathcal{H}_0$ holds.

The residuals provide a valuable tool for model checking. The residuals should be checked for individual large values. A large standardized or studentized residual is an indication of poor model fit for that point, and the reason for this outlying observation should be investigated. For observations obtained in a time-sequence the residuals should be checked for possible sequential correlation, seasonality, or non-stationarity (see, e.g., Madsen (2008) or Box and Jenkins (1970/1976)).

Also the distribution of the studentized residuals should be investigated and compared to the reference distribution (the t-distribution, or simply a normal distribution) by means of a qq-plot to identify possible anomalies. Section 3.10 outlines a series of useful plots for analyzing the residuals, and for pinpointing possible model deficiencies.

Influential observations, leverage

It follows from

$$\frac{\partial \widehat{\mu}}{\partial y} = H$$

that the i'th diagonal element h_{ii} of the hat-matrix H denotes the change in the fitted value $\widehat{\mu}_i$ induced by a change in the i'th observation y_i. In other words, the diagonal elements in the hat-matrix H indicate the "weight" with which the individual observation contributes to the fitted value for that data point.

DEFINITION 3.14 – LEVERAGE
The i'th diagonal element in the hat-matrix (3.34) is called the *leverage* of the i'th observation.

Example 3.8 – Leverage in a simple regression model
To illustrate the concept of leverage we shall consider a simple regression model with n data-points (x_i, y_i), $i = 1, 2, \ldots, n$. We parametrize the model as

$$E[Y_i] = \beta_0 + \beta_1(x_i - \overline{x}.)$$

with $\overline{x}. = \sum x_i / n$ denoting the average of the x-values. The corresponding design matrix is

$$X = \begin{pmatrix} 1 & (x_1 - \overline{x}.) \\ 1 & (x_2 - \overline{x}.) \\ \vdots & \vdots \\ 1 & (x_n - \overline{x}.) \end{pmatrix}.$$

We observe that the parametrization in terms of $(x_1 - \overline{x}.)$ has the effect that the two columns are orthogonal. Thus, one obtains

$$X^T X = \begin{pmatrix} n & 0 \\ 0 & \sum(x_i - \overline{x}.)^2 \end{pmatrix}$$

such that

$$(X^T X)^{-1} = \begin{pmatrix} 1/n & 0 \\ 0 & 1/\sum(x_i - \overline{x}.)^2 \end{pmatrix}.$$

Hence, the i'th diagonal term in the hat-matrix is

$$h_{ii} = \begin{pmatrix} 1 & (x_i - \overline{x}.) \end{pmatrix} (X^T X)^{-1} \begin{pmatrix} 1 \\ x_i - \overline{x}. \end{pmatrix}$$

$$= \frac{1}{n} + \frac{(x_i - \overline{x}.)^2}{\sum(x_j - \overline{x}.)^2}.$$

Thus, it is seen that h_{ii} is the Euclidean distance between the point $(1, x_i)$ corresponding to the i'th model vector, and the center of gravity $(1/n, \overline{x}.)$ for all model vectors.

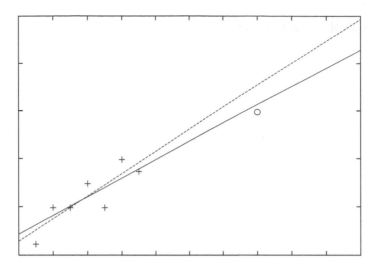

Figure 3.6: *A data point ∘ with large leverage. The full line indicates the regression line corresponding to all observations. The dashed line indicates the regression line estimated without the point marked ∘ with large leverage.*

For a regression model with k explanatory variables, one finds analogously that the i'th diagonal element in the hat-matrix is

$$h_{ii} = \frac{1}{n} + \frac{(x_{i1} - \overline{x}_{\cdot 1})^2}{\sum (x_{j1} - \overline{x}_{\cdot 1})^2} + \cdots + \frac{(x_{ik} - \overline{x}_{\cdot k})^2}{\sum (x_{jk} - \overline{x}_{\cdot k})^2} \, . \qquad (3.70)$$

Again, h_{ii} is the Euclidean distance between the i'th model vector and the center of gravity $(1/n, \overline{x}_{\cdot 1}, \ldots, \overline{x}_{\cdot k})$ for all model vectors.

The quantity (3.70) is also termed *Mahalanobis distance* between the i'th model vector and the cloud of points corresponding to all model vectors.

Thus, the diagonal elements in the hat-matrix indicate how much the corresponding model vector deviates from the bulk of values of the explanatory variables. When the model vector is in the center of gravity for the explanatory variables, we have $h_{ii} = 1/n$. The larger the value of h_{ii} the more extreme is the model vector.

Note that leverage is a property associated with the model and the *design*, i.e., the values of the explanatory variables. As we have seen in Section 3.4 on page 50, the consequences of a large leverage are of importance as well for the variance of the residual (see (3.37)), as for the variance on the fitted value (see (3.35)).

One should pay special attention to observations with large leverage. Not because it necessarily is undesirable that an observation has a large leverage. If an observation with large leverage is in agreement with the other data, that point would just serve to reduce the uncertainty of the estimated parameters, see Figure 3.6. When a model is reasonable, and data are in agreement with the model, then observations with large leverage will be an advantage.

Observations with a large leverage might however be an indication that the model does not represent the data. It is well known that one should be cautious by extrapolating an empirically established relation beyond that region in the "design space" where it has been fitted. A linear model might provide a good description of data in a region of the design space, but it need not provide a good description outside that region.

3.10 Analysis of residuals

Previously in Section 2.8 asymptotic results for the variance of the ML estimator were presented, and it was shown how these results can be used to provide estimates of the uncertainty of the individual parameters and for test of an individual parameter using, e.g., the Wald statistic. Also in Section 2.11 the likelihood ratio test was introduced as a tool for model selection.

In order to finally conclude on model sufficiency the task called *analysis of the residuals* is a very important step, and a part of what is called *model validation*.

For a sufficient model the (standardized) residuals must be approximatively independent and identically distributed. If any systematic variation is detected this indicates a *lack-of-fit* and indicates a potential direction for model improvement. For observations obtained as a time-sequence the residuals should be checked for possible autocorrelation, hidden harmonics, trends, etc., see, e.g., Madsen (2008).

In this section we will consider typically observed *scatter plots* of the residuals against various variables, and we will briefly explain how these plots can be used for pinpointing model deficiencies. Figure 3.7 on the next page shows such typical scatter plots, and it should be noticed that 1) always corresponds to a situation which is considered acceptable.

i) *Plot of residuals against time*

   1) Acceptable.

   2) The variance grows with time. Do a weighted analysis.

   3) Lack of term of the form $\beta \cdot \mathbf{time}$.

   4) Lack of term of the form $\beta_1 \cdot \mathbf{time}^2$.

ii) *Plot of residuals against independent variables*

   1) Acceptable.

   2) The variance grows with x_i. Perform weighted analysis or transform the Y's (e.g., with the logarithm or equivalent).

   3) Error in the computations.

   4) Lack of quadratic term in x_i.

iii) *Plot of residuals against fitted values*

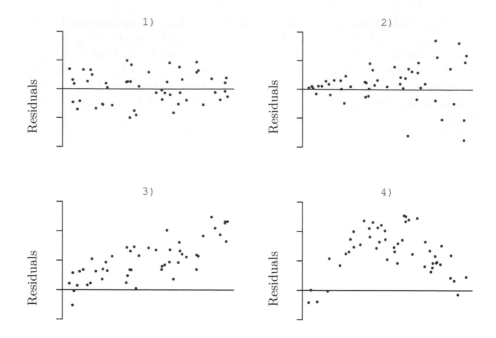

Figure 3.7: *Residual plots.*

1) Acceptable.

2) The variance grows with $\hat{\mu}_i$. Do a weighted analysis or transform the Y's.

3) Lack of constant term. The regression is possibly erroneously forced trough zero.

4) Bad model, try transforming the Y's.

▸ **Remark 3.24**

It should be noticed that plot of the residuals against the observations will, even for a validated model, almost always show a dependence structure like the one illustrated as No. 3 in Figure 3.7, i.e., the residuals and the observations are positively correlated. This is simply due to the fact that

$$\text{Cor}[Y_i, r_i] = 1 - R^2 \tag{3.71}$$

where R is the coefficient of determination defined in (3.45) on page 57.   ◂

3.11   Representation of linear models

As we have seen, classical GLMs may be represented geometrically in terms of vector spaces and subspaces. Other representations are by means of design matrices and parameters or, in terms of parametric expressions for the mean value as linear functions of values of the explanatory variables, like $E[Y_i] = \beta_0 + \beta_1 x_i$. Yet another representation is used in statistical software, see Section 3.12 on page 81.

Parametrization by contrasts

Factors entering in models normally produce more coefficients than can be estimated. Thus, all the dummy incidence variables u_{ij} for any factor adds to 1 for each observation i. This is identical to the variable used implicitly to fit the intercept term. Hence, the resulting $n \times (r+1)$ dimensional design matrix

$$\boldsymbol{X}_A = \{\boldsymbol{1}, \boldsymbol{u}_1, \ldots, \boldsymbol{u}_r\} \tag{3.72}$$

corresponding to a factor with r levels does only have rank r.

To resolve this problem, a linear restriction is sometimes introduced between the parameters $\beta_1, \ldots, \beta_r$, or the model is reparametrized (implicitly, or explicitly) to yield an identifiable parameterization.

If it is considered advantageous to operate with a design matrix of full rank, one may choose to reparametrize the model to obtain a $n \times r$ dimensional matrix

$$\boldsymbol{X}_A = \{\boldsymbol{1}, \boldsymbol{c}_1, \ldots, \boldsymbol{c}_{r-1}\}$$

of full rank, spanning the same space Ω_A as (3.72).

The columns $\boldsymbol{c}_j$ define a set of so-called *estimable contrasts* among the r parameters $\beta_1, \ldots, \beta_r$.

DEFINITION 3.15 – CONTRAST AMONG EFFECTS
A contrast among the effects $\beta_1, \ldots, \beta_r$ is a linear combination $\sum_i c_i \beta_i$ where the c_i's are known and $\sum_i c_i = 0$.

Clearly, there are a number of different possibilities for defining contrasts between r levels of a factor. In the following example we shall illustrate some of these possibilities in the case $r = 4$.

Example 3.9 – Contrasts among $r = 4$ levels

a) **Helmert-transformation** Parametrization by the so-called Helmert-transformation corresponds to the model

$$\boldsymbol{\mu} = \boldsymbol{X}_H \begin{pmatrix} \beta_0 \\ \gamma_1 \\ \gamma_2 \\ \gamma_3 \end{pmatrix}$$

with the design matrix $\boldsymbol{X}_H$

$$\boldsymbol{X}_H = \begin{pmatrix} 1 & -1 & -1 & -1 \\ 1 & 1 & -1 & -1 \\ 1 & 0 & 2 & -1 \\ 1 & 0 & 0 & 3 \end{pmatrix}.$$

In this parametrization the first contrast denotes the difference $\beta_2 - \beta_1$ between the results corresponding to level 2 and level 1. The second contrast denotes the difference $\beta_3 - (\beta_1 + \beta_2)/2$ between the result corresponding to level 3 and the average of level 1 and level 2. The j'th contrast denotes the difference $\beta_{j+1} - (\beta_1 + \cdots + \beta_j)/j$ between the result corresponding to level $j + 1$ and the average of levels 1 to j.

The Helmert-transformation assures that the individual columns $\boldsymbol{c}_j$ are mutually orthogonal, and further that they are orthogonal to $\boldsymbol{1}$.

b) Sum-coding Another possibility is the so-called 'sum-coding'

$$\boldsymbol{\mu} = \boldsymbol{X}_S \begin{pmatrix} \beta_0 \\ \gamma_1 \\ \gamma_2 \\ \gamma_3 \end{pmatrix}$$

with the design matrix $\boldsymbol{X}_S$

$$\boldsymbol{X}_H = \begin{pmatrix} 1 & 1 & 0 & 0 \\ 1 & 0 & 1 & 0 \\ 1 & 0 & 0 & 1 \\ 1 & -1 & -1 & -1 \end{pmatrix},$$

corresponding to using level r as a reference value (the intercept term).

The first contrast denotes the difference between β_1 and β_r, second contrast denotes the difference between β_2 and β_r, etc. Also these contrasts are mutually orthogonal and orthogonal to $\boldsymbol{1}$.

c) Treatment-coding Often a coding is used corresponding to the representation

$$\boldsymbol{\mu} = \boldsymbol{X}_T \begin{pmatrix} \beta_0 \\ \gamma_1 \\ \gamma_2 \\ \gamma_3 \end{pmatrix} \tag{3.73}$$

with the design matrix $\boldsymbol{X}_T$

$$\boldsymbol{X}_T = \begin{pmatrix} 1 & 0 & 0 & 0 \\ 1 & 1 & 0 & 0 \\ 1 & 0 & 1 & 0 \\ 1 & 0 & 0 & 1 \end{pmatrix}.$$

The coding is sometimes denoted 'treatment'-coding.

The reparametrization (3.73) corresponds to using the first level β_0 as reference level. The parameters $\gamma_1, \ldots, \gamma_{r-1}$ corresponding to this coding are not contrasts, as the columns do not sum to zero, and hence, the columns are not orthogonal to **1**.

3.12   General linear models in R

Model formulae

In statistical software models are often specified using a symbolic notation suggested by Wilkinson and Rogers (1973). A *model formulae* is a symbolic description of the mean value structure in a linear model using a combination of symbols for variables and for operators. Variables are referred to by *variable names*. When a variable plays the role of a *response variable*, it is placed on the left hand side in the model formulae and on the right hand side if it plays the role of a *explanatory variable*. A model formulae in R has the form:

response variable $\sim$ explanatory variable(s)

where the *tilde* symbol $\sim$ can be read as 'is modeled as a function of.' The variables can also appear as transformations, powers or polynomials.

The following data types can appear as a *term* in a formula:

- the number '1' indicating the intercept term (default in R, i.e., '-1' indicates no intercept)

- a numeric vector, implying a simple coefficient

- a factor (or ordered factor) implying one coefficient for each level

- a matrix, implying a coefficient for each column

Symbols are used differently in model formulae than in arithmetic expressions. In particular:

+	inclusion of an explanatory variable in the model
-	deletion of an explanatory variable in the model
:	inclusion of interaction of explanatory variables in the model
*	inclusion of interaction and explanatory variables in the model, i.e., A*B = A + B + A:B

Model fitting

Before fitting a model in R it is important that the variables are of the correct type, e.g., that categorical variables are defined as *factors*. A useful function to check this is the str function that displays the internal structure of an R object. If, for example, a categorical variable is not defined as factor, it needs to be changed into one using the **factor** function.

The **lm** function fits a linear model with normal errors and constant variance. This function is generally used for regression analysis using continuous explanatory variables. For models including only categorical explanatory variables or mix of categorical and continuous variables the **aov** function is generally used. The syntax for the two functions is similar. A typical call to the, e.g., **lm** function is:

```
> fit<-lm(y ~ x)
```

After a model has been fitted using **lm** or **aov**, a number of *generic* functions exists that can be used to get information about the fitted model. Some useful generic functions are:

```
> summary(fit)
```

The form of the output returned by the function depends on the class of its argument. When the argument is an **lm** object, the function computes and returns a list of summary statistics of the fitted linear model including parameter estimates and standard errors. If the argument is an **aov** object an analysis of variance table is returned. Possible to choose **summary.aov** or **summary.lm** to get the alternative form of output.

```
> update(fit)
```

Function to use to modify and re-fit a model.

```
> plot(fit)
```

Produces diagnostic plots for model checking.

```
> model.matrix(fit)
```

Extracts the model matrix/design matrix from a model formula.

```
> anova(fit)
```

Computes analysis of variance (or deviance) tables for one or more fitted model objects. When given a single argument it returns a table which tests whether the model terms are significant but when given a sequence of objects the function tests the models against one another.

```
> coef(fit)
```

Extracts model coefficients from the estimated model.

```
> fitted(fit)
```

Extracts fitted values, predicted by the model for the values of the explanatory variable(s) included.

```
> predict(fit)
```

Produces predicted values, standard errors of the predictions and confidence or prediction intervals.

```
> residuals(fit)
```

Extracts residuals from the model object.

```
> rstandard(fit)
```

Extracts standardized residuals from the model object.

```
> rstudent(fit)
```

Extracts studentized residuals from the model object.

```
> hatvalues(fit)
```

Returns the diagonal elements of the hat matrix.

Contrasts

Functions contrasts may be constructed by the functions, `contr.treatment`, `contr.helmert`, `contr.poly`, `contr.sum` with the number of levels as arguments. The default contrast in R is the treatment contrast which contrasts each level with the baseline level (specified by `base`). `cont.helmert` returns Helmert contrasts, `contr.poly` returns contrasts based on orthogonal polynomials and `contr.sum` uses "sum to zero contrasts." See also Example 3.9.

3.13   Problems

Please notice that some of the problems are designed for being solved by hand calculations while others require access to a computer.

Exercise 3.1
Consider the generalized loss function (deviance) defined in Definition 2.12. Show that for $(Y_1, Y_2, \ldots, Y_n)^T$ being a sequence of i.i.d. normally distributed variables the deviance becomes the classical sum of squared error loss function.

Exercise 3.2
Consider the regression model

$$Y_t = \beta x_t + \varepsilon_t$$

where $E[\{\varepsilon_t\}] = 0$. Suppose that n observations are given.

Question 1 Assume that $\text{Var}[\varepsilon_t] = \sigma^2 / x_t^2$ but that the elements of the sequence $\{\varepsilon_t\}$ are mutually uncorrelated. Consider the unweighted least squares estimator $(\widehat{\beta^*})$.

- Is the estimator unbiased?
- Calculate the variance of the estimator.

Question 2 Calculate the variance of the weighted least squares estimator $(\widehat{\beta})$.

Table 3.6: *Measured contamination with coal*

Method No. I	Volume (x) (in 10 cm³)	1	2	3
	Total coal contamination (in g)	0.22	0.38	0.72
Method No. II	Volume (x) (in 10 cm³)	1	2	3
	Total coal contamination (in g)	0.31	0.66	0.99

Question 3 Compare the variances of $(\widehat{\beta}^*)$ and $(\widehat{\beta})$.

Question 4 Assume now that $\mathrm{Var}[\varepsilon_t] = \sigma^2$, but that the elements of the sequence $\{\varepsilon_t\}$ are mutually correlated. We assume that

$$\rho[\varepsilon_t, \varepsilon_{t-1}] = \rho, \quad \text{for } k = 1$$
$$\rho[\varepsilon_t, \varepsilon_{t-k}] = 0, \quad \text{for } k \geq 2$$

Consider the unweighted least square estimator.

- Is the estimator unbiased?
- Calculate the variance of the estimator.

Exercise 3.3
Assume that $\boldsymbol{H}$ is a projection matrix. Prove that then the matrix $(\boldsymbol{I} - \boldsymbol{H})$ is a projection matrix.

Exercise 3.4
For the production of a certain alloy two different methods can be used. Both methods imply that there is some unwanted contamination of coal in the alloy.

The purpose of this exercise is to investigate whether the two methods lead to the same contamination of the alloy. For each of the two methods 3 blocks of the alloy of the size 10, 20 and 30 cm³ have been produced. The results are shown in Table 3.6.

The total coal contamination for method No. I is based on an analysis of a small part of the volume, and the total contamination is then determined by an up-scaling to the given volume. This method leads to the assumption that the variance on a measurement is proportional to the squared volume, i.e.,

$$\mathrm{Var}[Y|x] = \sigma_1^2 x^2$$

For Method No. II the coal is found based on analysis of the entire block, and therefore it is assumed that the variance on a single measurement is proportional to the volume, i.e.,

$$\mathrm{Var}[Y|x] = \sigma_2^2 x$$

Let us for simplicity assume that $\sigma_1^2 = \sigma_2^2 = \sigma^2$.

Table 3.7: *The six measurements of pH at the two factories*

Factory A	1.33	1.11	-0.73	0.99	-0.29	1.43
Factory B	-1.05	-0.29	1.29	0.55	0.09	1.83

Question 1 Find an estimate of the coal contamination of the alloy (in gram/10 cm^3) for each of the two methods.

Question 2 Find an estimate for the common variance, σ^2.

Question 3 Is it reasonable to assume that the contamination with coal is the same for both methods?

Exercise 3.5
In an investigation it is assumed that the relationship between measurements of a certain phenomena, Y, and the explanatory variable, x, can be described by

$$Y = ax^2 - bx + \epsilon$$

where ϵ is i.i.d. and normally distributed. The following measurements are available:

x:	0	1	1	2
y:	1/2	0	-1/2	-2

Question 1 Find estimates of a and b and state an estimate for the related uncertainty on the estimates.

Question 2 Now we want a successive testing of whether the relation can be described by the following equations

i) $ax^2 - bx + c$

ii) $ax^2 - bx$

iii) $-bx$

Question 3 Calculate an estimate of the uncertainty of the estimates in each of the above three cases.

Exercise 3.6
A company using a certain liquid which can be produced at two different factories is interested in an analysis of the pH value of the produced liquid at the two factories. The measurements of the pH value are encumbered with a rather large uncertainty, and therefore the pH-value is determined six times at each factory, as shown in Table 3.7.

The equipment for obtaining pH measurements at Factory B is much older that that of Factory A, and it seems reasonable to assume that the variance of the measurements taken at Factory B is twice the variance of the measurements taken at Factory A.

Now the company wants to analyze whether the pH values can be assumed to be the same for both factories, and secondarily whether such a common value can be assumed to be zero. Formulate appropriate models for this analysis and conduct the appropriate tests.

Exercise 3.7

Here we will consider Anscombe's artificial data (Anscombe (1973)). In R the data frame `anscombe` is made available by

```
> data(anscombe)
```

This contains four artificial datasets, each of 11 observations of a continuous response variable y and a continuous explanatory variable x.

All the usual summary statistics related to classical analysis of the fitted models are identical across the four datasets. In particular, the Multiple R^2: 0.6662, and R^2_{adj}: 0.6292 are identical for the four datasets.

Plot the data and discuss whether you consider R^2 to be a good measure of model fit.

CHAPTER 4

Generalized linear models

In the previous chapter we briefly summarized the theory of general linear models (classical GLM). The assumptions underlying a classical GLM are that (at least to a good approximation) the errors are normally distributed, the error variances are constant and independent of the mean, and the systematic effects combine additively, for short: normality, homoscedasticity and linearity.

When data for a continuous characteristic cover only a limited range of values, these assumptions may be justifiable. However, as the following example shows, there are situations where these assumptions are far from being satisfied. Let us illustrate this using an example.

Example 4.1 – Assessment of developmental toxicity of chemical compound

In toxicology it is usual practice to assess developmental effects of an agent by administering specified doses of the agent to pregnant mice, and assess the proportion of stillborn as a function of the concentration of the agent. An example of data from such an experiment is shown in Table 4.1. The study is reported in Price et al. (1987).

The quantity of interest is the *fraction*, $Y_i = \frac{z_i}{n_i}$, of stillborn pups as a function of the concentration x_i of the agent. A natural distributional assumption is the binomial distribution $Y_i \sim \mathrm{B}(n_i, p_i)/n_i$.

For p_i close to 0 (or 1) the distribution of Y is highly skewed, and hence, the assumption of normality is violated. Also, the variance, $\mathrm{Var}[Y_i] = p_i(1 - p_i)/n_i$ depends not only on the (known) number n_i of fetuses, but also on the mean value p_i, which is the quantity we want to model. Thus, also the assumption of homoscedasticity is violated. A linear model of the form

$$p_i = \beta_1 + \beta_2 x_i$$

will violate the natural restriction $0 \le p_i \le 1$.

Finally, a model formulation of the form $y_i = p_i + \epsilon_i$ (mean plus noise) is not adequate; if such a model should satisfy $0 \le y_i \le 1$, then the distribution of ϵ_i would have to be dependent on p_i.

Thus, none of the assumptions underlying a classical GLM are – even approximately – satisfied in this case. We shall return to this example, and provide a more satisfactory model in Example 4.6.

Table 4.1: *Results of a dose-response experiment on pregnant mice. Number of stillborn fetuses found for various dose levels of a toxic agent.*

Index	Number of stillborn, z_i	Number of fetuses, n_i	Fraction still-born, y_i	Concentration [mg/kg/day], x_i
1	15	297	0.0505	0.0
2	17	242	0.0702	62.5
3	22	312	0.0705	125.0
4	38	299	0.1271	250.0
5	144	285	0.5053	500.0

The example is not unique to situations where the response is a ratio of counts as in the example. A similar problem occurs when the response takes the form of a set of counts, or when the response is measured on a ratio scale; for such data often none of the above assumptions will even be approximately true. That is, the errors will not be normal, the variance will change with the mean, and the systematic effects are likely to combine additively on some transformed scale different from the original one.

Previously a response to this state of affairs has been to transform the data, i.e., to replace the data y by $z = f(y)$, in the hope that the assumptions for the classical linear model will be satisfied. For binomially distributed data, the arcsine transformation has been recommended. However, it is asking quite a lot of a single transformation to satisfy all these assumptions simultaneously. For Poisson distributed observations, no such transformation exists. For approximate normality, one requires the transformation $z = y^{2/3}$, for approximately equal variances $z = y^{1/2}$, and for multiplicative effects $z = \log(y)$. In practice, the square root transformation has often been used for count data, choosing constant variance over normality and linearity.

With the increasing availability of computing resources, a more satisfactory approach to these problems in the framework of the so-called *Generalized Linear Models* was developed. In the years since the term was first introduced, Nelder and Wedderburn (1972), this approach has slowly become well known and widely used. Instead of transforming the observations, the main idea in this approach is to formulate linear models for a transformation of the mean value, the *link function*, and keep the observations untransformed, thereby preserving the distributional properties of the observations.

In this chapter we shall introduce the class of *Generalized Linear Models*. The theory of *Generalized Linear Model* is tied to a special class of distributions, the *exponential family* of distributions.

For single parameter densities the actual choice of distribution affects the assumptions on the variances, since the relation between the mean and the variance is known and in fact characterizes the distribution. The Poisson distribution, for instance, has the property $\mathrm{Var}[Y] = \mathrm{E}[Y]$.

It is useful to distinguish between the exponential family (or the *natural exponential family*) and the *exponential dispersion family*. In the exponential dispersion family an extra parameter, the so-called *dispersion parameter*, is introduced. This allows for a description of distributions without a unique relation between the mean and the variance. The normal distribution is such an example. The exponential dispersion family enables a possibility for separating the parameters in a structural parameter describing the mean value, and another parameter, which typically is an index, scale or precision parameter.

As in the previous chapter we shall build on the likelihood principles, and the underlying distributions all belong to the class of distributions called the exponential family.

Using the likelihood theory the concept of the Residual Sum of Squares (RSS) used for the Gaussian case (with $\Sigma = I$) is further generalized. The fit of a generalized linear model to data is assessed through the *deviance*, and the deviance is also used to compare nested models.

The theory of generalized linear models is described by McCullagh and Nelder (1989) and Lindsey (1997).

4.1   Types of response variables

In building statistical models for data the concept of a response variable is crucial. For the classical GLM considered in the previous chapter the response variable Y is often assumed to be quantitative, continuous and normally distributed. However, in practice several types of response variable are seen:

i) Count data (e.g. $y_1 = 57$, $y_2 = 67$, $y_3 = 54$, ..., $y_n = 59$ accidents), Poisson distribution, see Example 4.7 on page 123.

ii) Binary (or quantal) response variables (e.g., $y_1 = 0, y_2 = 0$, $y_3 = 1$, ..., $y_n = 0$), or proportion of counts (e.g. $y_1 = 15/297$, $y_2 = 17/242$, $y_3 = 2/312$, ..., $y_n = 144/285$), Binomial distribution, see Examples 4.6, 4.14, 4.11 and 4.10.

iii) Count data, waiting times, Negative Binomial distribution.

iv) Count data, multiple ordered categories "Very unsatisfied", "Unsatisfied", "Neutral", "Satisfied", "Very satisfied", Multinomial distribution, see Example 4.12 on page 133.

v) Count data, multiple categories.

vi) Continuous responses, constant variance, normal distribution (e.g., $y_1 = 2.567$, $y_2 = -0.453$, $y_3 = 4.808$, ..., $y_n = 2.422$).

vii) Continuous positive responses with constant coefficient of variation (e.g., empirical variances, $s_1^2 = 3.82$, $s_2^2 = 2.56$, ..., $s_n^2 = 2.422$), Gamma distribution, see Examples 4.8 and 4.13.

viii) Continuous positive highly skewed, Inverse Gaussian, see Example 4.9 on page 125.

The data shown as examples are assumed to be statistical relevant data – not just numbers. Note, for instance, that data given for different types of responses may look alike, but the (appropriate) statistical treatment is different!

4.2 Exponential families of distributions

Consider a univariate random variable Y with a distribution described by a family of densities $f_Y(y; \theta)$, $\theta \in \Omega$.

DEFINITION 4.1 – A NATURAL EXPONENTIAL FAMILY

A family of probability densities which can be written on the form

$$f_Y(y; \theta) = c(y) \exp(\theta y - \kappa(\theta)), \quad \theta \in \Omega \tag{4.1}$$

is called a *natural exponential family* of distributions. The function $\kappa(\theta)$ is called the *cumulant generator*. The representation (4.1) is called the *canonical parameterization* of the family, and the parameter θ is called the *canonical parameter*. The parameter space, Ω, is a subset of the real line.

A natural exponential family is characterized by its support (the set of y values for which $f_Y(y; \theta) > 0$) and the cumulant generator $\kappa(\cdot)$. As we shall see in Section 4.2 on page 92, mean and variance of Y are determined as functions of θ by the cumulant generator.

▸ **Remark 4.1**
The characterization of natural exponential families extends to multidimension families for vector valued $\boldsymbol{y}$ and $\boldsymbol{\theta}$ and with θy replaced by the inner product, $\boldsymbol{\theta}^T \boldsymbol{y}$.

Exponential families were first introduced by Fisher, Fisher (1922). They play an important role in mathematical statistics because of their simple inferential properties. For exponential families the sum of the observations is a sufficient statistic, which is a property that is unique for the exponential families.   ◀

DEFINITION 4.2 – AN EXPONENTIAL DISPERSION FAMILY

A family of probability densities which can be written on the form

$$f_Y(y; \theta) = c(y, \lambda) \exp(\lambda \{\theta y - \kappa(\theta)\}) \tag{4.2}$$

is called an *exponential dispersion family* of distributions. The parameter $\lambda > 0$ is called the *precision parameter* or *index parameter*.

The basic idea behind the dispersion family is to separate the mean value related distributional properties described by the *cumulant generator* $\kappa(\theta)$

from such dispersion features as sample size, common variance, or common over-dispersion that are not related to the mean value. The latter features are captured in the precision parameter, λ. In some cases the precision parameter represents a known number of observations as for the binomial distribution, or a known shape parameter as for the gamma (or χ^2) distribution. In other cases the precision parameter represents an unknown dispersion like for the normal distribution, or an over-dispersion that is not related to the mean.

Formally, an exponential dispersion family is just an indexed set of natural families, indexed by the precision parameter λ, and therefore the family shares fundamental properties with the exponential family.

In some situations the exponential dispersion family of models (refeq:dexp) for Y is obtained by an addition of related natural exponential family random variables. In this case the exponential dispersion family is often called an *additive exponential dispersion model*, as also demonstrated in some of the following examples.

The corresponding family of densities for $Z = Y/\lambda$ is called the *reproductive exponential dispersion family*.

Example 4.2 – Poisson distribution
Consider $Y \sim \text{Pois}(\mu)$. The probability function for Y is

$$f_Y(y; \mu) = \frac{\mu^y e^{-\mu}}{y!}$$

$$= \exp\{y \log(\mu) - \mu - \log(y!)\}. \tag{4.3}$$

Comparing with (4.1) it is seen that $\theta = \log(\mu)$ which means that $\mu = \exp(\theta)$. Thus the Poisson distribution is a special case of a natural exponential family with canonical parameter $\theta = \log(\mu)$, cumulant generator $\kappa(\theta) = \exp(\theta)$ and $c(y) = 1/y!$. The canonical parameter space is $\Omega = \mathbb{R}$.

Example 4.3 – Normal distribution
Consider $Y \sim N(\mu, \sigma^2)$. The probability function for Y is

$$f_Y(y; \mu, \sigma^2) = \frac{1}{\sqrt{2\pi}\sigma} \exp\left[-\frac{(y-\mu)^2}{2\sigma^2}\right]$$

$$= \frac{1}{\sqrt{2\pi}\sigma} \exp\left\{\frac{1}{\sigma^2}\left(\mu y - \frac{\mu^2}{2}\right) - \frac{y^2}{2\sigma^2}\right\}. \tag{4.4}$$

Thus the normal distribution belongs to the exponential dispersion family with $\theta = \mu$, $\kappa(\theta) = \theta^2/2$ and $\lambda = 1/\sigma^2$. The canonical parameter space is $\Omega = \mathbb{R}$.

Example 4.4 – Binomial and Bernoulli distribution

The probability function for a binomially distributed random variable, Z, may be written as

$$
\begin{aligned}
f_Z(n,p) &= \binom{n}{z} p^z (1-p)^{n-z} \\
&= \binom{n}{z} \exp\left(z \log\left(\frac{p}{1-p}\right) + n\log(1-p)\right).
\end{aligned}
\tag{4.5}
$$

For $n = 1$ we have the Bernoulli distribution. By putting $\theta = \log(\frac{p}{1-p})$, i.e., $p = \frac{\exp(\theta)}{1+\exp(\theta)}$, it is seen that the Bernoulli distribution is a special case of a natural exponential family, i.e., of the form (4.1), with $\theta = \log(\frac{p}{1-p})$, and $\kappa(\theta) = \log(1+\exp(\theta))$.

The Binomial distribution is recognized as being an exponential dispersion model of the form described in (4.2) with $\theta = \log(\frac{p}{1-p})$, $\kappa(\theta) = \log(1+\exp(\theta))$, and $\lambda = n$.

By considering n in the Binomial distribution as an indexing parameter, the distribution of $Y = Z/n$ belongs to the additive exponential dispersion family, which is written on the form $f_Y(y;\theta) = c(y,\lambda)\exp(\lambda\{\theta y - \kappa(\theta)\})$ where $\theta = \log(\frac{p}{1-p})$, $\kappa(\theta) = \log(1+\exp(\theta))$ and $\lambda = n$. In this case the precision or index parameter λ represents the (known) number of observations.

Mean and variance

The properties of the exponential dispersion family are mainly determined by the cumulant generator $\kappa(\cdot)$. It is rather easily shown that if Y has a distribution given by (4.1) then

$$
\mathrm{E}[Y] = \kappa'(\theta) \tag{4.6}
$$

$$
\mathrm{Var}[Y] = \frac{\kappa''(\theta)}{\lambda} \tag{4.7}
$$

Hence, the function

$$
\tau(\theta) = \kappa'(\theta) \tag{4.8}
$$

is seen to be monotone, and the function defines a one to one mapping $\mu = \tau(\theta)$ of the parameter space, Ω, for the canonical parameter θ on to a subset, $\mathcal{M}$, of the real line, called the *mean value space*. Roughly speaking, the mean value space is the convex hull of the support of the distribution. For some distributions the mean value space is only a subset of the real line. For the binomial distribution $0 < \mu < 1$.

DEFINITION 4.3 – (UNIT) VARIANCE FUNCTION

By introducing the mean value parameter in (4.7) we may express the variance as a function, the *(unit) variance function*, of the mean value in the

distribution as
$$V(\mu) = \kappa''(\tau^{-1}(\mu)) \tag{4.9}$$

▸ **Remark 4.2 – Variance operator and variance function**
Note the distinction between the variance *operator*, $\text{Var}[Y]$, which calculates
the variance in the probability distribution of a random variable, Y, and the
variance *function*, which is a function, $V(\mu)$, that describes the variance as a
function of the mean value for a given family of distributions.   ◂

The parameterization by the mean value, μ facilitates the comparison of
observations, y, with parameter values according to some model. As a mean
for comparing observations, y, with parameter values, μ, we define the *unit
deviance*:

DEFINITION 4.4 – THE UNIT DEVIANCE
As a mean for comparing observations, y, with parameter values, μ, we define
the *unit deviance* as

$$d(y;\mu) = 2 \int_{\mu}^{y} \frac{y-u}{V(u)}\, du \ , \tag{4.10}$$

where $V(\cdot)$ denotes the variance function.

Examples for several distributions belonging to the exponential family distributions are found in Appendix B on page 257.

Thus, the unit deviance is an expression of the deviation of y from μ,
weighted by the reciprocal variance. Figure 4.1 on the next page shows the
unit deviance for observations $y = 2$, and $y = 8$ under various distributional
assumptions on the distribution of Y. It is seen that the deviance captures
curvature as well as asymmetry of the likelihood function.

▸ **Remark 4.3 – Unit deviance for the normal distribution**
It is readily found that in the case of a normal distribution with constant
variance, i.e., $V(\mu) = 1$, the unit deviance is simply $d(y;\mu) = (y - \mu)^2$.   ◂

▸ **Remark 4.4 – Density expressed by the unit deviance**
The density (4.2) for the exponential dispersion family may be expressed in
terms of the mean value parameter, μ as

$$g_Y(y;\mu,\lambda) = a(y,\lambda) \, \exp\left\{ -\frac{\lambda}{2}\, d(y;\mu) \right\} , \tag{4.11}$$

where $d(y;\mu)$ is the unit deviance in Equation (4.10).   ◂

Several examples are shown in Appendix B.

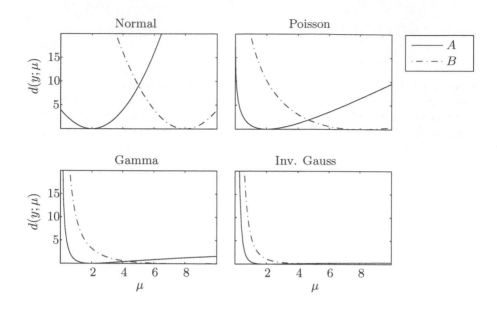

Figure 4.1: *Unit deviance for A and B corresponding to $y = 2$ and $y = 8$ respectively, under various distributional assumptions on Y.*

▶ **Remark 4.5 – Alternative definition of the deviance**

Let $\ell(y; \mu)$ denote the log-likelihood of the current model. Then apart from λ, the unit deviance may be defined as

$$d(y; \mu) = 2 \max_{\mu} \ell(\mu; y) - 2\ell(\mu; y) \,. \tag{4.12}$$

This definition is found in many textbooks. The definition corresponds to considering a *normalized, or relative likelihood* for μ corresponding to the observation y:

$$R(\mu; y) = \frac{L(\mu; y)}{\max_{\mu} L(\mu; y)}$$

Then $d(y; \mu) = -2 \log(R(\mu; y))$.   ◀

It can be noted that for the normal distribution with $\boldsymbol{\Sigma} = \boldsymbol{I}$, the deviance is just the residual sum of squares (RSS).

Notice, that there are two equivalent representations for an exponential dispersion family, namely

   i) by the cumulant generator, $\kappa(\cdot)$ and parametrized by the canonical (or natural) parameter, $\theta \in \Omega$, and the precision parameter λ, or

ii) by the variance function $V(\cdot)$ specifying the variance as a function of the mean value parameter, $\mu \in \mathcal{M}$, and further parametrized by the precision parameter λ.

The two parameterizations (canonical and mean value parameterization) supplement each other. While the parameterization in terms of the canonical parameter, θ has the advantage that the parameter space is the real line and therefore well suited for linear operations, the parameterization in terms of the mean value parameter, μ has the advantage that the fit of the model can be directly assessed as the mean value is measured in the same units as the observations, Y.

DEFINITION 4.5 – THE CANONICAL LINK FUNCTION
The relation between the two parameterizations is determined as $\mu = \tau(\theta)$ with $\tau(\cdot)$ given by (4.8). The inverse mapping

$$\theta = \tau^{-1}(\mu) \qquad (4.13)$$

is called the *canonical link function*.

In general, when considering generalized linear models, a *link function* is any monotone mapping $\eta = g(\mu)$ of the mean value space to (a subset of) the real line used to form the *linear predictor* (i.e., $\boldsymbol{X}\boldsymbol{\beta}$).

Table 4.2 lists the variance function, $V(\mu)$, the unit deviance, $d(y; \mu)$, and the precision parameter, λ for exponential dispersion family distributions with variance function that is at most a quadratic function of the mean[1], see Morris (1982).

To emphasize that a family is characterized by the variance function, $V(\cdot)$ and the precision parameter, λ, the notation $Y \sim \mathrm{ED}(\mu, V(\mu)/\lambda)$ is sometimes used to indicate that Y has density (4.11) with $d(y; \mu)$ specified by the variance function $V(\cdot)$, through (4.10). The notation is in analogy with the notation $\mathrm{N}(\mu, \sigma^2)$ to indicate normal distributions. The notation $Y \sim \mathrm{ED}(\mu, V(\mu)/\lambda)$ highlights the importance of the variance function.

Exponential family densities as a statistical model

Consider n independent observations $\boldsymbol{Y} = (Y_1, Y_2, \cdots, Y_n)^T$, and assume that they belong to the same exponential dispersion family with the cumulant generator, $\kappa(\cdot)$, and the precision parameter is a known weight, $\lambda_i = w_i$, and the density is on the form (4.2).

Then the joint density, using the canonical parameter, is

$$f(\boldsymbol{y}; \boldsymbol{\theta}) = \exp\left[\sum_{i=1}^{n} w_i(\theta_i y_i - \kappa(\theta_i))\right] \prod_{i=1}^{n} c(y_i, w_i) \qquad (4.14)$$

---

[1] We have included the inverse Gaussian distribution family because of its wide applicability.

Table 4.2: *Mean value space, unit variance function and unit deviance for exponential dispersion families.*

Family	$\mathcal{M}$	Var(μ)	unit deviance$(y;\mu)$	λ	θ
Normal	$(-\infty,\infty)$	1	$(y-\mu)^2$	$1/\sigma^2$	μ
Poisson	$(0,\infty)$	μ	$2\left[y\log\left(\frac{y}{\mu}\right) - (y-\mu)\right]$	$-$ [a]	$\log(\mu)$
Gamma	$(0,\infty)$	μ^2	$2\left[\frac{y}{\mu} - \log\left(\frac{y}{\mu}\right) - 1\right]$	α [b]	$1/\mu$
Bin	$(0,1)$	$\mu(1-\mu)$	$2\left[y\log\left(\frac{y}{\mu}\right) + (1-y)\log\left(\frac{1-y}{1-\mu}\right)\right]$	n [c]	$\log\left(\frac{\mu}{1-\mu}\right)$
Neg Bin	$(0,1)$	$\mu(1+\mu)$	$2\left[y\log\left(\frac{y(1+\mu)}{\mu(1+y)}\right) + \log\left(\frac{1+\mu}{1+y}\right)\right]$	r [d]	$\log(\mu)$
I Gauss	$(0,\infty)$	μ^3	$\frac{(y-\mu)^2}{y\mu^2}$		$1/\mu^2$
GHSS [e]	$(-\infty,\infty)$	$1+\mu^2$	$2y\left[\arctan(y) - \arctan(\mu)\right] + \log\left(\frac{1+\mu^2}{1+y^2}\right)$	$1/\sigma^2$	$\arctan(\mu)$

[a] The precision parameter λ can not be distinguished from the mean value.
[b] Gamma distribution with shape parameter α and scale parameter μ/α.
[c] $Y = Z/n$, where Z is the number of successes in n independent Bernoulli trials.
[d] $Y = Z/r$, where Z is the number of successes until the rth failure in independent Bernoulli trials.
[e] Generalized Hyperbolic Secant Distribution.

or, by introducing the mean value parameter, $\mu = \tau(\theta)$ cf. (4.8) we find, using (4.11), the equivalent joint density

$$
g(\boldsymbol{y}; \boldsymbol{\mu}) = \prod_{i=1}^{n} g_Y(y_i; \mu_i, w_i)
$$

$$
= \exp\left[-\frac{1}{2}\sum_{i=1}^{n} w_i d(y_i; \mu_i)\right] \prod_{i=1}^{n} c(y_i, w_i)
$$

$$(4.15)$$

Notice, the difference between the two parameterizations.

The likelihood theory

Assume that the joint density for $Y_1, Y_2, \ldots, Y_n$ can be described by either (4.14) or the equivalent form (4.15). Then the log-likelihood function in the two cases is

$$
\ell_\theta(\boldsymbol{\theta}; \boldsymbol{y}) = \sum_{i=1}^{n} w_i(\theta_i y_i - \kappa(\theta_i)) \tag{4.16}
$$

$$
\ell_\mu(\boldsymbol{\mu}; \boldsymbol{y}) = -\frac{1}{2}\sum_{i=1}^{n} w_i d(y_i; \mu_i) = -\frac{1}{2}\,\mathrm{D}(\boldsymbol{y}; \boldsymbol{\mu})\,, \tag{4.17}
$$

where

$$
\mathrm{D}(\boldsymbol{y}; \boldsymbol{\mu}) = \sum_{i=1}^{n} w_i d(y_i, \mu_i)\,. \tag{4.18}
$$

Note the similarity with the model for the normal density (3.5) and the related deviance $\mathrm{D}(\boldsymbol{y}; \boldsymbol{\mu})$ (3.4).

▶ **Remark 4.6**
The quantity $\mathrm{D}(\boldsymbol{y}; \boldsymbol{\mu})$ given by (4.18) is not a norm as such. However, $\mathrm{D}(\boldsymbol{y}; \boldsymbol{\mu})$ is very useful for measuring the distance between the observation and the mean value of the model.   ◀

The *score function* wrt. the canonical parameter, θ, is

$$
\frac{\partial}{\partial \theta_i}\,\ell_\theta(\boldsymbol{\theta}; \boldsymbol{y}) = w_i(y_i - \tau(\theta_i)) \tag{4.19}
$$

or in matrix form

$$
\frac{\partial}{\partial \boldsymbol{\theta}}\,\ell_\theta(\boldsymbol{\theta}; \boldsymbol{y}) = \mathrm{diag}(\boldsymbol{w})(\boldsymbol{y} - \boldsymbol{\tau}(\boldsymbol{\theta})) \tag{4.20}
$$

diag($\boldsymbol{w}$) denotes a diagonal matrix where the ith element is w_i, and

$$\boldsymbol{\tau}(\boldsymbol{\theta}) = \begin{Bmatrix} c\tau(\theta_1) \\ \tau(\theta_2) \\ \vdots \\ \tau(\theta_n) \end{Bmatrix}$$

is the result of the mean value mapping.

The observed information (2.25) wrt. $\boldsymbol{\theta}$ is

$$\begin{aligned} \boldsymbol{j}(\boldsymbol{\theta}; \boldsymbol{y}) &= \text{diag}\{\boldsymbol{w}\} \, \frac{\partial}{\partial \boldsymbol{\theta}} \boldsymbol{\tau}(\boldsymbol{\theta}) \\ &= \text{diag}\{w_i \kappa''(\theta_i)\} \\ &= \text{diag}\{w_i \, V(\tau(\theta_i))\} \end{aligned} \tag{4.21}$$

where $V(\tau(\theta))$ denotes the value of the variance function for $\mu = \tau(\theta)$.

▸ **Remark 4.7**
Note that, since the observed information wrt. the canonical parameter θ does not depend on the observation $\boldsymbol{y}$, the expected information is the same as the observed information.   ◂

▸ **Remark 4.8**
Note that the Hessian of the likelihood function depends on θ. In the normal case the Hessian was constant.   ◂

By using (2.32) we find the score function wrt. the mean value parameter

$$\ell'_{\boldsymbol{\mu}}(\boldsymbol{\mu}; \, \boldsymbol{y}) = \text{diag}\left\{\frac{w_i}{V(\mu_i)}\right\} (\boldsymbol{y} - \boldsymbol{\mu}) \tag{4.22}$$

which shows that the ML-estimate for $\boldsymbol{\mu}$ is $\widehat{\boldsymbol{\mu}} = \boldsymbol{y}$.

The observed information is

$$\boldsymbol{j}(\boldsymbol{\mu}; \boldsymbol{y}) = \text{diag}\left\{ w_i \left[\frac{1}{V(\mu_i)} + (y_i - \mu_i)\frac{V'(\mu_i)}{V(\mu_i)^2}\right]\right\}, \tag{4.23}$$

and then the expected information corresponding to the set $\boldsymbol{\mu}$ is

$$\boldsymbol{i}(\boldsymbol{\mu}) = \text{diag}\left\{\frac{w_i}{V(\mu_i)}\right\}, \tag{4.24}$$

which could have been seen directly using (2.34).

▸ **Remark 4.9**
Note that the observed information wrt. the mean value parameter $\boldsymbol{\mu}$ depends on the observation, $\boldsymbol{y}$.   ◂

4.3   Generalized linear models

DEFINITION 4.6 – THE GENERALIZED LINEAR MODEL
Assume that $Y_1, Y_2, \ldots, Y_n$ are mutually independent, and the density can be described by an exponential dispersion model (see equation (4.11)) with the same variance function $V(\mu)$.

A *generalized linear model* for $Y_1, Y_2, \ldots, Y_n$ describes an affine hypothesis for $\eta_1, \eta_2, \ldots, \eta_n$, where

$$\eta_i = g(\mu_i)$$

is a transformation of the mean values $\mu_1, \mu_2, \ldots, \mu_n$.

The hypothesis is of the form

$$\mathcal{H}_0 \ : \ \boldsymbol{\eta} - \boldsymbol{\eta}_0 \in L \,, \tag{4.25}$$

where L is a linear subspace $\mathbb{R}^n$ of dimension k, and where $\boldsymbol{\eta}_0$ denotes a vector of *known off-set values*.

▶ **Remark 4.10**
The differences from the general linear model for the normal distribution are

  i) The probability density of the observations belongs to a much more general family, namely an exponential dispersion family.

 ii) The deviation between the observations and the predictions is measured in accordance with the assumed distribution.

iii) The "mean value plus error" model for normally distributed observations is substituted by a "distribution model specified by the mean value".

 iv) The linear part of the model $\boldsymbol{X\beta}$ describes a function of the mean value (and not the mean value directly, as for the classical GLM).

  v) We are able to formally test the goodness of fit of the model.   ◀

However, many of the results from the classical GLM setup may be carried over to hold also for generalized linear models.

DEFINITION 4.7 – DIMENSION OF THE GENERALIZED LINEAR MODEL
The dimension k of the subspace L for the generalized linear model is the *dimension of the model*.

DEFINITION 4.8 – DESIGN MATRIX FOR THE GENERALIZED LINEAR MODEL
Consider the linear subspace $L = \text{span}\{x_1, \ldots, x_k\}$, i.e., the subspace is spanned by k vectors $(k < n)$, such that the hypothesis can be written

$$\boldsymbol{\eta} - \boldsymbol{\eta}_0 = \boldsymbol{X\beta} \ \text{ with } \boldsymbol{\beta} \in \mathbb{R}^k \,, \tag{4.26}$$

where $\boldsymbol{X}$ has full rank. The $n \times k$ matrix $\boldsymbol{X}$ is called the *design matrix*.

The ith row of the design matrix is given by the *model vector*

$$\boldsymbol{x}_i = \begin{pmatrix} x_{i1} \\ x_{i2} \\ \vdots \\ x_{ik} \end{pmatrix}, \tag{4.27}$$

for the ith observation.

▶ **Remark 4.11 – Specification of a generalized linear model**
The specification of a generalized linear model involves:

a) Distribution/Variance function Specification of the distribution – or the *variance function* $V(\mu)$.

b) Link function Specification of the *link function* $g(\cdot)$, which describes a function of the mean value which can be described linearly by the explanatory variables.

c) Linear predictor Specification of the linear dependency

$$g(\mu) = \eta = (\boldsymbol{x})^T \boldsymbol{\beta}$$

d) Precision (optional) If needed the precision is formulated as *known individual weights*, $\lambda_i = w_i$, or as a *common dispersion parameter*[2], $\lambda = 1/\sigma^2$, or a *combination* $\lambda_i = w_i/\sigma^2$.   ◀

DEFINITION 4.9 – THE LINK FUNCTION
The *link function*, $g(\cdot)$ describes the relation between the linear predictor η_i and the mean value parameter $\mu_i = \mathrm{E}[Y_i]$. The relation is

$$\eta_i = g(\mu_i) \tag{4.28}$$

The inverse mapping $g^{-1}(\cdot)$ thus expresses the mean value μ as a function of the linear predictor η:

$$\mu = g^{-1}(\eta)$$

that is

$$\mu_i = g^{-1}(\boldsymbol{x}_i^T \boldsymbol{\beta}) = g^{-1}\left(\sum_j x_{ij}\beta_j\right) \tag{4.29}$$

The concept of a *full*, or *saturated*, model is defined as for the classical GLM. Under the full model the estimate for μ_i is just the observation y_i.

---

[2] In some presentations of generalized linear models the dispersion parameter is symbolized by ϕ.

DEFINITION 4.10 – THE LOCAL DESIGN MATRIX
Consider a generalized linear model with the design matrix X and variance function $V(\cdot)$. The mappings

$$\mu_i = g^{-1}(\eta_i)$$

and the linear mapping (4.26)

$$\eta = X\beta$$

defines μ as a function of β. The function is a vector function $\mathbb{R}^k \to \mathbb{R}^n$. The matrix

$$X(\beta) = \frac{\partial \mu}{\partial \beta} = \left[\frac{d\mu}{d\eta}\right]^T \frac{\partial \eta}{\partial \beta} = \text{diag}\left\{\frac{1}{g'(\mu_i)}\right\} X \qquad (4.30)$$

is called the *local design matrix* corresponding to the parameter value β.

In (4.30), $\text{diag}\{1/g'(\mu_i)\}$ denotes a $n \times n$ dimensional diagonal matrix, whose i^{th} diagonal element is $1/g'(\mu_i)$, where μ_i is determined by (4.29), i.e., as the mean value of the ith observation under the hypothesis (4.26) corresponding to the parameter value β, i.e.,

$$\mu_i = \mu_i(\beta) = \mu(g^{-1}(\eta_i)) = \mu(g^{-1}(x_i\beta)), \qquad (4.31)$$

where x_i is the model vector for the ith observation given by (4.27).

For the canonical link we have $g(\cdot) = \tau^{-1}(\cdot)$.

In the case of the canonical link we have

$$g'(\mu) = \frac{1}{\tau'(\tau^{-1}(\mu))} = \frac{1}{V(\mu)}. \qquad (4.32)$$

Thus, the smaller the variance, the steeper is the graph of canonical link function, and the larger the variance, the flatter the canonical link, see Figure 4.2 on the following page.

▶ **Remark 4.12 – Interpretation of the local design matrix**
The local design matrix $X(\beta)$ expresses the local (corresponding to the value of β) rescaling of the design matrix, X, which is needed in order to account for the nonlinearity in the transformation from the linear predictor η to μ. ◀

▶ **Remark 4.13 – The local design matrix for the canonical link**
If the link function is the canonical link, i.e., $g(\mu) = \tau^{-1}(\mu)$ then

$$g'(\mu) = \frac{1}{V(\mu)},$$

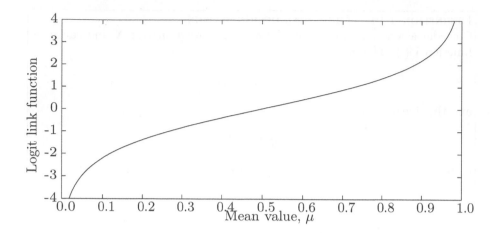

Figure 4.2: *The canonical link function for the binomial distribution. Note that the link function is steep when the variance function approaches 0, i.e., for $\mu \to 0$ and $\mu \to 1$.*

cf. (4.32), and hence

$$\left[\frac{d\mu}{d\eta}\right] = \text{diag}\left\{\frac{1}{g'(\mu_i)}\right\} = \text{diag}\left\{V(\mu_i)\right\}.$$

The local design matrix then becomes

$$X(\beta) = \text{diag}\{V(\mu_i)\}X. \tag{4.33}$$

◄

▶ **Remark 4.14**

Note that the link function is only used coordinate by coordinate.   ◄

Link functions

The most commonly used link functions, $\eta = g(\mu)$, are listed in Table 4.3. The canonical link is the function which transforms the mean to the canonical location parameter of the exponential dispersion family, i.e., it is the function for which $g(\mu) = \theta$. The canonical link function for the most widely considered densities is given in Table 4.4.

4.4   Maximum likelihood estimation

It may be shown that a generalized linear model with the canonical link by itself defines an exponential family. For a generalized linear model with the

Table 4.3: *Commonly used link functions.*

Name	Link function: $\eta = g(\mu)$	$\mu = g^{-1}(\eta)$
Identity	μ	η
logarithm	$\log(\mu)$	$\exp(\eta)$
logit	$\log(\mu/(1-\mu))$	$\exp(\eta)/[1+\exp(\eta)]$
reciprocal	$1/\mu$	$1/\eta$
power	μ^k	$\eta^{1/k}$
squareroot	$\sqrt{\mu}$	η^2
probit	$\Phi^{-1}(\mu)$	$\Phi(\eta)$
log-log	$\log(-\log(\mu))$	$\exp(-\exp(\eta))$
cloglog	$\log(-\log(1-\mu))$	$1-\exp(-\exp(\eta))$

Table 4.4: *Canonical link functions for some widely used densities.*

Density	Link: $\eta = g(\mu)$	Name
Normal	$\eta = \mu$	identity
Poisson	$\eta = \log(\mu)$	logarithm
Binomial	$\eta = \log[\mu/(1-\mu)]$	logit
Gamma	$\eta = 1/\mu$	reciprocal
Inverse Gauss	$\eta = 1/\mu^2$	power $(k=-2)$

canonical link it may be shown that the likelihood function is convex, and therefore, the maximum likelihood estimate is unique – if it exists. When the canonical parameter space is the real line, or an *open* subset thereof, it may happen that there is no maximum of the likelihood function, but the likelihood function is increasing for θ approaching the border of the parameter space.

Example 4.5 – Example of non-existing maximum likelihood estimate

Let $Z \sim \mathrm{B}(z; n, p)$ be a binomially distributed random variable. We saw in Example 4.4 on page 92 that the canonical parameter is $\theta = \log(p/(1-p))$ with log-likelihood function

$$\ell(\theta; z) = \log\binom{n}{z} + z\theta - n\log(1+\exp(\theta))$$

For $z = 0$ we find $\ell(\theta; 0) = C - n\log(1+\exp(\theta))$ which increases towards C for $\theta \to -\infty$. Thus, although the binomial distribution is well defined for $p = 0$, and the maximum-likelihood estimate for p is $\widehat{p} = 0$, the maximum-likelihood for the canonical parameter does not exist.

This is the price that has to be paid for the advantage of expanding the bounded parameter space $0 \leq \mu \leq 1$ for the mean value μ to the parameter space, $-\infty < \theta < \infty$ allowing for arbitrary linear transformation of the canonical parameter θe.

THEOREM 4.1 – ESTIMATION IN GENERALIZED LINEAR MODELS
Consider the generalized linear model in equation (4.25) for the observations $Y_1, \ldots Y_n$ and assume that $Y_1, \ldots Y_n$ are mutually independent with densities, which can be described by an exponential dispersion model with the variance function $V(\cdot)$, dispersion parameter σ^2, and optionally the weights w_i.

Assume that the linear predictor is parameterized with β corresponding to the design matrix X (4.26), then the maximum likelihood estimate $\widehat{\beta}$ for β is found as the solution to

$$[X(\beta)]^T i_\mu(\mu)(y - \mu) = 0 , \tag{4.34}$$

where $X(\beta)$ denotes the local design matrix (4.30), and $\mu = \mu(\beta)$ given by

$$\mu_i(\beta) = g^{-1}(x_i^T \beta) , \tag{4.35}$$

denotes the fitted mean values corresponding to the parameters β, and $i_\mu(\mu)$ is the expected information with respect to μ, (4.24).

Proof Disregarding the constant factor $1/\sigma^2$ we obtain the score function with respect to the mean value parameter μ as, cf. (4.22)

$$\ell'_\mu(\mu; y) = \text{diag}\left\{\frac{w_i}{V(\mu_i)}\right\}(y - \mu).$$

It follows from (2.32), that the score function with respect to β is

$$\ell'_\beta(\beta; y) = \left[\frac{\partial \mu}{\partial \beta}\right]^T \ell'_\mu(\mu; y) = [X(\beta)]^T \text{diag}\left\{\frac{w_i}{V(\mu_i)}\right\}(y - \mu(\beta)) . \tag{4.36}$$

By using (4.24) we get

$$\ell'_\beta(\beta; y) = \frac{\partial}{\partial \beta}\ell_\beta(\beta; y) = [X(\beta)]^T i_\mu(\mu)(y - \mu(\beta)) ,$$

which gives (4.34).   ∎

The score function (4.34) is nonlinear in β, so the ML-estimates have to be found by an iterative procedure. The *iteratively reweighted least squares method* is often used. We shall sketch this method in Section 4.4 on page 109.

▶ **Remark 4.15 – The score function for the canonical link**
For the canonical link the local design matrix $X(\beta)$ is $\text{diag}\{V(\mu_i)\}X$, and hence the score function (4.34) becomes

$$X^T \text{diag}\{V(\mu_i)\} \text{diag}\left\{\frac{w_i}{V(\mu_i)}\right\}(y - \mu(\beta)) ,$$

or

$$X^T \operatorname{diag}\{w_i\} \, y = X^T \operatorname{diag}\{w_i\} \, \mu. \tag{4.37}$$

Equation (4.37) is called the *mean value equation*, or *the normal equation* – compare with (3.26).

For an unweighted model the mean value equation simply becomes

$$X^T y = X^T \mu(\beta) \tag{4.38}$$

◄

▶ **Remark 4.16 – The ML-estimate minimizes the deviance**
It follows from (4.18) that the log-likelihood function for μ is

$$\ell_\mu(\mu; y) = - \frac{1}{2\sigma^2} \, D(y; \mu)$$

and hence maximizing the likelihood-function corresponds to minimizing the deviance.

◄

DEFINITION 4.11 – RESIDUAL DEVIANCE
Consider the generalized linear model in Theorem 4.1. The *residual deviance* corresponding to this model is

$$D(y; \mu(\widehat{\beta})) = \sum_{i=1}^{n} w_i d(y_i; \widehat{\mu}_i) \tag{4.39}$$

with $d(y_i; \widehat{\mu}_i)$ denoting the unit deviance corresponding the observation y_i and the fitted value $\widehat{\mu}_i$ (4.49), and where w_i denotes the weights (if present).

If the model includes a dispersion parameter σ^2, the *scaled* residual deviance is

$$D^*(y; \mu(\widehat{\beta})) = \frac{D(y; \mu(\widehat{\beta}))}{\sigma^2} . \tag{4.40}$$

It follows from (4.34), that:

▶ **Remark 4.17 – Fitted – or predicted values**
The fitted value, $\widehat{\mu}$ is determined as the projection (wrt. the information matrix) on to the local tangent plane of the model, and the residuals are orthogonal to this tangent plane (see Figure 4.3 on the next page). This is analogous to the result for a classical GLM in Remark 3.9 on page 50. ◄

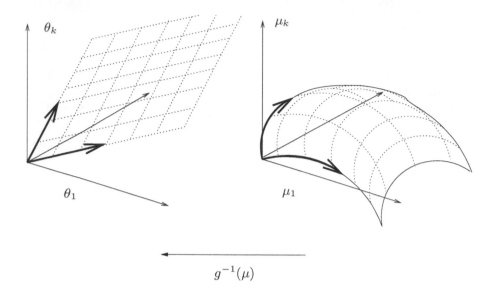

Figure 4.3: *The linear model space (left) and the mean value space (right).*

Properties of the ML estimator

THEOREM 4.2 – ASYMPTOTIC DISTRIBUTION OF THE ML ESTIMATOR
Under the hypothesis $\boldsymbol{\eta} = \boldsymbol{X}\boldsymbol{\beta}$ we have asymptotically

$$\frac{\widehat{\boldsymbol{\beta}} - \boldsymbol{\beta}}{\sqrt{\sigma^2}} \in \mathrm{N}_k(\boldsymbol{0}, \boldsymbol{\Sigma}) \,, \tag{4.41}$$

where the dispersion matrix $\boldsymbol{\Sigma}$ for $\widehat{\boldsymbol{\beta}}$ is

$$\mathrm{D}[\widehat{\boldsymbol{\beta}}] = \boldsymbol{\Sigma} = [\boldsymbol{X}^T \boldsymbol{W}(\boldsymbol{\beta}) \boldsymbol{X}]^{-1} \tag{4.42}$$

with

$$\boldsymbol{W}(\boldsymbol{\beta}) = \mathrm{diag}\left\{ \frac{w_i}{[g'(\mu_i)]^2 V(\mu_i)} \right\} . \tag{4.43}$$

▸ **Remark 4.18**
In the case of the canonical link, the weight matrix $\boldsymbol{W}(\boldsymbol{\beta})$ is

$$\boldsymbol{W}(\boldsymbol{\beta}) = \mathrm{diag}\left\{ w_i V(\mu_i) \right\} . \tag{4.44}$$

◂

Linear prediction and fitted values

Consider the generalized linear model from Theorem 4.1. Assume that $\widehat{\boldsymbol{\beta}}$ is the maximum likelihood estimate for the parameter $\boldsymbol{\beta}$. Let us then introduce:

Definition 4.12 – Linear prediction for the generalized linear model

The linear prediction $\widehat{\boldsymbol{\eta}}$ is defined as the values

$$\widehat{\boldsymbol{\eta}} = \boldsymbol{X}\widehat{\boldsymbol{\beta}} \tag{4.45}$$

with the linear prediction corresponding to the i'th observation is

$$\widehat{\eta}_i = \sum_{j=1}^{k} x_{ij}\widehat{\beta}_j = (\boldsymbol{x}_i)^T \widehat{\boldsymbol{\beta}} . \tag{4.46}$$

The linear predictions $\widehat{\boldsymbol{\eta}}$ are approximately normally distributed with

$$\mathrm{D}[\widehat{\boldsymbol{\eta}}] \approx \widehat{\sigma}^2 \boldsymbol{X}\boldsymbol{\Sigma}\boldsymbol{X}^T \tag{4.47}$$

with $\boldsymbol{\Sigma}$ given by (4.42).

Definition 4.13 – Fitted values for the generalized linear model

The fitted values are defined as the values

$$\widehat{\boldsymbol{\mu}} = \boldsymbol{\mu}(\boldsymbol{X}\widehat{\boldsymbol{\beta}}) , \tag{4.48}$$

where the ith value is given as

$$\widehat{\mu}_i = g^{-1}(\widehat{\eta}_i) \tag{4.49}$$

with the fitted value $\widehat{\eta}_i$ of the linear prediction given by (4.46).

The fitted values $\widehat{\boldsymbol{\mu}}$ are approximately normally distributed with

$$\mathrm{D}[\widehat{\boldsymbol{\mu}}] \approx \widehat{\sigma}^2 \left[\frac{\partial \boldsymbol{\mu}}{\partial \boldsymbol{\eta}}\right]^2 \boldsymbol{X}\boldsymbol{\Sigma}\boldsymbol{X}^T \tag{4.50}$$

with $\boldsymbol{\Sigma}$ given by (4.42).

▶ Remark 4.19

If the canonical link is used, then the canonical parameter θ is used as linear predictor, i.e., instead of (4.46) we use

$$\widehat{\theta}_i = \sum_{j=1}^{k} x_{ij}\widehat{\beta}_j = (\boldsymbol{x}_i)^T \widehat{\boldsymbol{\beta}} . \tag{4.51}$$

◀

Residuals

Residuals for generalized linear models may be defined in several different
ways some of which will be presented below.

DEFINITION 4.14 – RESPONSE RESIDUAL

Consider the generalized linear model from Theorem 4.1 on page 104 for the
observations $Y_1, \ldots, Y_n$.

The *response residual* (or just the residual) is defined as the values

$$r_i^R = r_R(y_i; \widehat{\mu}_i) = y_i - \widehat{\mu}_i \, , \, i = 1, 2, \ldots, n \, , \tag{4.52}$$

where $\widehat{\mu}_i$ denotes the fitted value (4.49) corresponding to the ith observation.

DEFINITION 4.15 – DEVIANCE RESIDUAL

Consider the generalized linear model from Theorem 4.1 on page 104 for the
observations $Y_1, \ldots, Y_n$.

The *deviance residual* for the i'th observation is defined as

$$r_i^D = r_D(y_i; \widehat{\mu}_i) = \text{sign}(y_i - \widehat{\mu}_i)\sqrt{w_i d(y_i, \widehat{\mu}_i)} \tag{4.53}$$

where $\text{sign}(x)$ denotes the *sign function* $\text{sign}(x) = 1$ for $x > 0$ and $\text{sign}(x) = -1$ for $x < 0$, and with w_i denoting the weight (if relevant), $d(y; \mu)$ denoting
the unit deviance and $\widehat{\mu}_i$ denoting the fitted value corresponding to the i'th
observation.

As the deviance is non-negative, it does not indicate whether the observation
y_i is larger or smaller than the fitted value $\widehat{\mu}_i$. As the sign of the deviation
between the observation and the fitted value may be of interest in analyses
of residuals, this sign has been introduced in the definition of the deviance
residual.

Assessments of the deviance residuals is in good agreement with the likelihood approach as the deviance residuals simply express differences in log-
likelihood.

DEFINITION 4.16 – PEARSON RESIDUAL

Consider again the generalized linear model from Theorem 4.1 on page 104
for the observations $Y_1, \ldots, Y_n$. The *Pearson residuals* are defined as the
values

$$r_i^P = r_P(y_i; \widehat{\mu}_i) = \frac{y_i - \widehat{\mu}_i}{\sqrt{V(\widehat{\mu}_i)/w_i}}. \tag{4.54}$$

The Pearson residual is thus obtained by scaling the response residual with
$\sqrt{\text{Var}[Y_i]}$. Hence, the Pearson residual is the response residual normalized
with the estimated standard deviation for the observation.

▸ **Remark 4.20 – The squared Pearson residual approximates the deviance**

It follows from (4.10) that the Taylor expansion of the unit deviance $d(y;\mu)$ around $y = \mu$ is

$$d(y;\mu) \approx \frac{(y-\mu)^2}{V(\mu)} \,, \tag{4.55}$$

which is the squared Pearson residual.

As the squared deviance residual is produced by multiplying the unit deviance with the weight (if relevant) it follows that

$$r_D(y;\widehat{\mu})^2 \approx \frac{w(y-\widehat{\mu})^2}{V(\widehat{\mu})} = r_P(y;\widehat{\mu})^2 \,. \tag{4.56}$$

◂

DEFINITION 4.17 – WORKING RESIDUAL

Consider the generalized linear model from Theorem 4.1 on page 104 for the observations $Y_1,\ldots Y_n$. The *working residual* for the i'th observation is defined as

$$r_i^W = (y_i - \widehat{\mu}_i)\frac{\partial\widehat{\eta}_i}{\partial\widehat{\mu}_i} \,. \tag{4.57}$$

These residuals are the difference between the working response and the linear predictor at the final iteration of the iteratively reweighted least squares algorithm that often is used in the estimation procedure.

Iteratively reweighted least squares solution

We recall the *score equations* (4.36)

$$\left[\frac{\partial\mu}{\partial\beta}\right]^T \ell'_\mu(\boldsymbol{\mu};\boldsymbol{y}) = [\boldsymbol{X}(\boldsymbol{\beta})]^T \operatorname{diag}\left\{\frac{w_i}{V(\mu_i)}\right\} (\boldsymbol{y} - \boldsymbol{\mu}(\boldsymbol{\beta}))$$

with

$$\boldsymbol{X}(\boldsymbol{\beta}) = \operatorname{diag}\left\{\frac{1}{g'(\mu_i)}\right\}\boldsymbol{X} \,.$$

For the canonical link the local design matrix $\boldsymbol{X}(\boldsymbol{\beta})$ simplifies to $\boldsymbol{X}(\boldsymbol{\beta}) = \operatorname{diag}\{V(\mu_i)\}\boldsymbol{X}$, and the score function is

$$S(\boldsymbol{\beta}) = \boldsymbol{X}^T \operatorname{diag}\{V(\mu_i)\} \operatorname{diag}\left\{\frac{w_i}{V(\mu_i)}\right\} (\boldsymbol{y}-\boldsymbol{\mu}(\boldsymbol{\beta})) = \boldsymbol{X}^T \operatorname{diag}\{w_i\}(\boldsymbol{y}-\boldsymbol{\mu}(\boldsymbol{\beta}))$$

$$S(\boldsymbol{\beta}) = \boldsymbol{X}^T\boldsymbol{\Sigma}^{-1}\left[\frac{\partial\eta}{\partial\mu}\right](\boldsymbol{y} - \boldsymbol{\mu}(\boldsymbol{\beta}))$$

$$\boldsymbol{i}_\beta(\widehat{\boldsymbol{\beta}}) = \boldsymbol{X}^T\boldsymbol{\Sigma}^{-1}\boldsymbol{X}$$

with

$$\Sigma = \left[\frac{\partial \eta}{\partial \mu}\right]^2 V(\mu_i)/w_i = \text{diag}\left\{\frac{1}{w_i V(\mu_i)}\right\}$$

so that

$$i_\beta(\widehat{\beta}) = X^T \text{diag}\{w_i V(\mu_i)\} X .$$

The score equation is of the form $S(\beta) = 0$. We shall use the Newton-Raphson procedure to determine the solution by iteration. At each step, ν, the updating formula is

$$\widehat{\beta}_{\nu+1} = \widehat{\beta}_\nu - [S'(\widehat{\beta}_\nu)]^{-1} S(\widehat{\beta}_\nu)$$
$$= \widehat{\beta}_\nu + [i_\beta(\widehat{\beta})]^{-1} S(\widehat{\beta}_\nu).$$

But using the expression for $i_\beta(\widehat{\beta})$ we have

$$\widehat{\beta}_{\nu+1} = \widehat{\beta}_\nu + (X^T \Sigma^{-1} X)^{-1} X^T \Sigma^{-1} \left[\frac{\partial \eta}{\partial \mu}\right] (y - \mu(\widehat{\beta}))$$
$$= (X^T \Sigma^{-1} X)^{-1} X^T \Sigma^{-1} \left\{ X\widehat{\beta}_\nu + \left[\frac{\partial \eta}{\partial \mu}\right] (y - \mu(\widehat{\beta})) \right\} \qquad (4.58)$$
$$= (X^T \Sigma^{-1} X)^{-1} X^T \Sigma^{-1} Z,$$

where

$$Z = X\beta_\nu + \left[\frac{\partial \eta}{\partial \mu}\right] (y - \mu(\beta)) \qquad (4.59)$$

is called the *working response*.

The iteration continues by updating the working response (4.59), and then calculates the next iterate of $\widehat{\beta}$ by (4.58). The iteration continues until successive estimates are sufficiently close to each other.

The method is called the *iteratively reweighted least squares* method because (4.58) is just a weighted least squares estimate.

We derived the algorithm as the Newton Raphson algorithm for solving the score equations (4.36) for a canonical link function. When the algorithm is used for other link functions it is called the *Fisher scoring algorithm*. (It corresponds to using the expected Fisher information instead of the observed information.)

Quasi-deviance and quasi-likelihood

We recall the score function (4.36). The estimate $\widehat{\beta}$ is determined by setting the score function equal to 0, i.e., as the solution to

$$\left[\frac{\partial \mu}{\partial \beta}\right]^T \text{diag}\left\{\frac{w_i}{V(\mu_i)}\right\} (y - \mu(\beta)) = 0 \qquad (4.60)$$

The objective function corresponding to (4.60) is the deviance

$$D(\boldsymbol{\mu}(\boldsymbol{\beta}); \boldsymbol{y}) = \sum_{i=1}^{n} \lambda_i d(y_i; \mu_i)$$

with $d(y; \mu)$ defined by

$$d(y; \mu) = 2 \int_{\mu}^{y} \frac{y - u}{V(u)} \, du \ . \tag{4.61}$$

Now, consider a model, specified only by the first two moments in the distribution of Y

$$E[Y_i] = \mu_i(\boldsymbol{\beta})$$
$$\mathrm{Var}[Y_i] = \lambda_i V(\mu_i)$$

for known functions $\mu_i(\cdot)$ and $V(\cdot)$ of an unknown regression parameter $\boldsymbol{\beta}$. Then (4.60) with $d(y; \mu)$ defined by (4.61) provides an *estimating equation* for $\boldsymbol{\beta}$.

However, this estimating equation does not necessarily correspond to a full likelihood solution because in general the normalizing constant in the probability distribution of the observations would also depend on the unknown parameters. Therefore, in this context, the function (4.61) is called a *quasi-deviance*, and the objective function associated with (4.60) is called a *quasi-likelihood*.

The advantage of using such a quasi-likelihood approach is that it allows for using user-defined variance functions, and the estimating equations assure that the observations are precision weighted in the estimating procedure.

4.5   Likelihood ratio tests

The approximative normal distribution of the ML-estimator implies that many distributional results from the classical GLM-theory are carried over to generalized linear models as approximative (asymptotic) results.

An example of this is the likelihood ratio test in Theorem 3.6 on page 54. In the classical GLM case it was possible to derive the exact distribution of the likelihood ratio test statistic (the F-distribution), but for generalized linear models, this is not possible, and hence we shall use the asymptotic results for the logarithm of the likelihood ratio in Theorem 2.5 on page 26.

THEOREM 4.3 – LIKELIHOOD RATIO TEST
Consider the generalized linear model from Theorem 4.1 on page 104.
  Assume that the model

$$\mathcal{H}_1 : \ \boldsymbol{\eta} \in L \subset \mathbb{R}^k$$

holds with L parameterized as $\boldsymbol{\eta} = \boldsymbol{X}_1 \boldsymbol{\beta}$, and consider the hypothesis

$$\mathcal{H}_0 : \ \boldsymbol{\eta} \in L_0 \subset \mathbb{R}^m$$

where $\boldsymbol{\eta} = \boldsymbol{X}_0\boldsymbol{\alpha}$ and $m < k$, and with the alternative $\mathcal{H}_1 : \boldsymbol{\eta} \in L\backslash L_0$.
Then the likelihood ratio test for $\mathcal{H}_0$ has the test statistic

$$-2\log\ell = \mathrm{D}\left(\boldsymbol{\mu}(\widehat{\boldsymbol{\beta}}); \boldsymbol{\mu}(\boldsymbol{\beta}(\widehat{\boldsymbol{\alpha}}))\right) \tag{4.62}$$

where the deviance statistic $\mathrm{D}\left(\boldsymbol{\mu}(\widehat{\boldsymbol{\beta}}); \boldsymbol{\mu}(\boldsymbol{\beta}(\widehat{\boldsymbol{\alpha}}))\right)$ is given by (4.39).

When $\mathcal{H}_0$ is true, the test statistic $\mathrm{D}\left(\boldsymbol{\mu}(\widehat{\boldsymbol{\beta}}); \boldsymbol{\mu}(\boldsymbol{\beta}(\widehat{\boldsymbol{\alpha}}))\right)$ will asymptotically
follow a $\chi^2(k-m)$ distribution.

If the model includes a dispersion parameter, σ^2, then $\mathrm{D}\left(\boldsymbol{\mu}(\widehat{\boldsymbol{\beta}}); \boldsymbol{\mu}(\boldsymbol{\beta}(\widehat{\boldsymbol{\alpha}}))\right)$
will asymptotically follow a $\sigma^2\chi^2(k-m)$ distribution.

▶ **Remark 4.21 – Partitioning of residual deviance**
In analogy with (3.43) we have the useful Pythagorean relation

$$\mathrm{D}\left(\boldsymbol{y}; \boldsymbol{\mu}(\boldsymbol{\beta}(\widehat{\boldsymbol{\alpha}}))\right) = \mathrm{D}\left(\boldsymbol{y}; \boldsymbol{\mu}(\widehat{\boldsymbol{\beta}})\right) + \mathrm{D}\left(\boldsymbol{\mu}(\widehat{\boldsymbol{\beta}}); \boldsymbol{\mu}(\boldsymbol{\beta}(\widehat{\boldsymbol{\alpha}}))\right), \tag{4.63}$$

which can be used to calculate $\mathrm{D}\left(\boldsymbol{\mu}(\widehat{\boldsymbol{\beta}}); \boldsymbol{\mu}(\boldsymbol{\beta}(\widehat{\boldsymbol{\alpha}}))\right)$ using the residual deviances
$\mathrm{D}\left(\boldsymbol{y}; \boldsymbol{\mu}(\widehat{\boldsymbol{\beta}})\right)$ and $\mathrm{D}\left(\boldsymbol{y}; \boldsymbol{\mu}(\boldsymbol{\beta}(\widehat{\boldsymbol{\alpha}}))\right)$.

When $\mathcal{H}_1$ is true, then $\mathrm{D}\left(\boldsymbol{y}; \boldsymbol{\mu}(\widehat{\boldsymbol{\beta}})\right)$ is approximately distributed like
$\chi^2(n-k)$. When also $\mathcal{H}_0$ is true, then $\mathrm{D}\left(\boldsymbol{y}; \boldsymbol{\mu}(\boldsymbol{\beta}(\widehat{\boldsymbol{\alpha}}))\right)$ is approximately distributed like $\chi^2(n-m)$, and by Cochran's theorem (Theorem 3.4 on page 51).
$\mathrm{D}\left(\boldsymbol{\mu}(\widehat{\boldsymbol{\beta}}); \boldsymbol{\mu}(\boldsymbol{\beta}(\widehat{\boldsymbol{\alpha}}))\right)$ is approximately distributed like $\chi^2(k-m)$ and independent of $\mathrm{D}\left(\boldsymbol{y}; \boldsymbol{\mu}(\widehat{\boldsymbol{\beta}})\right)$.   ◀

Initial test for goodness of fit

In analogy with classical GLMs one often starts with formulating a rather
comprehensive model, and then uses Theorem 4.3 on the previous page to
reduce the model by successive tests. In contrast to classical GLMs we may
however test the goodness of fit of the initial model. The test is a special case
of Theorem 4.3 on the preceding page.

▶ **Remark 4.22 – Test for model "sufficiency"**
Consider the generalized linear model from Theorem 4.1 on page 104, and
assume that the dispersion $\sigma^2 = 1$. Let $\mathcal{H}_{\mathrm{full}}$ denote the *full*, or *saturated*
model allowing each observation to have its own, freely varying mean value, i.e.,
$\mathcal{H}_{\mathrm{full}}: \boldsymbol{\mu} \in \mathbb{R}^n$ (restricted only by restrictions on mean values), and consider
the hypothesis
$$\mathcal{H}_0 : \boldsymbol{\eta} \in L \subset \mathbb{R}^k$$

with L parameterized as $\boldsymbol{\eta} = \boldsymbol{X}_0\boldsymbol{\beta}$.

Then, as the residual deviance under $\mathcal{H}_{\mathrm{full}}$ is 0, the test statistic is the
residual deviance $\mathrm{D}\left(\boldsymbol{\mu}(\widehat{\boldsymbol{\beta}})\right)$. When $\mathcal{H}_0$ is true, the test statistic is distributed
as $\chi^2(n-k)$. The test rejects for large values of $\mathrm{D}\left(\boldsymbol{\mu}(\widehat{\boldsymbol{\beta}})\right)$.   ◀

When the hypothesis is rejected, this indicates that the model cannot be maintained. This is often taken to mean that the distributional assumptions on Y are inadequate. We shall return to possible directions to pursue in such situations in Section 4.5 and Chapter 6 on page 225.

▶ **Remark 4.23 – Residual deviance measures goodness of fit**
In itself, the residual deviance $D\left(y; \mu(\widehat{\beta})\right)$ that was introduced in (4.39) is a reasonable measure of the goodness of fit of a model. The above remark shows that the residual deviance is a test statistic for the goodness of fit of the model $\mathcal{H}_0$. Therefore, when referring to a hypothesized model $\mathcal{H}_0$, we shall sometimes use the symbol $G^2(\mathcal{H}_0)$ to denote the residual deviance $D\left(y; \mu(\widehat{\beta})\right)$.

Using that convention, the partitioning of residual deviance in (4.63) may be formulated as

$$G^2(\mathcal{H}_0|\mathcal{H}_1) = G^2(\mathcal{H}_0) - G^2(\mathcal{H}_1)$$
$$= D\left(\mu(\widehat{\beta}); \mu(\beta(\widehat{\alpha}))\right) \qquad (4.64)$$

with $G^2(\mathcal{H}_0|\mathcal{H}_1)$ interpreted as the goodness fit test statistic for $\mathcal{H}_0$ conditioned on $\mathcal{H}_1$ being true, and $G^2(\mathcal{H}_0)$ and $G^2(\mathcal{H}_1)$, denoting the unconditional goodness of fit statistics for $\mathcal{H}_0$ and $\mathcal{H}_1$, respectively. ◀

Analysis of deviance table

The initial test for goodness of fit of the initial model is often represented in an *analysis of deviance table* in analogy with the table for classical GLMs, Table 3.2 on page 57.

In the table the goodness of fit test statistic corresponding to the initial model $G^2(\mathcal{H}_1) = D\left(y; \mu(\widehat{\beta})\right)$ is shown in the line labeled "Error". The statistic should be compared to percentiles in the $\chi^2(n - k)$ distribution.

The table also shows the test statistic for $\mathcal{H}_{null}$ under the assumption that $\mathcal{H}_1$ is true. This test investigates whether the model is necessary at all, i.e., whether at least some of the coefficients differ significantly from zero.

The table is analogous to Table 3.2. Note, however that in the case of a generalized linear model, we can start the analysis by using the residual (error) deviance to test whether the model may be maintained, at all.

This is in contrast to the classical GLMs in the previous chapter where the residual sum of squares around the initial model $\mathcal{H}_1$ served to estimate σ^2, and therefore we had no reference value to compare with the residual sum of squares. But in generalized linear models the variance is a known function of the mean, and therefore in general there is no need to estimate a separate variance.

Overdispersion

It may happen that even if one has tried to fit a rather comprehensive model (i.e., a model with many parameters), the fit is not satisfactory, and the

Table 4.5: *Initial assessment of goodness of fit of a model* $\mathcal{H}_1$. $\mathcal{H}_{\text{null}}$ *and* $\widehat{\mu}_{\text{null}}$ *refer to the minimal model, i.e., a model with all observations having the same mean value.*

Source	f	Deviance	Mean deviance	Goodness of fit interpretation
Model $\mathcal{H}_{\text{null}}$	$k-1$	$\mathrm{D}\left(\mu(\widehat{\beta}); \widehat{\mu}_{\text{null}}\right)$	$\dfrac{\mathrm{D}\left(\mu(\widehat{\beta}); \widehat{\mu}_{\text{null}}\right)}{k-1}$	$G^2(\mathcal{H}_{\text{null}}\vert\mathcal{H}_1)$
Residual (Error)	$n-k$	$\mathrm{D}\left(y; \mu(\widehat{\beta})\right)$	$\dfrac{\mathrm{D}\left(y; \mu(\widehat{\beta})\right)}{n-k}$	$G^2(\mathcal{H}_1)$
Corrected total	$n-1$	$\mathrm{D}\left(y; \widehat{\mu}_{\text{null}}\right)$		$G^2(\mathcal{H}_{\text{null}})$

residual deviance $\mathrm{D}\left(y; \mu(\widehat{\beta})\right)$ is larger than what can be explained by the χ^2-distribution.

An explanation for such a poor model fit could be an improper choice of linear predictor, or of link or response distribution.

If the residuals exhibit a random pattern, and there are no other indications of misfit, then the explanation could be that the variance is larger than indicated by $V(\mu)$. We say that the data are *overdispersed*, and a more appropriate model might be obtained by including a *dispersion parameter*, σ^2, in the model, i.e., a distribution model of the form (4.11) with $\lambda_i = w_i/\sigma^2$, and σ^2 denoting the overdispersion, $\text{Var}[Y_i] = \sigma^2 V(\mu_i)/w_i$.

As the dispersion parameter only would enter in the score function as a constant factor, this does not affect the estimation of the mean value parameters β. However, because of the larger error variance, the distribution of the test statistics will be influenced.

If, for some reasons, the parameter σ^2 had been known beforehand, one would include this known value in the weights, w_i. Most often, when it is found necessary to choose a model with overdispersion, σ^2 shall be estimated from the data.

▶ **Remark 4.24 – Approximate moment estimate for the dispersion parameter**

For the normal distribution family, the dispersion parameter is just the variance σ^2; the estimation in this case has been described in the chapter on classical GLM.

In the case of a gamma distribution family, the shape parameter α acts as dispersion parameter; and for this family ML estimation of the shape parameter is not too complicated.

For other exponential dispersion families, ML estimation of the dispersion parameter is more tricky. The problem is that the dispersion parameter enters

Table 4.6: *Example of deviance table in the case of overdispersion. It is noted that the scaled deviance is equal to the model deviance scaled by the error deviance according to Table 4.5 on page 113.*

Source	f	Deviance	Scaled deviance
Model $\mathcal{H}_{\text{null}}$	$k-1$	$D\left(\boldsymbol{\mu}(\widehat{\boldsymbol{\beta}}); \widehat{\boldsymbol{\mu}}_{\text{null}}\right)$	$\frac{D(\boldsymbol{\mu}(\widehat{\boldsymbol{\beta}}); \widehat{\boldsymbol{\mu}}_{\text{null}})/(k-1)}{D(\boldsymbol{y}; \boldsymbol{\mu}(\widehat{\boldsymbol{\beta}}))/(n-k)}$
Residual (Error)	$n-k$	$D(\boldsymbol{y}; \boldsymbol{\mu}(\widehat{\boldsymbol{\beta}}))$	
Corrected total	$n-1$	$D\left(\boldsymbol{y}; \widehat{\boldsymbol{\mu}}_{\text{null}}\right)$	

in the likelihood function, not only as a factor to the deviance, but also in the normalizing factor $a(y_i, w_i/\sigma^2)$, see (4.15). And it is necessary to have an explicit expression for this factor as function of σ^2 (as in the case of the normal and the gamma distribution families) in order to perform the maximum likelihood estimation.

It is therefore common practice in these cases to use the residual deviance $D(\boldsymbol{y}; \boldsymbol{\mu}(\widehat{\boldsymbol{\beta}}))$ as basis for the estimation of σ^2 and use the result in Remark 4.21 on page 112 that $D(\boldsymbol{y}; \boldsymbol{\mu}(\widehat{\boldsymbol{\beta}}))$ is approximately distributed as $\sigma^2\chi^2(n-k)$. It then follows that

$$\widehat{\sigma}^2_{dev} = \frac{D(\boldsymbol{y}; \boldsymbol{\mu}(\widehat{\boldsymbol{\beta}}))}{n-k} \tag{4.65}$$

is asymptotically unbiased for σ^2.

Alternatively, one would utilize the corresponding Pearson goodness of fit statistic

$$X^2 = \sum_{i=1}^{n} w_i \frac{(y_i - \widehat{\mu}_i)^2}{V(\widehat{\mu}_i)}$$

which likewise follows a $\sigma^2\chi^2(n-k)$-distribution, and use the estimator

$$\widehat{\sigma}^2_{Pears} = \frac{X^2}{n-k} \tag{4.66}$$

◄

Table 4.6 shows a table of deviance in the case of overdispersion. In this case the *scaled deviance*, D^* (4.40), i.e., deviance divided by $\widehat{\sigma}^2$ should be used in the tests instead of the crude deviance. For calculation of p-values etc. the asymptotic χ^2-distribution of the scaled deviance is used.

In Chapter 6 on page 225 we shall demonstrate another way of modeling overdispersion in generalized linear models.

4.6   Test for model reduction

The principles for model reduction in generalized linear models are essentially the same as the principles outlined in Section 3.6 on page 58 for classical

GLMs. In classical GLMs the deviance is calculated as a (weighted) sum of squares, and in generalized linear models the deviance is calculated using the expression (4.10) for the unit deviance.

Besides this, the major difference is that instead of the exact F-tests used for classical GLMs the tests in generalized linear models are only approximate tests using the χ^2-distribution.

In particular, the principles of successive testing in hypotheses chains use a type I or type III partition of the deviance carry over to generalized linear models. In Example 4.14 on page 140 we shall show an example of hypotheses chains with two different paths of type I partition.

Also, the various tools for model diagnostics, estimable contrasts, etc. carry over from classical GLMs.

4.7 Inference on individual parameters

Test

THEOREM 4.4 – TEST OF INDIVIDUAL PARAMETERS β_j
A hypothesis $\mathcal{H}:$ $\beta_j = \beta_j^0$ *related to specific values of the parameters is tested by means of the test statistic*

$$u_j = \frac{\hat{\beta}_j - \beta_j^0}{\sqrt{\widehat{\sigma^2}\widehat{\sigma}_{jj}}} \,, \tag{4.67}$$

where $\widehat{\sigma^2}$ *indicates the estimated dispersion parameter (if relevant), and* $\widehat{\sigma}_{jj}$ *denotes the j'th diagonal element in* $\widehat{\boldsymbol{\Sigma}}$, *obtained from (4.42) by inserting* $\hat{\boldsymbol{\beta}}$.

Under the hypothesis u_j *is approximately distributed as a standardized normal distribution. The test statistic is compared with quantiles of a standardized normal distribution (some software packages use a $t(n-k)$ distribution). The hypothesis is rejected for large values of* $|u_j|$. *The p-value is found as* $p = 2\big(1 - \Phi(|u_j|)\big)$.

In particular is the test statistic for the hypothesis $\mathcal{H}:$ $\beta_j = 0$

$$u_j = \frac{\hat{\beta}_j}{\sqrt{\widehat{\sigma^2}\widehat{\sigma}_{jj}}} \,. \tag{4.68}$$

An equivalent test is obtained by considering the test statistic

$$z_j = u_j^2 \tag{4.69}$$

and reject the hypothesis for for $z_j > \chi^2_{1-\alpha}(1)$.

Proof Omitted. See Section 2.8 on page 22.   ∎

The test is often termed a *Wald test*.

Confidence intervals

Confidence intervals for individual parameters

Using again Theorem 4.2 on page 106 it is seen that a Wald-type approximate $100(1-\alpha)\%$ confidence interval is obtained as

$$\widehat{\beta}_j \pm u_{1-\alpha/2}\sqrt{\widehat{\sigma^2\widehat{\sigma}_{jj}}}$$

Likelihood ratio based confidence interval

Using Theorem 2.5 one finds that likelihood ratio based approximate $100(1-\alpha)\%$ confidence interval for β_j is determined by

$$\{\beta_j \;:\; \tilde{l}(\beta_j) \geq l_0\}$$

with

$$l_0 = l(\widehat{\boldsymbol{\beta}}) - 0.5\chi^2_{1-\alpha}(1)$$

and $\tilde{l}(\beta_j)$ denoting the profile log-likelihood for β_j.

The interval has to be determined by iteration. The interval reflects the shape of the likelihood function, and therefore it is not necessarily symmetric. Statistical software packages such as R have an option for producing likelihood ratio confidence intervals.

Confidence intervals for fitted values

An approximate $100(1-\alpha)\%$ confidence interval for the linear prediction is obtained as

$$\widehat{\eta}_i \pm u_{1-\alpha/2}\sqrt{\widehat{\sigma^2\widehat{\sigma}_{ii}}} \qquad (4.70)$$

with $\widehat{\eta}_i$ given by (4.46) and $\widehat{\sigma}_{ii}$ denoting the i'th diagonal element in $\boldsymbol{X\Sigma X}^T$ where $\boldsymbol{\Sigma}$ given by (4.42).

The corresponding interval for the fitted value $\widehat{\mu}_i$ is obtained by applying the inverse link transformation $g^{-1}(\cdot)$ to the confidence limits (4.70) in analogy with (4.49).

4.8 Examples

In this section we shall present a number of examples of generalized linear models. The examples serve to illustrate the versatility of this class of models rather than to demonstrate technicalities about computing, etc. However, in Example 4.14 on page 140 we report the various options in more detail. Examples will refer to the R software.

Applied statistics is a data-driven discipline. This is also reflected in the use of generalized linear models. When addressing a problem, after having identified response, and explanatory variables, then the next question to be considered is the type of response variable (Section 4.1), and the modeling of the random component in that type of response. The important implication of the choice of response distribution is that this determines the *variance structure*, i.e., how the variance varies with the mean, and, consequently, how the observations are weighted in the analysis. Example 4.9 on page 125 demonstrates this analytically.

Having selected a relevant distribution model, the next step is to select the *link function*. Selecting a distribution implies an associated *canonical link* that generally maps the (bounded) mean value space to the unbounded real line. In many cases the canonical link will be a good choice with good statistical properties (simple estimation by fitting mean values), but in some cases, the problem at hand suggests the use of another link function.

In the examples we shall illustrate a "linear regression model", $\eta = \beta x$, for the canonical link for the various exponential dispersion families. The Binomial response distribution is considered in Example 4.6, ordered categorical response in Example 4.12 on page 133, Poisson distribution in Example 4.7 on page 123, the χ^2-distribution of empirical variances in Example 4.13, and Gamma distribution in Example 4.8 on page 125. In Example 4.10 we consider a "regression model" for a Binomial response distribution and using various link functions, and Example 4.11 on page 131 shows an example of Bernoulli-filtered Poisson-data.

Further, in Example 4.14 on page 140, we shall illustrate a two-way analysis for binomially distributed data and canonical link, and in Example 4.15 on page 148 we show a two-way analysis for Poisson distributed data. Example 4.16 on page 151 extends this analysis to a situation with overdispersed Poisson data.

Logistic regression model for binary response

Example 4.6 – Assessment of toxicity
In Example 4.1 we introduced the data from a study of developmental toxicity of a chemical compound. The study is reported in Price et al. (1987). In the study a specified amount of the ether was dosed daily to pregnant mice, and after 10 days all fetuses were examined. The size of each litter and the number of stillborns were recorded, see Table 4.1 on page 88.

Below we shall introduce the *logistic regression model* which is often used to assess dose-response relationships for such studies. The logistic regression model is a special case of the generalized linear model.

Let Z_i denote the number of stillborns at dose concentration x_i. We shall assume $Z_i \sim B(n_i, p_i)$ (a binomial distribution corresponding to n_i independent trials (fetuses), and the probability, p_i, of stillbirth being the same for all n_i fetuses). We want to model $Y_i = Z_i/n_i$, and in particular we want a model for $E[Y_i] = p_i$.

We shall use a linear model for a function of p, the *link function*. The canonical link for the binomial distribution is the *logit transformation*

$$g(p) = \log\left(\frac{p}{1-p}\right),$$

and we will formulate a *linear model* for the transformed mean values

$$\eta_i = \log\left(\frac{p_i}{1-p_i}\right), \ i = 1, 2, \ldots, 5.$$

The *linear model* is

$$\eta_i = \beta_1 + \beta_2 x_i, \ i = 1, 2, \ldots, 5. \tag{4.71}$$

The inverse transformation, which gives the probabilities, p_i, for stillbirth is the *logistic function*

$$p_i = \frac{\exp \beta_1 + \beta_2 x_i}{1 + \exp(\beta_1 + \beta_2 x_i)}, \ i = 1, 2, \ldots, 5. \tag{4.72}$$

The mean value equation (4.37) for determination of β_1 and β_2 is

$$\begin{pmatrix} \sum_{i=1}^{5} n_i \frac{\exp(\beta_1+\beta_2 x_i)}{1+\exp(\beta_1+\beta_2 x_i)} \\ \sum_{i=1}^{5} x_i n_i \frac{\exp(\beta_1+\beta_2 x_i)}{1+\exp(\beta_1+\beta_2 x_i)} \end{pmatrix} = \begin{pmatrix} \sum_{i=1}^{5} z_i \\ \sum_{i=1}^{5} x_i z_i \end{pmatrix}$$

where we have used the weights[3] $w_i = n_i$ and utilized that $w_i y_i = z_i$, when $y_i = z_i/n_i$. The solution has to be determined by iteration.

Assume that the data is stored in the R object mice with mice$conc, mice$alive mice$stillb denoting the concentration, the number of live and the number of stillborn respectively, and let

```
> mice$resp <- cbind(mice$stillb,mice$alive)
```

denote the response variable composed by the vector of the number of stillborns, z_i, and the number of live fetuses, $n_i - z_i$. Then the commands

```
> mice.glm <- glm(formula = resp ~ conc, family = binomial(link =
    logit), data= mice)
> anova(mice.glm)
```

will give the output

```
Analysis of Deviance Table
Model: binomial, link: logit
Response: resp
```

---

[3]The weight parameter is relevant when you perform the calculations yourself. In standard statistical software, the weighting of the unit deviance is automatically determined when you specify the binomial distribution and the number of observations.

Table 4.7: *Deviance table.*

Source	f	Deviance	Goodness of fit interpretation
conc	1	253.330	$G^2(\mathcal{H}_M\|\mathcal{H}_1)$
Residual (Error)	3	5.777	$G^2(\mathcal{H}_1)$
Corrected total	4	259.107	$G^2(\mathcal{H}_M)$

```
Terms added sequentially (first to last)
    Df Deviance Resid. Df Resid. Dev
NULL                    4      259.107
conc  1   253.330       3        5.777
```

corresponding to Table 4.7.

The goodness of fit statistic is $D(y; \mu(\widehat{\beta})) = 5.777$ and as

```
> pval <- 1-pchisq(5.777,3)
```

results in a *p*-value, *pval* = 0.1229783, we have that $P[\chi(3)^2 \geq 5.78] = 0.12$. Thus, the goodness of fit statistic is not significant for significance levels less than 12%, and hence, there is no evidence contradicting the model assumption of a binomial distribution with p_i specified by (4.72).

The command > **summary(mice.glm)** results in the output

```
Call:
glm(formula = resp ~ conc, family = binomial(link = logit),
data = mice)

Deviance Residuals:
      1        2        3        4        5
 1.1317   1.0174  -0.5968  -1.6464   0.6284

Coefficients:
             Estimate Std. Error z value Pr(>|z|)
(Intercept) -3.2479337  0.1576602  -20.60   <2e-16
conc         0.0063891  0.0004348   14.70   <2e-16

(Dispersion parameter for binomial family taken to be 1)

    Null deviance: 259.1073  on 4  degrees of freedom
Residual deviance:   5.7775  on 3  degrees of freedom
AIC: 35.204

Number of Fisher Scoring iterations: 4
```

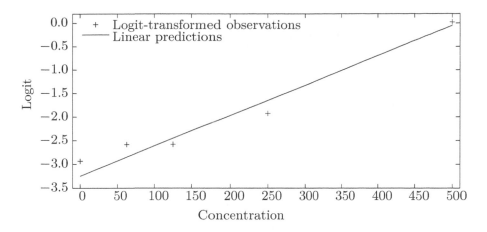

Figure 4.4: *Logit-transformed observations and corresponding linear predictions for dose response assay.*

Thus, β_2 as well as β_1 are significant, so all terms in the model are necessary. The command > **predict**(mice.glm) results in the linear predictions (4.46)

```
      1           2           3           4           5
-3.24793371 -2.84861691 -2.44930011 -1.65066652 -0.05339932
```

The observed logits and the linear predictions are shown in Figure 4.4. The empirical logits seem to be reasonably well fitted by the logit line. The command > **fitted**(mice.glm) results in the fitted values (4.49)

```
      1          2          3          4          5
0.03740121 0.05475285 0.07948975 0.16101889 0.48665334
```

The observed fractions and the fitted values are shown in Figure 4.5 on the following page. The logistic curve seems to fit the observed fractions reasonably well. The visual check is supported by the numerical check of the deviance residuals. The command > **residuals**(mice.glm) results in the output

```
     1          2          3          4          5
 1.1316578  1.0173676 -0.5967859 -1.6464253  0.628428
```

As none of the deviance residuals exceed ± 2 there is no indication of outliers. Thus, the logistic model is not contradicted and the estimated model may be used for specifying "safe levels."

Table 4.8 on the next page sums up the various types of residuals.

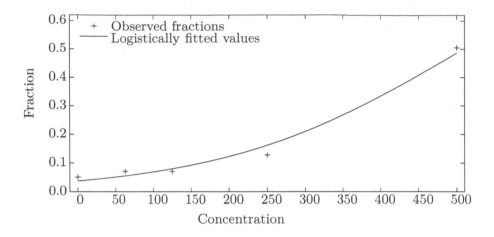

Figure 4.5: *Observed fraction stillborn and corresponding fitted values under logistic regression for dose response assay.*

Table 4.8: *Various types of residuals.*

Obs no. i	1	2	3	4	5
Conc. [mg/kg/day]	0.0	62.5	125.0	250.0	500.0
# fetuses w_i	297	242	312	299	285
Frac. stillb. y_i	0.0505	0.0703	0.0705	0.1271	0.5053
$\tau^{-1}(y_i)$	-2.9339	-2.5829	-2.5788	-1.9269	0.0211
$\widehat{\theta}_i$	-3.2480	-2.8486	-2.4493	-1.6507	-0.0534
$\widehat{p}_i$	0.0374	0.0548	0.0795	0.1610	0.4867
$w_i d(y_i; \widehat{p}_i)$	1.2812	1.0356	0.3557	2.7090	0.3960
$r_r(y_i; \widehat{p}_i)$	0.0131	0.0155	-0.0090	-0.0339	0.0186
$r_D(y_i; \widehat{p}_i)$	1.1317	1.0174	-0.5968	-1.6464	0.6284
$r_P(y_i; \widehat{p}_i)$	1.1904	1.0598	-0.5859	-1.5957	0.6294
$r_W(y_i; \widehat{p}_i)$	1.1856	1.0567	-0.5854	-1.5907	0.6293

Table 4.9: *The number of accidents with personal injury during daylight hours in January quarter during the years 1987 to 1990.*

Year	1987	1988	1989	1990
Accidents, y_i	57	67	54	59
Traf.index, x_i	100	111	117	120

Regression models, $\mu_i = \beta x_i$

Example 4.7 – Poisson "regression", use of offset

Table 4.9 shows the number of accidents with personal injury during daylight hours in January quarter during the years 1987 to 1990. Also shown is the traffic index for the same periods. The traffic index is a measure for the amount of traffic on Danish roads (Index for 1987 was set to 100).

It is standard practice in road research to model accident counts Y_i as Poisson distributed random variables, $Y_i \sim \text{Pois}(\mu_i)$, corresponding to a random distribution of the accidents over time and space.

Also, we shall assume that the accident count is proportional to the amount of traffic,

$$\mu_i = \gamma x_i \tag{4.73}$$

The canonical link for the Poisson distribution is

$$g(\mu) = \log(\mu) \,,$$

therefore, in terms of the canonical link the model (4.73) is

$$\eta_i - \log(x_i) = \beta_1$$

with $\beta_1 = \log(\gamma)$, or

$$\eta_i = \beta_1 + \log(x_i)$$

viz. a model with *offset* equal to $\log(x_i)$, and only one term, the intercept, β_1. The accident *rate* $\gamma = \exp(\beta_1)$.

Since

$$\mu_i(\beta_1) = \exp\left(\beta_1 + \log(x_i)\right) \,,$$

the mean value equation (4.38) becomes

$$\mathbf{1}^T \mathbf{y} = \mathbf{1}^T \begin{pmatrix} \exp\left(\beta_1 + \log(x_1)\right) \\ \vdots \\ \exp\left(\beta_1 + \log(x_4)\right) \end{pmatrix}$$

i.e.,

$$\sum_{i=1}^{4} y_i = \sum_{i=1}^{4} \exp(\beta_0) x_i$$

and, hence

$$\widehat{\gamma} = \exp(\widehat{\beta_1}) = \frac{\sum_{i=1}^{4} y_i}{\sum_{i=1}^{4} x_i} = \frac{237}{448} = 0.529 \tag{4.74}$$

Assuming that the data from the table above is in a dataset acc with the further variable lninx calculated as the logarithm of the traffic index then the R command

```
glm(formula = Acc ~ offset(lninx), family = poisson(link = log),
        data = acc)
```

results in the output

```
Coefficients:
(Intercept)
   -0.6367


Degrees of Freedom: 3 Total (i.e. Null);  3 Residual
Null Deviance:       2.803
Residual Deviance: 2.803           AIC: 28.48
```

The goodness of fit statistic $D(\boldsymbol{y}; \boldsymbol{\mu}(\widehat{\boldsymbol{\beta}})) = 2.80$ with $P[\chi(3)^2 \geq 2.80] = 0.42$. Thus, the data do not contradict the assumptions underlying the model.

▶ **Remark 4.25 – Simple interpretation of the estimator**
The estimator in (4.74) could have been derived without resorting to the theory of generalized linear models.

Observing that $Y_i \sim \text{Pois}(\gamma x_i)$ implies that $\sum Y_i \sim \text{Pois}(\gamma \sum x_i)$ reflecting that under the proportionality assumption the total number of accidents follows a Poisson distribution corresponding to the total amount of traffic.

Hence, the accident rate is estimated by the total number of accidents divided by the total amount of traffic. ◀

▶ **Remark 4.26 – Interpretation of the estimator as a weighted average**
Consider the problem above as a problem of determining the slope of the regression line

$$E[Y_i] = \gamma x_i.$$

For each data-point x_i, y_i we may determine an individual estimate $\widehat{\gamma}_i = y_i/x_i$ of the slope. We want to investigate how these individual estimates are combined to form the overall estimate (4.74).

Inserting the individual estimates $\widehat{\gamma}_i$ into (4.74) we find

$$\widehat{\gamma} = \frac{\sum x_i \widehat{\gamma}_i}{\sum x_i},$$

which shows that the ML-estimate $\widehat{\gamma}$ is a *weighted average* of the individual estimates $\widehat{\gamma}_i$ with the traffic indexes, x_i as weights.

Noting that

$$\text{Var}[\widehat{\gamma}_i] = \frac{\text{Var}[Y_i]}{x_i^2} = \frac{\gamma x_i}{x_i^2} = \frac{\gamma}{x_i}$$

we observe that the weights x_i represent the precision (reciprocal variance) of the individual estimates, and hence the ML-estimate is nothing but a *precision weighted* average of the individual slope estimates. ◀

Example 4.8 – Gamma regression

Let $Y_1, Y_2, \ldots, Y_n$ be mutually independent random variables with $Y_i \in G(\alpha, \mu_i/\alpha_i)$, where the shape parameters α_i are known. We recall from Table 4.2 that the shape parameter acts as a *precision parameter* λ.

The *Gamma regression model* is of the form

$$\mu_i = \gamma\, x_i, \quad i = 1, 2, \ldots, n\,, \tag{4.75}$$

where $\mu_i = \mathrm{E}[Y_i]$ and $x_1, x_2, \ldots, x_n$ are known values of a quantitative covariate.

As the canonical link function for the gamma distribution is the reciprocal link, $\eta = 1/\mu$, we find that the model (4.75) corresponds to

$$\eta_i = \frac{1}{\gamma x_i}\,, \tag{4.76}$$

i.e.,

$$\eta_i = \frac{1}{\gamma}\, t_i$$

with $t_i = 1/x_i$.

The mean value equation (4.37) becomes

$$\boldsymbol{t}^T \operatorname{diag}\{\alpha_i\}\boldsymbol{y} = \boldsymbol{t}^T \operatorname{diag}\{\alpha_i\} \begin{pmatrix} \gamma x_1 \\ \vdots \\ \gamma x_n \end{pmatrix}$$

with the solution

$$\widehat{\gamma} = \frac{\sum_{i=1}^n \alpha_i y_i/x_i}{\sum_{i=1}^n \alpha}\,. \tag{4.77}$$

When the shape parameter α_i is the same for all observations the estimate is simply the crude average of the individual slope estimates

$$\widehat{\gamma} = \frac{1}{n} \sum_{i=1}^n \frac{y_i}{x_i}\,. \tag{4.78}$$

Observing that $\mathrm{Var}[Y_i/x_i] = \alpha_i \mu_i^2/x_i^2 = \alpha_i \gamma^2$ we find again that the ML-estimate $\widehat{\gamma}$ (4.77) is a *precision weighted average* of the individual estimates $\widehat{\gamma}_i$.

Example 4.9 – Simple regression models under various distributional assumptions

We consider the regression model $\mathrm{E}[Y_i] = \beta x_i$, and let $\widehat{\beta}_i = y_i/x_i$ denote the individual slope estimate based on the i'th observation.

Below we shall supplement the examples above and express the ML-estimate $\widehat{\beta}$ as a precision weighted average of the individual slope estimates $\widehat{\beta}_i$ under various distributional assumptions on Y.

a) Normal distribution $Y \in N(\beta x, \sigma^2)$; $\mathrm{Var}[Y] = \sigma^2$.

$$\widehat{\beta} = \frac{\sum y_i x_i}{\sum x_i^2} = \frac{\sum x_i^2 \widehat{\beta}_i}{\sum x_i^2}$$

b) Poisson distribution $Y \in \mathrm{Pois}(\beta x)$; $\mathrm{Var}[Y] = \beta x$.

$$\widehat{\beta} = \frac{\sum y_i}{\sum x_i} = \frac{\sum x_i \widehat{\beta}_i}{\sum x_i}$$

c) Gamma distribution $Y \in G(\alpha, \beta x / \alpha)$; $\mathrm{Var}[Y] = (\beta x)^2$.

$$\widehat{\beta} = \frac{\alpha \sum y_i / x_i}{\sum \alpha} = \frac{\sum \widehat{\beta}_i}{k}$$

d) Inverse Gaussian distribution $Y_i \sim \mathrm{IGau}(\mu)$; $\mathrm{Var}[Y] = (\beta x)^3$.

$$\widehat{\beta} = \frac{\sum (y_i / x_i) / x_i}{\sum 1 / x_i} = \frac{\sum \widehat{\beta}_i / x_i}{\sum 1 / x_i}$$

The example shows that the distributional assumptions really make a difference, even under the same model for the mean value structure. Also, we have illustrated that the ML-solution takes the precision of the observations into account.

Link functions for binary response regression

Example 4.10 – "Dose-response" curve for electrical insulation
In Example 4.6 on page 118 we considered the *logistic regression model* corresponding to the canonical link for the binomial distribution. In the present example we shall investigate various other regression models for binary response.

Table 4.10 shows the result of an experiment testing the insulation effect of a gas (SF$_6$). In the experiment a gaseous insulation was subjected to 100 high voltage pulses with a specified voltage, and it was recorded whether the insulation broke down (spark), or not. After each pulse the insulation was reestablished. The experiment was repeated at twelve voltage levels from 1065 kV to 1135 kV.

We shall assume that the data is stored in R with the names Volt, Breakd, Trials, and moreover that the model formula has been specified as

```
model <- cbind(Breakd,Trials-Breakd) ~ Volt
```

Table 4.10: *The insulation effect of a gas (SF₆).*

Voltage (kV)	1065	1071	1075	1083	1089	1094
Breakdowns	2	3	5	11	10	21
Trials	100	100	100	100	100	100

Voltage (kV)	1100	1107	1111	1120	1128	1135
Breakdowns	29	48	56	88	98	99
Trials	100	100	100	100	100	100

As the insulation was restored after each voltage application it seems reasonable to assume that the trials were independent. At each trial the response is binary (Breakdown/Not), and therefore it seems appropriate to use a binomial distribution model for the experiment.

Let Z_i denote the number of breakdowns at the i'th trial at the voltage x_i. We shall then use the model $Z_i \sim \mathrm{B}(n_i, p_i)$ with $n_i = 100$, and $p_i = p(x_i)$, where $p(x)$ is some suitable dose-response function.

The logit transformation, logistic regression

The logistic regression is of the form

$$g(p) = \log\left(\frac{p}{1-p}\right) = \beta_1 + \beta_2 x$$

$$p(x) = \frac{\exp(\eta)}{1+\exp(\eta)} = \frac{\exp(\beta_1 + \beta_2 x)}{1+\exp(\beta_1 + \beta_2 x)} \tag{4.79}$$

The curve is symmetric around $x = 0$

In R we get

```
> logist.glm<-glm(model,binomial(logit))
```

```
Coefficients:
(Intercept)          Volt
 -127.7002        0.1155
```

```
Degrees of Freedom: 11 Total (i.e. Null);  10 Residual
Null Deviance:      783.1
Residual Deviance: 21.02          AIC: 70.61
```

The p-value corresponding to the goodness of fit statistic $D(\boldsymbol{y}; \boldsymbol{\mu}(\widehat{\boldsymbol{\beta}})) = 21.02$ is assessed by calculating

```
> pval<-1-pchisq(21.02,10)
```

leading to pval = 0.02095454. Thus, $\mathcal{H}_{logist}$ is rejected at any significance
level greater than 2%. Also, a look at the *deviance residuals*:

```
> residuals(logist.glm)
         1          2          3          4          5          6
 1.0162934  0.8554846  1.2245599  1.5869759 -0.7922914  0.1608718
         7          8          9         10         11         12
-1.0302561 -1.0832120 -1.7571526  1.2046593  2.3465774  1.5170429
```

They indicate underestimation in the tails, and overestimation in the central
part of the curve.

The probit transformation

The transformation

$$g(p) = \Phi^{-1}(p) = \beta_1 + \beta_2 x$$
$$p(x) = \Phi(\eta) = \Phi(\beta_1 + \beta_2 x) \tag{4.80}$$

with $\Phi(\cdot)$ denoting the cumulative distribution function for the standardized
normal distribution is termed the *probit*-transformation.

The function is symmetric around $x = 0$. The function tends towards
0 and 1 for $x \to \mp\infty$, respectively. The convergence is faster than for the
logistic transformation.

There is a long tradition in biomedical literature for using the probit
transformation, mainly because tables of the normal distribution were available.

In R we get

```
> probit.glm<-glm(model,binomial(probit))

Coefficients:
(Intercept)          Volt
  -71.10502       0.06432

Degrees of Freedom: 11 Total (i.e. Null);   10 Residual
Null Deviance:        783.1
Residual Deviance: 26.22          AIC: 75.81
```

The p-value corresponding to the goodness of fit statistic $D(\boldsymbol{y}; \boldsymbol{\mu}(\widehat{\boldsymbol{\beta}})) = 26.22$
is assessed by calculating

```
> pval<-1-pchisq(26.22,10)
```

leading to pval = 0.003455298. Thus, $\mathcal{H}_{probit}$ is rejected at any significance
level greater than 0.3%. The fit is not satisfactory. Again, a look at the
deviance residuals:

```
> residuals(probit.glm)
         1          2          3          4          5          6
```

1.6661628 1.2389805 1.3964999 1.2603399 -1.3579378 -0.5152912
        7   8   9   10   11   12
-1.5628465 -1.2161132 -1.6530131 1.4925659 2.3946024 1.2806034

They indicate systematic underestimation in both tails, and overestimation in the central part.

Complementary log-log

The transformation

$$g(p) = \log(-\log(1-p)) = \beta_0 + \beta_1 x$$
$$p(x) = 1 - \exp[-\exp(\beta_0 + \beta_1 x)] \tag{4.81}$$

is termed the *complementary log-log* transformation.

Note that in some software packages this link function is termed the *log-log link function*. Sometimes the function (4.81) is called the *Gompit*-regression model named after the Gompertz-distribution with cumulative distribution function given by (4.81).

The response function is *asymmetrical*; it increases slowly away from 0, whereas it approaches 1 in a rather steep manner.

Consider two values, x_1 and x_2, of the explanatory variable. Then it follows from the model (4.81), that

$$\frac{\log[1 - p(x_2)]}{\log[1 - p(x_1)]} = \exp[\beta_1(x_2 - x_1)]$$

or, that

$$1 - p(x_2) = [(1 - p(x_1)]^{\exp[\beta_1(x_2 - x_1)]}$$

i.e., when the explanatory variable x is increased by one unit, then the probability of a "negative response" will be raised to the power $\exp(\beta_1)$.

```
> cloglog.glm<-glm(model,binomial(cloglog))

Coefficients:
(Intercept)        Volt
   -91.1063      0.0819

Degrees of Freedom: 11 Total (i.e. Null);  10 Residual
Null Deviance:        783.1
Residual Deviance: 5.671        AIC: 55.27
```

The p-value corresponding to the goodness of fit statistic $D(y; \mu(\widehat{\beta})) = 5.671$ is assessed by calculating `pval<-1-pchisq(5.671,10)`, leading to `pval = 0.8421057`. Thus, data do not provide any evidence against the `cloglog` model. This is further supported by the deviance residuals,

```
> residuals(cloglog.glm)
          1           2          3          4           5          6
-0.02716413 -0.17608904 0.20605221 0.82313318 -1.11484977 0.28323667
          7           8          9         10          11         12
-0.29861616  0.12373696 -0.60303934 1.00951550 0.49449585 -1.36539500
```

There is no systematic pattern in the residuals, and all residuals are in the interval ± 2.

log-log link

If, instead the pattern had been a steep increase away from zero, we could have used the response function

$$g(p) = \log(-\log(p)) = \beta_1 + \beta_2 x,$$
$$p(x) = \exp[-\exp(\beta_1 + \beta_2 x)]. \tag{4.82}$$

This model is sometimes called a *Gumbel*-regression, as the probability distribution with cdf expressed by (4.82) is termed a Gumbel (extreme value) distribution.

If $p(x)$ may be described by the model (4.81), then $1 - p(x)$ may be described by (4.82) and converse. Thus, a complementary log-log link for $p(x)$ is a log-log link for $1 - p(x)$.

The R commands

```
> model2<-cbind(Trials-Breakd,Breakd)~Volt
> loglog.glm<-glm(model2,binomial(cloglog))
```

result in

```
Coefficients:
(Intercept)         Volt
   65.12371      -0.05934


Degrees of Freedom: 11 Total (i.e. Null);  10 Residual
Null Deviance:      783.1
Residual Deviance: 80.13          AIC: 129.7
```

The goodness of fit statistic $D(y; \mu(\widehat{\beta})) = 80.13$ is highly significant. The p-value is found by 1-pchisq(80.13,10) resulting in

```
p =  4.733991e-13
```

The deviance residuals are

```
> residuals(loglog.glm)
         1          2          3          4          5         6
-2.8310667 -1.8595999 -1.5871556 -0.5050439 2.5248632 1.8926860
         7          8          9         10         11        12
 2.7621188 1 .7519484 1.6327615 -2.8212448 -4.4638996 -3.8307331
```

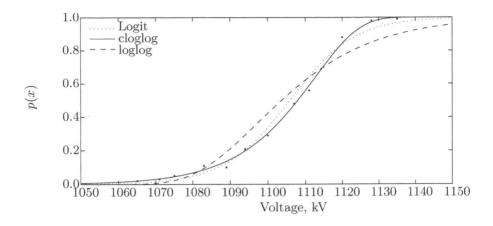

Figure 4.6: *Probability of breakdown for an insulator as function of applied pulse voltage. The curves correspond to different assumptions on the functional form of the relation.*

clearly indicating an unsatisfactory fit. The graph in Figure 4.6 illustrates the fit by these three models. The example illustrates the importance of having observations in the whole region of interest for the explanatory variable x, if one wants to distinguish between different models. If, e.g., we only had observations corresponding to voltage levels $x \leq 1090$ [kV], then we would probably not have been able to distinguish between these three models, as all three may be fit reasonably well to the response pattern in a limited interval for p.

Non-canonical link for binary response

Example 4.11 – Determination of bacterial density from dilution assay

In bacteriology a so-called *dilution assay* is often used to assess the concentration of bacteria and other microorganisms in a suspension. The method is used in situations where it is not possible to perform a simple count, e.g., because the growing bacteria form indistinguishable colonies. The method consists primarily in preparing several different dilutions of the original suspension, and determining the presence or absence of the bacteria in question in each sample, by recording whether there was evidence of bacterial growth or not. This method is considered to be one of the earliest methods in bacteriology.

The earliest published attempt to deduce an estimate from some considerations connected with statistical modeling seems to be due to McCrady

Table 4.11: *Results found in an assay with the 3-fold dilutions 1/50, 1/250 and 1/1250 ml.*

Volume [ml]	0.0008	0.0040	0.0200
No. of samples	5	5	5
No. of not fertile	5	1	0

(1915). The comprehensive survey paper by Eisenhart and Wilson (1943) provides an account of McCradys and other approaches up to 1943.

Let x_i denote the known concentration of the suspension in the i'th dilution. When the bacteria are randomly spread in the suspension (without clusters), it follows that the number of bacteria in a given volume will follow a Poisson distribution. Hence, the number of bacteria in a plate with the i'th dilution may be described by a Poisson distributed random variable with mean γx_i where γ denotes the unknown density of bacteria in the original suspension. The probability, p_i that a plate from the i'th solution will not exhibit growth is simply the probability that there are no bacteria in the plate,

$$p_i = P[P(\lambda x_i) = 0] = \exp(-\gamma x_i) \,. \tag{4.83}$$

Let now Z_i denote the number of plates from the i'th dilution that did not exhibit growth. It then follows that $Z_i \sim B(n_i, p_i)$, a binomial distribution with p_i given by (4.83).

Let $Y_i = Z_i/n_i$. Then $E[Y_i] = p_i$, and the model for p_i is

$$\log p_i = -\gamma x_i \,, \tag{4.84}$$

The model is a generalized linear model for the fraction Y_i of plates without growth with $Y_i \sim B(n_i, p_i)/n_i$. The expression (4.84) shows that we are looking for a *linear predictor* for the *logarithm* to the mean value. Thus, the distribution is *binomial* and the link function is "log".

In an assay with the 3-fold dilutions 1/50, 1/250 and 1/1250 ml, the results shown in Table 4.11 were found. In R, a model formula could be

```
baktmod <- cbind(nonfert, samples-nonfert) ~ -1+Vol
```

and

```
> glm(baktmod, binomial(log))
```

These commands specify a binomial distribution with log-link and no intercept term.

In R one finds the coefficient to vol is -282.27 corresponding to 282 bacteria/ml. The residual deviance is $D(\boldsymbol{y}; \boldsymbol{\mu}(\widehat{\beta})) = 2.6723$ with 2 degrees of freedom, indicating a quite satisfactory fit.

Table 4.12: *Customer satisfaction*

Delay min	V. diss.	Diss.	response, number(%) Neutral	Satis.	V. satis.	Total
0	234(2.3)	559(5.4)	1157(11.2)	5826(56.4)	2553(24.7)	10329
2	41(3.6)	100(8.9)	145(12.9)	602(53.5)	237(21.1)	1125
5	42(7.9)	76(14.3)	89(16.7)	254(47.7)	72(13.5)	533
7	35(14.3)	48(19.7)	39(16.0)	95(38.9)	27(11.1)	244

The maximum likelihood solution (termed *the most probable number*, MPN) was suggested by R.A. Fisher in connection with an extensive experiment carried out in the protozoological laboratory at Rothamstead, and reported in his general paper on principles of statistical inference, Fisher 1922.

Logistic regression with ordered categorical response

Example 4.12 – Customer satisfaction for bus passengers
A widespread method for assessing customer satisfaction is to present a questionnaire to customers. The questionnaire contains a series of items each representing a quality feature of the product in question, and the respondent is asked to indicate his satisfaction by selecting one of the options:

Very dissatisfied	☐
Dissatisfied	☐
Neutral	☐
Satisfied	☐
Very satisfied	☐

Often this ordered categorical scale is termed a Likert scale named after the originator, Likert (1932). A Likert-type scale consists of a series of declarative statements. The respondent is asked to indicate whether he agrees or disagrees with each statement. Commonly, five options are provided: 'strongly agree,' 'agree,' 'undecided,' 'disagree,' and 'strongly disagree.' Other Likert-type scales include four or six steps rather than five, excluding the undecided position.

Questionnaires were distributed to 12,233 passengers in randomly selected buses. One of the questions was:

How satisfied are you with the punctuality of this bus?

The delay of the bus was also recorded, and classified as 0, 2, 5 or 7 minutes delay. The results are shown in Table 4.12.

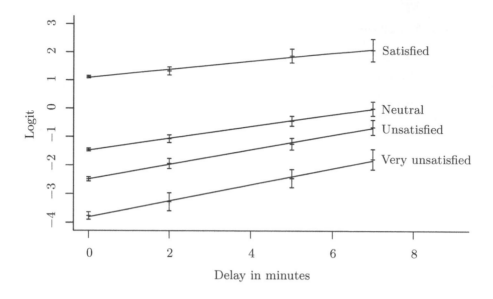

Figure 4.7: *The logit of the observed cumulative fractions, $\widehat{\Pi}_j$ with confidence limits for Π_j determined from the binomial distribution, and transformed to the logit scale.*

For a given delay may the number of responses in different categories be described by a *Multinomial distribution*, $\mathrm{Mult}(n, p_1, p_2, \ldots, p_5)$ with $\sum_i p_i = 1$.

As the categories are *ordered*, we may meaningfully introduce the *cumulative probabilities*,

$$\Pi_j = p_1 + p_2 + \cdots + p_j \ , \quad j = 1, 2, 3, 4$$

of responding in category j, or lower. For other parameterizations, e.g., continuation logit see Agresti (2002).

Consider the *logistic model* for the cumulative probabilities

$$\log \left(\frac{\Pi_j(t)}{1 - \Pi_j(t)} \right) = \alpha_j + \beta_j t \ , j = 1, \ldots, 4$$

corresponding to

$$\Pi_j(t) = \frac{\exp(\alpha_j + \beta_j t)}{1 + \exp(\alpha_j + \beta_j t)} \tag{4.85}$$

Each cumulative logit includes data from all 5 categories The graph in Figure 4.7 shows the logit of the observed cumulative fractions, $\widehat{\Pi}_j$ with confidence limits for Π_j determined from the binomial distribution, and transformed to the logit scale. The hypothesis of linearity is not rejected. A hypothesis of a common slope β is also not rejected, see Figure 4.8. Consider

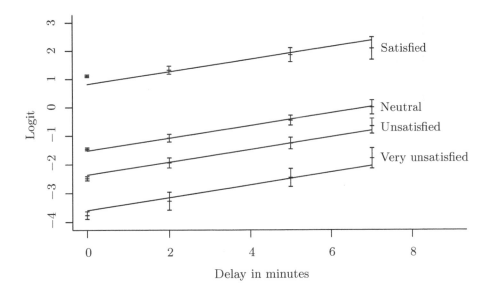

Figure 4.8: *Testing a hypothesis of a common slope β.*

two delay times t_1 and t_2. The model with common slope implies for any category, j, that

$$\log\left(\frac{\Pi_j(t_1)/(1-\Pi_j(t_1))}{\Pi_j(t_2)/(1-\Pi_j(t_2))}\right) = (t_1 - t_2)\beta$$

log odds-ratio is proportional to the difference in delay time, but does not depend on the category, j.

The common slope of the logit is estimated to be $\widehat{\beta} = 0.213$. For any category, j, is the cumulative odds, $\Pi_j/(1 - \Pi_j)$ multiplied by $\exp(0.213) = 1.24$ per minute delay. The model is illustrated in Figure 4.9 in terms of probabilities for the different response categories (extrapolated to 20 minutes). Consider a variable, R with a logistic distribution over the population of respondents

$$f_R(r) = \frac{\exp(-r)}{(1+\exp(-r))^2}$$

(mean 0, variance $\pi^2/3$). Define the categorical variable J as $J = j$ if $\alpha_{j-1} < R \leq \alpha_j$ with $\alpha_0 = -\infty$. For an undelayed bus, the passenger responds in category j if his *latent satisfaction parameter* R is in the interval $\alpha_{j-1} < R \leq \alpha_j$. If the value of the latent parameter is reduced by β for each minute the bus is delayed, then the probability of a response at most in category j is given by (4.85). This interpretation is illustrated in Figure 4.10 on page 137.

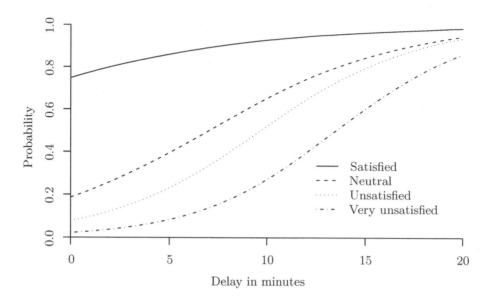

Figure 4.9: *The model in terms of probabilities for the different response categories.*

Example 4.13 – Empirical variances for normally distributed observations

An experiment aiming at the determination of washing power for a detergent consists in uniformly staining sheets of cloth, and subsequently washing them with the detergent and afterward the cleanness is determined by means of measurements of the reflection.

In the experiment 3, 5, and 7 sheets were washed simultaneously. Table 4.13 on page 138 shows the empirical variances for each of these trials. It is of interest to assess whether the variance decreases inverse proportional to the number of sheets (in the interval between 3 and 7 sheets).

Assuming that the variation of the cleanness may be described by a normal distribution with a mean, and a variance that only depends on the number of sheets, it follows that the empirical variances, $S_i^2 \in G\big(f_i/2, \; \sigma_i^2/(f_i/2)\big)$ with $E[s_i^2] = \sigma_i^2$.

Although the variable we want to model is symbolized by σ^2, it is the *mean value* in the distribution of S_i^2.

It is of interest to assess the model

$$\sigma_i^2 = \frac{\gamma}{x_i}, \quad i = 1, 2, 3,$$

i.e., the regression model (4.76) for the canonical link $\eta_i = 1/\sigma_i^2 = (1/\gamma)x_i$.

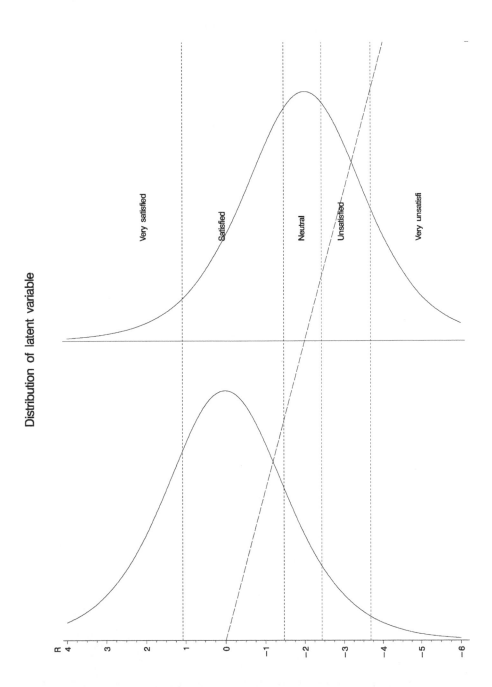

Figure 4.10: *The model interpreted in terms of latent satisfaction parameter.*

Table 4.13: *The empirical variances for each of the trials.*

Number of sheets, x_i	3	5	7
Sum of squares	2.5800	4.8920	4.9486
Degrees of freedom, f_i	2	4	6
Empirical variance, s_i^2	1.2900	1.2230	0.8248

Table 4.14: *The fitted values and the deviance residuals, determined from the unit deviance for the Gamma distribution.*

Number of sheets, x_i	3	5	7
Empirical variance, s_i^2	1.2900	1.2230	0.8248
Fitted value, $\widehat{\sigma}_i^2$	1.8566	1.1140	0.7957
unit dev. $d(s_i^2; \widehat{\sigma}_i^2)$	0.11785	0.00900	0.00130
$(f_i/2)d(s_i^2; \widehat{\sigma}_i^2)$	0.11785	0.01800	0.00391
$r_D(s_i^2; \widehat{\sigma}_i^2)$	−0.34329	0.13416	0.06254

Therefore we find the ML estimate as (4.77)

$$\sum_{i=1}^{3} \frac{f_i}{2} s_i^2 x_i = \widehat{\gamma} \sum_{i=1}^{3} \frac{f_i}{2}$$

$$\widehat{\gamma} = \frac{\sum_{i=1}^{3} \frac{f_i}{2} s_i^2 x_i}{\sum_{i=1}^{3} \frac{f_i}{2}} = 5.57$$

Table 4.14 shows the fitted values and the deviance residuals, determined from the unit deviance for the Gamma distribution (cf. Table 4.2 on page 96)

$$d(y; \mu) = 2 \times \left(\frac{y}{\mu} - \log\left(\frac{y}{\mu}\right) - 1 \right)$$

One finds the goodness of fit test statistic $D(s^2; \widehat{\sigma}^2) = 0.1376$, which should be compared with the quantiles in a $\chi^2(2)$-distribution. As $P[\chi^2(2) \geq 0.140] = 0.93$ there is no evidence against the hypothesis.

The model may be analyzed in R by choosing a Gamma-distribution, and *weights* $f_i/2$, canonical link and explanatory variable x equal to number of sheets (corresponds to $1/\sigma_i^2 = x_i/\gamma$). The program would then estimate $1/\gamma$, but the fitted values, etc. would be like in the table above.

```
> detergent.glm<-glm(formula = var ~-1 + nsheets,
data=detergent,family=Gamma(link=inverse),weights=wgt)


> summary(detergent.glm)
```

```
Deviance Residuals:
       1         2         3
-0.34334   0.13407   0.06251

Coefficients:
        Estimate Std. Error t value Pr(>|t|)
nsheets  0.17953    0.01767   10.16  0.00955

(Dispersion parameter for Gamma family taken to be 0.05815066)

    Null deviance:     NaN  on 3  degrees of freedom
Residual deviance: 0.13976  on 2  degrees of freedom
AIC: -1.3479

Number of Fisher Scoring iterations: 4

> anova(detergent.glm)

Analysis of Deviance Table
Model: Gamma, link: inverse
Response: var
Terms added sequentially (first to last)


          Df Deviance Resid. Df Resid. Dev
NULL                        3
nsheets  1                  2      0.13977
```

▸ **Remark 4.27 – Bartlett's test for homogeneity of variances**
Some textbooks quote a test for homogeneity of variances for normally distributed variables, *Bartlett's test*. That test corresponds to a hypothesis of total homogeneity in a generalized linear model for n gamma-distributed observations. The test is named after the British statistician M.S. Bartlett, who published the derivation of that test in Bartlett (1937). Bartlett did not only derive the test statistic, but he also derived a correction factor,

$$c = 1 + \frac{1}{3(n-1)} \left(\sum_{i=1}^{n} \frac{1}{f_i} - \frac{1}{f} \right), \qquad (4.86)$$

to the test statistic $D(s^2; \hat{\sigma}^2)$, such that the distribution of

$$\frac{1}{c} D(s^2; \hat{\sigma}^2)$$

Table 4.15: *Number of cuttings alive for different factor combinations in plum root propagation experiment. The symbols Meag, Medium and Thick denote thin, medium and thick cuttings, respectively.*

Obs no.	Length	Thickness	Number of cuttings	Number alive	Number dead
1	Long	Meag	20	6	14
2	Long	Medium	20	14	6
3	Long	Thick	20	18	2
4	Short	Meag	20	4	16
5	Short	Medium	20	10	10
6	Short	Thick	20	11	9

is better approximated by the $\chi^2(n-1)$-distribution. The principle for deriving this correction factor may be generalized to the likelihood ratio test statistic (2.46). Such a correction to the likelihood ratio test-statistic is termed the "Bartlett-correction." ◄

▶ **Remark 4.28 – Previous approaches to models for empirical variances**
Until the development of generalized linear models and the identification of the distribution of S^2 as an exponential dispersion family, practicing statisticians used to model the distribution of $\log(S_i^2)$ by a normal distribution, and then use classical GLMs on the distribution of $\log(S_i^2)$. This approach has been suggested and discussed by Bartlett and Kendall (1946). ◄

Example 4.14 – Experiment with propagation of plum rootstocks
In an experiment with propagation of plum rootstocks from root cuttings (reported by Hoblyn and Palmer (1934)) cuttings were made of two different *lengths*, 6 [cm] and 12 [cm], respectively, and three *thicknesses*, 3-6 [mm], 6-9 [mm] and 9-12 [mm], respectively. In October, 20 cuttings of each combination of length and thickness were planted, and the following spring the condition of the cuttings was assessed. The result is shown in Table 4.15. The model consists of two explanatory variables, length and thickness. We will consider both variables as classification variables (factors) with two levels of length, {Short, Long} corresponding to 6 [cm] and 12 [cm]. Thickness has the three levels {Meag, Medium, Thick} corresponding to 3-6 [mm], 6-9 [mm], and 9-12 [mm], respectively.

In the parametric representation we shall respect the factorial structure of the explanatory variables, and we shall therefore parametrize the observations in accordance with Table 4.16. We will denote the observations by Z_{ij}, $i = 1, 2$, $j = 1, 2, 3$, where index i denotes index for length, and j the index for thickness. We choose to model the number, Z_{ij}, of surviving cuttings

Table 4.16: *Number of surviving/number of cuttings in plum root propagation experiment. (Factorial representation)*

		Thickness	
Length	Meag	Medium	Thick
Short	4/20	10/20	11/20
Long	6/20	14/20	18/20

Table 4.17: *Parametrization of the linear predictor (logit) for the probability of survival in plum root propagation experiment. (Factorial representation)*

		Thickness	
Length	Meag	Medium	Thick
Short	μ	$\mu + \beta_2$	$\mu + \beta_3$
Long	$\mu + \alpha_2$	$\mu + \alpha_2 + \beta_2$	$\mu + \alpha_2 + \beta_3$

by a binomial distribution, $Z_{ij} \sim B(n_{ij}, p_{ij})$ where the number of cuttings, $n_{ij} = 20$.

The family of distributions of $Y_{ij} = Z_{ij}/n_{ij}$ is an exponential dispersion family with mean p_{ij}, canonical link function

$$\eta = \log\left(\frac{p}{1-p}\right)$$

variance function

$$V(p) = p(1 - p)$$

and precision parameter (weight)[4] $w_{ij} = n_{ij}$. We select the canonical link, and formulate a hypothesis of no interaction in analogy with the model for a two-factor experiment with normally distributed observations:

$$\eta_{ij} = \mu + \alpha_i + \beta_j \ , i = 1, 2; \ j = 1, 2, 3 \tag{4.87}$$

As it is, this parameterized model is overparametrized, and therefore we need to impose some restrictions to make sure that the model matrix is of full rank.

If we choose a *treatment coding*, the default in R, then the first level for each factor is chosen as reference, i.e., $\alpha_1 = 0$ and $\beta_1 = 0$.

---

[4]The weight parameter is relevant when you perform the calculations yourself. In standard statistical software, the weighting of the unit deviance is automatically determined when you specify the binomial distribution and the number of observations.

Model fit

Since we have selected the canonical link, the parameters are estimated by solving the *mean value equations* (4.37):

$$\sum_{i=1}^{2} y_{ij} = \sum_{i=1}^{2} \widehat{p}_{ij}\,, \quad j = 1, 2, 3$$

$$\sum_{j=1}^{3} y_{ij} = \sum_{j=1}^{3} \widehat{p}_{ij}\,, \quad i = 1, 2$$

with

$$\widehat{p}_{ij} = \frac{\exp(\widehat{\mu} + \widehat{\alpha}_i + \widehat{\beta}_j)}{1 + \exp(\widehat{\mu} + \widehat{\alpha}_i + \widehat{\beta}_j)}$$

The equations are solved by iteration.

In R, the response corresponding to a binomial distribution is a *vector* giving the number of "successes" (corresponding to p) and the number of "failures" (corresponding to $1 - p$). Note that although we are modeling p which is the mean of the *fraction*, the response vector is a vector of integer *counts*. We construct the vector resp by

```
plumroot$resp <- cbind(plumroot$alive,plumroot$dead)
```

and invoke **glm** by

```
plumroot.glm1 <- glm(formula = resp ~ Length + Thick,
    family = binomial(link =logit), data = plumroot )
```

The result is

```
Coefficients:
            Estimate Std. Error z value Pr(>|z|)
(Intercept)  -0.6342     0.4048  -1.567  0.11717
LengthShort  -1.0735     0.4211  -2.549  0.01079
ThickMedium   1.6059     0.5082   3.160  0.00158
ThickThick    2.2058     0.5335   4.135 3.55e-05

(Dispersion parameter for binomial family taken to be 1)

    Null deviance: 28.9152  on 5  degrees of freedom
Residual deviance: 1.8854  on 2  degrees of freedom
AIC: 28.983

Number of Fisher Scoring iterations: 4
```

Thus, we can construct the deviance table (Table 4.18) for the initial goodness of fit, as in Table 4.5. As $P[\chi^2(2) > 1.8854] = 0.39$ there is no strong evidence

Table 4.18: *Deviance table.*

Source	f	Deviance	Goodness of fit
Model $\mathcal{H}_1$	3	27.0298	$G^2(\mathcal{H}_M|\mathcal{H}_1)$
Residual (Error)	2	1.8854	$G^2(\mathcal{H}_1)$
Corrected total	5	28.9152	$G^2(\mathcal{H}_M)$

Table 4.19: *Estimated fraction of surviving cuttings using additive model for logits.*

		Thickness	
Length	Meag	Medium	Thick
Short	0.15345	0.47455	0.62200
Long	0.34655	0.72545	0.82800

against $\mathcal{H}_1$, and the hypothesis of binomial distribution of numbers of cuttings alive may be maintained as well as the hypothesis of no interaction between length and thickness under the logistic link. The deviance residuals are

```
> residuals(plumroot.glm1)
         1          2          3          4          5          6
-0.4425440 -0.2526784  0.9123295  0.5564751  0.2277588 -0.6571325
```

No residuals exceed ± 2, and there are no signs of systematic pattern in the residuals.

Comparing this analysis with the classical GLM analysis of a two-factor experiment we note that in this case of a binomial distribution we were able to test the assumption of no interaction by comparing the error deviation with fractiles in the χ^2-distribution. In the classical GLM case with only one observation in each cell, we had to assume that there was no interaction, and then we could use the error sum of squares as an estimate of the unknown error variance, σ^2.

In order to identify the parameter estimates, we might request to output the design matrix X

```
> model.matrix(plumroot.glm1)
  (Intercept) LengthShort ThickMedium ThickThick
1           1           0           0          0
2           1           0           1          0
3           1           0           0          1
4           1           1           0          0
5           1           1           1          0
6           1           1           0          1
```

Comparing the design matrix to the list of data in Table 4.15 we identify that the parameter Length corresponds to α_2, and Medium and Thick correspond to β_2 and β_3, respectively, in the parametric representation (4.87).

Test for reduction of model

In order to investigate whether the model may be reduced by dropping the effect of length or thickness, we request a Type I partitioning of the model deviance.

```
> anova(plumroot.glm1,test='Chisq')

Analysis of Deviance Table
Model: binomial, link: logit
Response: resp
Terms added sequentially (first to last)

        Df Deviance Resid. Df Resid. Dev P(>|Chi|)
NULL                       5    28.9152
Length 1   5.6931          4    23.2221    0.0170
Thick  2  21.3367          2     1.8854 2.327e-05
```

As the contribution from thickness (when controlled for length) is strongly significant, we cannot delete the thickness term in the model. In order to assess whether length may be deleted from the model, we formulate the model with the terms in reverse order:

```
> plumroot.glm2<-glm(formula=resp ~ Thick + Length,
    family = binomial(link=logit),data=plumroot)

> anova(plumroot.glm2,test='Chisq')
Analysis of Deviance Table
Model: binomial, link: logit
Response: resp
Terms added sequentially (first to last)

        Df Deviance Resid. Df Resid. Dev P(>|Chi|)
NULL                       5    28.9152
Thick  2  20.1740          3     8.7412 4.162e-05
Length 1   6.8558          2     1.8854    0.0088
```

Thus, the contribution from length (when controlled for thickness) is also significant, so we cannot reduce the model by dropping some of the terms. These conclusions could also have been achieved by applying a Type III partitioning of the model deviance.

The inclusion diagram in Figure 2.6 on page 30 illustrates the two hypotheses chains that can be constructed for this experiment, the chains going from a full model with interaction to the null model with total homogeneity.

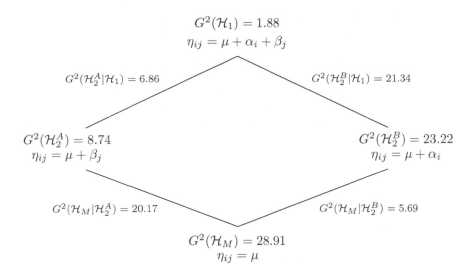

Figure 4.11: *Goodness of fit statistics corresponding to the two hypotheses chains in the plum root propagation example. $\mathcal{H}_2^A$ corresponds to only thickness influence, $\mathcal{H}_2^B$ to only length influence.*

The sequence of tests obtained by choosing different paths in the inclusion diagram, and their goodness of fit interpretation is illustrated in Figure 4.11. As we have accepted the hypotheses of no interaction (i.e., we have adopted the model Length + Thick), then the test designated on the left hand path in Figure 2.6 on page 30 will correspond to investigating whether there is a significant effect of the thickness when it is recognized that the cuttings are of differing length, and corrects for this effect. Analogously, the test on the right hand path in the diagram will correspond to investigating whether there is a significant effect when corrected for the possible effect of the different thickness of the cuttings. The sequence of tests reflects the order in which the terms are presented in the model-formula.

Test of hypothesis for individual parameters

The method **summary()** provides information about the estimated parameters

```
> summary(plumroot.glm1)

Coefficients:
              Estimate Std. Error z value Pr(>|z|)
(Intercept)    -0.6342     0.4048  -1.567  0.11717
LengthShort    -1.0735     0.4211  -2.549  0.01079
```

```
ThickMedium    1.6059      0.5082   3.160  0.00158
ThickThick     2.2058      0.5335   4.135 3.55e-05
```

We might have interest in investigating whether the survival for thickness "Medium" differs significantly from the reference level "Meag" (with $\beta_1 = 0$). The parameter β_2 corresponding to Medium was estimated as $\widehat{\beta}_2 = 1.6059$ with the estimated standard deviation $\sqrt{\sigma_{33}} = 0.5082$.

From (4.67) we obtain the test statistic for the hypothesis $\mathcal{H}_3 \; : \; \beta_2 = 0$

$$u_3 = \frac{1.6059 - 0}{0.5082} = 3.16$$

In the output from R this value is output under the heading z value.

As $P[N(0,1) \geq 3.16] = 0.0016$ we find that the hypothesis is rejected at any significance level greater than 1.6%.

Confidence intervals for individual parameters

The assessment of whether survival for thickness "Medium" differs significantly from the reference level "Meag" may also be made using a confidence interval approach, viz. by investigating whether the value 0 is included in the confidence interval for β_2.

The quantities to be used for the Wald confidence interval are found in the output. The parameter β_2 was estimated by $\widehat{\beta}_2 = 1.6059$ with the estimated standard deviation $\sqrt{\widehat{\sigma}_{33}} = 0.5082$. The Wald confidence interval corresponding to 95% confidence probability is therefore found as

$$\widehat{\beta}_2 \pm 1.96 \sqrt{\widehat{\sigma}_{33}} \,,$$

as $z_{0.975} = 1.96$. Thus, we obtain the interval

$$1.6059 \pm 1.96 \times 0.5082 = \begin{cases} 0.6098 \\ 2.6020. \end{cases}$$

The interval does not include the value 0; therefore we cannot claim that the survival for "Medium" is the same as for "Meag".

The estimated parameters may be viewed by using the command **coef()**. To view the confidence intervals for the parameter estimates one can either use the command confint.**default**() which gives the Wald confidence interval, or the command confint() which prints the more accurate likelihood ratio confidence interval.

```
> coef(plumroot.glm1)
(Intercept) LengthShort ThickMedium  ThickThick
 -0.6342497  -1.0735261   1.6058969   2.2058015


> confint.default(plumroot.glm1)
```

```
                    2.5 %        97.5 %
(Intercept) -1.4276662    0.1591669
LengthShort -1.8988646   -0.2481876
ThickMedium  0.6098162    2.6019776
ThickThick   1.1602494    3.2513537

> confint(plumroot.glm1)
Waiting for profiling to be done...
                    2.5 %        97.5 %
(Intercept) -1.4672057    0.1378372
LengthShort -1.9266835   -0.2659578
ThickMedium  0.6393492    2.6439706
ThickThick   1.2012131    3.3050806
```

We observe that the likelihood ratio based 95% confidence interval corresponding to Medium is $(0.6393; 2.6440)$. The interval is not symmetric around the maximum likelihood estimate $\hat{\beta}_2 = 1.6059$, but the interval does not deviate much from the simpler Wald confidence interval.

▸ **Remark 4.29 – Interactions in tables of counts**
Bartlett (1935) introduced the concept of interactions in contingency tables (e.g., $2 \times 2 \times 2$ tables of counts). Bartlett suggested to use the additive model (4.87) for the logits to mean *no interaction*. A major argument for this choice of additivity of the logits was that this model is symmetrical with respect to the classification of the response (Alive vs Dead), and hence the result of the analysis would be unchanged if, instead we had considered logits of dead plants.

Looking at Hoblyn and Palmer's original presentation of the experimental result, we observe that they had analyzed the model

$$\mathcal{H}_1^{log}: \quad \log(p_{ij}) = \mu + \alpha_i + \beta_j \ , i = 1,2; \ , j = 1,2,3 \qquad (4.88)$$

i.e., an additive model using logarithmic link. The model corresponds to a multiplicative model for the survival probabilities,

$$\mathcal{H}_1^{log}: \quad p_{ij} = \zeta \rho_i \tau_j \ , i = 1,2; \ , j = 1,2,3 \qquad (4.89)$$

This model has an analogy in reliability engineering. Consider a two-component system with components A and B, and assume that the components are coupled in *series*. Then, for the system to function, it is necessary that both components are functioning. Assume that component A is functioning with probability p_A, and B is functioning with probability p_B, and that the two components fail independently, then the probability of the system functioning is

$$P[A \cap B] = p_A p_B$$

Table 4.20: *Number of surviving/number of cuttings in plum root propagation experiment. Bartlett's data.*

Length	Time of planting At once	In spring
Short	107/240	31/240
Long	. 156/240	84/240

which is analogous to (4.89). The probability that the system is *not functioning* is the probability that at least one of the components is not functioning, i.e.,

$$1 - P[A \cap B] = P[\neg A \cup \neg B]$$
$$= (1 - p_A) + (1 - p_B) - (1 - p_A)(1 - p_B)$$
$$= 1 - p_A p_B$$

We find that the model (4.88) is more appropriate for these data, not only because of the better fit, but also because of this interpretation in terms of a serial system.

In Table 4.20 we have reproduced the subset of Hoblyn and Palmer's experimental data that Bartlett used in his paper suggesting the logit transformation as a basis for a claim of no interaction.   ◄

Example 4.15 – Two-factor model, Poisson distribution
Table 4.21 shows the quarterly accident counts for accidents with personal injury, involving vehicles on the same road going in opposite directions during daylight hours during the years 1987-89 in Denmark. In analogy with Example 4.7 on page 123 we shall choose to model the random component of the counts by a Poisson distribution. The cyclic pattern of the year suggest to present the data in a two-way layout, see Table 4.22. The canonical link for the Poisson distribution is the logarithm, $\eta_{pq} = \log(\lambda_{pq})$.

In order to assess graphically whether an additive model

$$\eta_{i,j} = \alpha_i + \beta_j \tag{4.90}$$

for the logarithms is appropriate, we plot a "profile plot" for $\log(y_{i,j})$ as shown in Figure 4.12. A model without interaction would be represented by parallel graphs on the plot of $\log(y_{i,j})$ vs j for $i = 1, \ldots, 3$.

To a fair approximation the connected lines are parallel. It is, however, not easy to assess how large leeway there should be allowed due to random fluctuations. Therefore, we shall assess the hypothesis (4.90) by a numerical test.

The *full model*, that assigns an individual mean value to each observation $y_{i,j}$ is of the form

$$\eta_{i,j} = \alpha_i + \beta_j + \gamma_{i,j}$$

Table 4.21: *Quarterly accident counts.*

Index, i	Number of accidents, y_i	Year, x_{i1}	Quarter, x_{i2}
1	128	1987	1
2	95	1987	2
3	100	1987	3
4	75	1987	4
5	94	1988	1
6	85	1988	2
7	119	1988	3
8	71	1988	4
9	82	1989	1
10	81	1989	2
11	98	1989	3
12	72	1989	4

Table 4.22: *Number of accidents with personal injury in the years 1987-1989. Category "opp.direction."*

Year	Quarter 1	2	3	4	Total
1987	128	95	100	75	398
1988	94	85	119	71	369
1989	82	81	98	72	333
Total	304	261	317	218	1100

or, in the Wilkinson Rogers notation: Acc = Year+ Quart+ Year.Quart. We choose to fit the model $\mathcal{H}_1$ (4.90) formulated as Acc = Year+ Quart and choose Poisson distribution and canonical link:

```
> glmacc<-glm(formula=Numacc~Year+Quarter,
           family=poisson(link=log),data=accdat)
> deviance(glmacc)
[1] 8.953269
```

The residual deviance, $D\left(y; \mu(\widehat{\beta})\right) = 8.95$ corresponds to the 82% fractile in a $\chi^2(6)$-distribution. Thus, data do not contradict the hypothesis (4.90) of no interaction. Also, the deviance residuals do not contradict the hypothesis. Therefore, we will proceed by investigating whether both terms, Year and Quarter are necessary.

The Type III analysis of deviance table as given by R is:

```
> drop1(glmacc,test='Chisq')
```

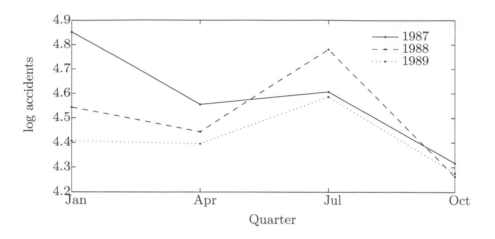

Figure 4.12: *Profile plot for the logarithm to quarterly accident counts for the years 1987-1989. Accident counts for each year are connected with straight lines.*

```
Single term deletions

Model:
Numacc ~ Year + Quarter
          Df Deviance     AIC      LRT    Pr(Chi)
<none>          8.953  97.049
Year       2   14.763  98.859   5.810   0.05476
Quarter    3   31.473 113.569  22.520 5.085e-05
```

The table shows that when corrected for the effect of quarters, there is not a significant effect of the years. When corrected for the possible effect of the years the effect of the quarters is strongly significant. Therefore, the first step in a model reduction could be to drop the term `Year` from the model.

We observe that in this case with Poisson distributed data is the partitioning of the contributions from the factors *orthogonal*. The Type I decomposition of the model deviance is

```
> anova(glmacc,test='Chisq')

Analysis of Deviance Table
Model: poisson, link: log
Response: Numacc
Terms added sequentially (first to last)

          Df Deviance Resid. Df Resid. Dev P(>|Chi|)
NULL                        11       37.283
Year       2    5.810        9       31.473     0.055
```

Table 4.23: *Number of injured persons in accidents with personal injury in the years 1987-1989. Category "opp.direction".*

		Quarter			
Year	1	2	3	4	Total
1987	217	144	177	131	669
1988	156	163	205	130	654
1989	149	137	189	117	592
Total	522	444	571	378	1915

```
Quarter  3   22.520          6       8.953 5.085e-05
```
where the (uncorrected) Year contribution is the same as the Type III Year contribution found in the previous output table.

The additive model (4.90) for $\eta_{i,j}$ corresponds to a *multiplicative* model for $\mu_{i,j} = \mathrm{E}[Y_{i,j}]$, viz. a model of the form

$$\mu_{i,j} = \alpha_i^* \beta_j^*. \tag{4.91}$$

Thus, the model without interaction term for the canonical parameter has the consequence that in all years there is the same ratio between accident counts in two selected quarters. Hence, a statement like "there are 50% more accidents in July quarter than in October quarter" makes sense, as well for all years as a whole, as for the individual years.

Example 4.16 – Overdispersion, Poisson distribution

As an example of a situation with overdispersion, we consider the same source of data as in Example 4.15 on page 148.

Now, instead of studying the accident count (of accidents with personal injury), we study the number of injured persons at accidents with personal injury. The data are shown in Table 4.23. Proceeding as in the example with the number of accidents, we formulate a model, $Y_{i,j} \sim \mathrm{Pois}(\mu_{i,j})$. Also, we shall formulate the hypothesis (4.90) of no interaction for the logarithmic link $\eta_{i,j} = \log(\mu_{i,j})$.

```
> glmacc<-glm(formula=Numacc ~ Year + Quarter,
          family = poisson(link=log),data=accdat2)
> deviance(glmacc)
[1] 15.74113
```

As $\mathrm{P}[\chi^2(6) \geq 15.7411] = 0.015$ the hypothesis of Poisson distribution with no interaction is rejected at all levels larger than 1.5%.

We have no reasons to believe that there is interaction between years and quarters for the number of injured, and therefore we choose to interpret the test results as indicating *overdispersion*.

This assumption does not contradict the assumption of a multiplicative Poisson model (no interaction) for accident *counts*. The count of injured persons is compounded with the number of accidents in the sense that for each accident there is at least one injured person, but the number of injured persons may vary from 1 to some upper limit. Therefore, it is meaningful to assume that the number of injured persons may be described by a Poisson distributed random variable with some constant overdispersion σ^2.

To estimate σ^2, we use the residual deviance, $D\left(y; \mu(\widehat{\beta})\right) = 15.74$ with 6 degrees of freedom, leading to $\widehat{\sigma}^2 = 2.6235$. For testing purposes, we shall use *scaled deviances* in analogy with (4.40).

Rescaling the residual deviances by division by $\widehat{\sigma}^2 = 2.6235$ we obtain the following table of Type III partitioning of the scaled deviances:

```
> OD <- 1/2.6235
> w <- rep(OD,12)

> glmacc2<-glm(formula = Numacc ~ Year + Quarter,
          weights = w, family=poisson(link=log),
          data=accdat2)

> anova(glmacc2,test='Chisq')

Analysis of Deviance Table
Model: poisson, link: log
Response: Numacc
Terms added sequentially (first to last)

        Df Deviance Resid. Df Resid. Dev P(>|Chi|)
NULL                     11    25.5392
Year     2   2.0123       9    23.5270    0.3656
Quarter  3  17.5269       6     6.0000    0.0006
```

4.9 Generalized linear models in R

Model fitting

Generalized linear models are analyzed by means of the function **glm**. Most of the features of a **lm** object are also present in a **glm** object. The first arguments to the **glm** function are:

- A *linear model* specified by a model formulae.

- A **family** object specifying the distribution. Valid family names are **binomial**, **gaussian**, **Gamma**, **inverse.gaussian**, **poisson**, **quasi**, quasibinomial and quasipoisson.

- A link function given as an argument to the family.

The model formulae are specified the same way as in the **lm** function, see Section 3.12 on page 81. A typical call to the **glm** function is:

```
\item fit<-glm(y ~ x, family=poisson(link = log))
```

After a model has been fitted using **glm**, some of the same *generic* functions that are used for the general linear model can by used. A list of some useful functions is given in Section 3.12 on page 81.

The linear prediction and the fitted values are extracted from a **glm** object (e.g., `fit`) by using the method **predict.glm**. A simple way of extracting the linear prediction and the fitted values is to use the expressions **predict.glm(**`fit`**)** and **fitted(**`fit`**)**, respectively.

Residuals

The residuals may be extracted from a **glm**-object by using the **residuals()** method with an argument `type =` specifying the type. The deviance residuals is the default option in the **residuals** method. They may also be extracted by specifying `type='`**deviance**`'`. The response residuals are produced by specifying `type='`**response**`'`, the Pearson residuals by specifying `type='pearson'` and the the working residuals are produced from a **glm**-object by specifying `type='`**working**`'` in the **residuals** extractor.

Dispersion parameter

The **summary** reports the value for the dispersion parameter σ^2. For **binomial** and Poisson families, the dispersion parameter is 1, and is not estimated in these cases. For the **Gamma** and Gaussian families, σ^2 is estimated by the Pearson X^2 statistic (4.66). The dispersion parameter is used in the computation of the reported standard errors and z-values for the individual coefficients. These defaults can be overridden by specifying the value for the dispersion parameter using the `dispersion =` argument to the **summary** method. Specification of `dispersion = 0` will result in the Pearson estimate, irrespective of the family of the object.

In case of overdispersion, `family = `**quasi** can be used which allows for the choice of a quasi-distribution where the user specifies the variance function $V(\cdot)$. For quasibinomial and quasipoisson errors, `family = quasibinomial` or `family = quasipoisson` can be used, respectively.

4.10   Problems

Please notice that some of the problems are designed for being solved by hand calculations while others require access to a computer.

Exercise 4.1

Consider the two observations, $y_1 = 4$, and $y_2 = 14$, respectively. For each of these two observations sketch a graph of the deviance

$$\lambda d(y; \mu)$$

where $d(y; \mu)$ denotes the unit deviance, and λ is the precision parameter. Sketch the graph for each of the following distribution models:

1. Normal distribution, $Y \in N(\mu, \sigma^2)$ with $\sigma^2 = 1$

2. Gamma distribution $Y \in G(\alpha, \mu/\alpha)$ for $\alpha = 1$

3. The inverse Gaussian distribution, $(IG)(\mu, \lambda)$ for $\lambda = 1$

4. The Poisson distribution, $Y \in P(\lambda)$

For each of these distribution models give the variance in the distribution of Y for $\mu = 4$ and $\mu = 14$.

Exercise 4.2

In order to monitor the general health condition in society, governmental statistical agencies record a number of demographical variables. The dataset stillborn gives for each of the years 1963-1976 the number of births for mothers who were married at the time of birth, and the number of births for mothers who were unmarried. Moreover the number of stillbirths for each category has been recorded. (Source: Statistics Denmark)

Question 1 Plot the fraction of stillborn vs year. What is the general tendency?

Question 2 Modify the plot from Question 1 by using different plot-symbols for the two categories. What do you see?

Question 3 Set up a model describing the development of the fraction of stillborn during that period. In order to avoid too many decimals in the regression coefficients it might be wise to choose a new origin for the year-axis. An interpretation of the analysis should be made in light of the societal norms during that period – and not the present days norms. The development of the number of births for the two groups clearly indicates the change in society's attitude towards unwed mothers. A model suitable for extrapolation beyond 1976 would let two curves approach each other.

Exercise 4.3

The night of January 27, 1986, there was a three-hour teleconference among people at NASA, Kennedy Space Center and Morton Thiokol (a solid rocket motor manufacturer). The discussion was about the effect of low temperature on O-ring performance and whether to postpone launch time of the space shuttle Challenger scheduled the next morning. The predicted temperature next morning was 31 degrees Fahrenheit. It was decided not to change the time of launch. The Challenger crashed the next morning because of an O-ring failure in the right rocket booster.

The Challenger had two solid rocket motors manufactured by Morton Thiokol. Each of the motors was shipped in four pieces and assembled at the Kennedy Space Center. The O-rings were used to seal the motor field joints.

After each launch the rocket motors are recovered from the ocean for inspection. There were 24 launches prior to the Challenger. For one launch the rocket motors were lost at sea. The dataset challenger_data has data from the 23 earlier flights available with the variables:

n number of O-rings
damage number of O-rings showing thermal distress
temp temperature at launch
pres leak-check pressure
date date of launch.

This data has of course been analysed frequently, see for example http://www-pao.ksc.nasa.gov/shuttle/missions/51-l/mission-51-l.html,

Question 1 Look at a plot of number of O-rings showing thermal distress vs temperature at launch and leak-check pressure. Do you see any dependence?

Question 2 Fit a model for the relationship between proportion of O-ring showing thermal distress to temperature and leak-check pressure. Use the logit link. Reduce the model if possible.

Question 3 Predict the risk of thermal distress at temperatures from 31 to 85 degrees Fahrenheit. Plot the available data for proportion O-rings showing thermal distress and the predictions vs. temperature.

Question 4 How much does the odds of failure increase with decrease in temperature of 10 degrees?

Exercise 4.4
The data shown in Table 4.20 on page 148 is given in the file barlett.

Question 1 Verify Bartlett's finding that there is no evidence of interaction under the logit transformation.

Question 2 Verify Hoblyn's finding that the benefit of planting at once is greater for short roots than for long.

Exercise 4.5
The dataset Metal has the results from an experiment with the purpose to determine the toxicity of nickel. Five groups of 25-28 female rats each were used in the experiment. The rats were given nickel orally for ten weeks before pairing and during the pregnancy in the following doses 0, 1, 2.5, 5 and 10 mg/kg/day, respectively. After giving birth the size of the litter for every rat was noted and the number of dead newborns in the litter counted (until four days after the birth). It is of interest to assess the possible relationship between the given doses and number of dead newborns.

Question 1 Fit a normal logistic regression model to the data. Look at the goodness of fit, is the fit satisfactory?

Question 2 Fit a new model to the data, now taking the overdispersion into account. The overdispersion can possibly be explained by the fact that even for the same doses there is some variation between the survival probability in the litters.

CHAPTER 5

Mixed effects models

In this chapter we will at first consider a general class of models for the analysis of *grouped data*. It is assumed that the (possibly experimental) conditions within groups are the same, whereas the conditions vary between groups. For most experimental conditions it is reasonable to talk about *repetitions* for the observations within groups.

We will initially represent the observations using the following table

Group	Observations
1	$Y_{11}, Y_{12}, \ldots, Y_{1n_1}$
2	$Y_{21}, Y_{22}, \ldots, Y_{2n_2}$
$\vdots$	$\vdots$, $\vdots$
k	$Y_{k1}, Y_{k2}, \ldots, Y_{kn_k}$

which corresponds to a so-called *classification* in k groups (cells) with n_i, $(i = 1, 2, \ldots, k)$ observations in each group. In the case $n_1 = n_2 = \cdots = n_k = n$, i.e., the same number of observations or repetitions is available for each of the k homogeneous *groups*, we say that the experiment is *balanced*. The grouping may be the result of one or several *factors*, and each set of factor levels defines a single (homogeneous) situation or *treatment*, often called a *cell*. The possible values of the factors are often called *levels*. If the factor is "sex," the levels are "male" and "female."

For the so-called *factorial experiment* outlined above all the explanatory variables are *categorical*, and often called factors. Later on in this chapter situations will be considered where such variables are combined with for instance continuous quantitative variables.

In the chapters about general and generalized linear models, i.e., Chapters 3 and 4, methods for estimating the structure of the mean value function, $\boldsymbol{\mu}$ have been considered. For *grouped data* a typical parameterization consists of a parameter or level specific for each group, and these so-called *fixed effects models* were introduced in Chapter 3. However, as briefly discussed while we introduced for instance the one-way ANOVA in Example 3.2 on page 48 it might be more reasonable instead to consider the level as an outcome of picking a number of groups in a *large population*, where only the variation between groups within this population is of interest and not the specific level for each group as for the fixed effects model. For the one-way ANOVA case this leads to a simple *hierarchical model* where the levels are considered as random variables, and this gives rise to the so-called *random effects models*. Models containing both fixed and random effects are called *mixed effects models* or just *mixed models*. The hierarchical structure arises here from the fact that the so-called first stage model describes the observations given the random effects, and the second stage model is a model for these random effects.

In a general setting mixed models describes dependence between observations within and between groups by assuming the existence of one or more unobserved *latent variables* for each group of data. The latent variables are assumed to be random and hence referred to as *random effects*. Hence a mixed model consists of both fixed model parameters θ and random effects U, where the random effects are described by another model and hence another set of parameters describing the assumed distribution for the random effects.

The hierarchical structure of the models implies that the mixed models are a powerful class of models used for the analysis of *correlated data*. As it will be demonstrated in this chapter, the grouping structure induces a correlation structure of the data even in the classical case of independent data within the groups.

Mixed effects models can handle more general correlation structures than simply correlations between groups. Examples include simple correlation in time between the observations as for time series data; see e.g., Madsen (2008) and problems with missing data.

Examples of correlated data include grouped or clustered data, repeated observations, longitudinal data, multiple dependent variables, spatial data or population based time series data. Key references to the analysis of longitudinal data are Diggle et al. (2002) and Verbeke and Molenberghs (2000). Methods for mixed effect modeling based on time series data have developed extensively during the past decade and are extensively used for drug development as further outlined in the following remark.

▶ **Remark 5.1 – Methods for population based time series data**
An important example related to population based time series data are population pharmacokinetic/pharmacodynamic (PK/PD) studies. Considering PK/PD studies the variation in time can be described explicitly as for the longitudinal data, or by ordinary differential equations (ODEs) or stochastic differential equations (SDEs). The ODE based mixed models for analysis of PK/PD data were first introduced in Beal and Sheiner (1980) and implemented in the software tool NONMEM (Beal and Sheiner 2004) and R (Tornøe et al. 2004). A method using stochastic differential equations is first reported in Tornøe, Jacobsen, and Madsen (2004) and later implemented in software as described in Mortensen et al. (2007) and Klim et al. (2009). These methods are related to the methods for estimating parameters in stochastic differential equations described in Kristensen, Madsen, and Jørgensen (2004).   ◀

A key feature of mixed models is that, by introducing random effects in addition to fixed effects, they allow you to address multiple sources of variation, e.g., in the longitudinal study they allow you to take into account both within and between subject or group variation.

As in the previous chapters we will adapt a likelihood framework both for introducing the concepts and for making statistical inference. In the final section about nonlinear mixed effects models there will be no constraints on the assumed distribution for the random effects, but apart from that the presentation and almost all of the examples in this chapter are based on an assumed Gaussian distribution which implies that we consider models for the mean value directly. In the next chapter, i.e., in Chapter 6 we shall consider hierarchical generalized models where we will consider the concept of random effects and mixed effects for various exponential family distributions.

In this chapter we will outline the theory for both *linear* and *nonlinear* mixed effects models. First the difference between linear and nonlinear Gaussian mixed models is outlined in Section 5.1. Then in Section 5.2 the concepts of random effects models are introduced followed by an extensive presentation of the important one-way model with random effects where a number of analytic results is available. Further examples of hierarchical models are briefly mentioned in Section 5.3. Now a rather comprehensive description of the linear (and Gaussian) mixed effects models are given in Section 5.4 followed by a Bayesian interpretation of the linear mixed effects models in Section 5.5. The chapter is concluded with an introduction to the theory behind nonlinear mixed effects models, and some remarks and hints on software are provided.

Important references for mixed effects models are McCulloch and Searle (2001), Pinheiro (2000), Searle, Casella, and McMulloch (1992), Lee and Nelder (2001), and Lee and Nelder (1996).

5.1   Gaussian mixed effects model

It is useful to understand the important difference between linear and nonlinear models, and this difference will be illustrated by introducing a general formulation of the Gaussian mixed model.

DEFINITION 5.1 − GAUSSIAN MIXED MODEL
A general formulation of the *Gaussian mixed model* is

$$Y|U = u \sim N(\mu(\beta, u), \Sigma(\beta)) \qquad (5.1)$$
$$U \sim N(0, \Psi(\psi)) \qquad (5.2)$$

where the dimension of U and therefore Ψ might be large, but ψ is generally small.

It is clearly seen that the model is also a so-called *hierarchical model* where (5.1) and (5.2) are the first and second stage model, respectively.

It is seen that the conditional distribution of Y given the outcome, u of the *random effect*, U is Gaussian.

Assuming that the random effects are scalar and independent and that the residuals within groups are independent, then

$$Y_{ij}|U_i = u_i \sim N(\mu_{ij}(\boldsymbol{\beta}, u_i), \sigma^2) \tag{5.3}$$

$$U_i \sim N(0, \sigma_u^2), \; i = 1, \ldots, k; j = 1, \ldots, n_i \tag{5.4}$$

where k is the number of groups and n_i is the number of observations within group i.

Due to the within group independence of the residuals this is equivalently written

$$Y_i|U_i = u_i \sim N(\boldsymbol{\mu}_i(\boldsymbol{\beta}, u_i), \sigma^2 \boldsymbol{I}) \tag{5.5}$$

$$U_i \sim N(0, \sigma_u^2), \; i = 1, \ldots, k \tag{5.6}$$

Using the mean value function here, where the random effect is scalar, this case is equivalently written

$$\boldsymbol{Y}_i = \boldsymbol{\mu}_i(\boldsymbol{\beta}, U_i) + \boldsymbol{\epsilon}_i \tag{5.7}$$

$$\boldsymbol{\epsilon}_i|U_i = u_i \sim N(\mathbf{0}, \sigma^2 \boldsymbol{I}) \; U_i \sim N(0, \sigma_u^2) \tag{5.8}$$

where $\boldsymbol{\mu}_i(\boldsymbol{\beta}, u_i)$ is the *mean value function*.

Let us now extend the discussion to the vector case of the random effects. Then if $\boldsymbol{\mu}(\boldsymbol{\beta}, \boldsymbol{U})$ is nonlinear in $\boldsymbol{\beta}$ then we have a *nonlinear mixed model*, whereas a model with the mean value function

$$\boldsymbol{\mu} = \boldsymbol{X\beta} + \boldsymbol{ZU} \tag{5.9}$$

with $\boldsymbol{X}$ and $\boldsymbol{Z}$ denoting known matrices, is called a *linear mixed model*. The useful case of linear Gaussian mixed models will be further described in Section 5.4. Notice how the mixed effect linear model in (5.9) is a linear combination of *fixed effects*, $\boldsymbol{X\beta}$ and *random effects*, $\boldsymbol{ZU}$.

Since we have so far, in the previous chapters, focused on fixed effects models we shall now focus on the random effects.

5.2 One-way random effects model

Motivation

We shall start by an example to supplement the abstract presentation.

Example 5.1 – Balanced data, ANalysis Of VAriance table
Unprocessed (baled) wool contains varying amounts of fat and other impurities that need to be removed prior to further processing. The price – and the value of the baled wool depends on the amount of pure wool that is left

Table 5.1: *The purity in % pure wool for four samples from each of seven bales of Uruguayan wool. The data marked with a bullet will be excluded later on in order to illustrate methods for unbalanced data.*

				Bale No.			
Sample	1	2	3	4	5	6	7
1	52.33	56.99	54.64	54.90	59.89	57.76	60.27
2	56.26	58.69	57.48	60.08	57.76	59.68	60.30
3	62.86•	58.20•	59.29•	58.72	60.26	59.58	61.09
4	50.46•	57.35•	57.51•	55.61	57.53	58.08	61.45
Bale average	55.48	57.81	57.23	57.33	58.86	58.78	60.78

Table 5.2: *ANOVA table for the baled wool data.*

Variation	Sum of Squares		f	$s^2 = \text{SS}/f$	F-value	Prob > F
Between bales	SSB	65.9628	6	10.9938	1.76	0.16
Within bales	SSE	131.4726	21	6.2606		
Total	SST	197.4348	27			

after removal of fat and impurities. The purity of the baled wool is expressed as the mass percentage of pure wool in the baled wool. The data are from Cameron (1951).

As part of the assessment of different sampling plans for estimation of the purity of a shipment of several bales of wool U.S. Customs Laboratory, Boston had selected seven bales at random from a shipment of Uruguayan wool, and from each bale, four samples were selected for analysis.

The sample results are shown in Table 5.1.

Model with fixed effects

We could formulate a one-way model as discussed in Chapter 3 on page 48:

$$\mathcal{H}_1: \quad Y_{ij} \sim \text{N}(\mu_i, \sigma^2) \quad i = 1, 2, \ldots, k; \quad j = 1, 2, \ldots, n_i$$

and obtain the ANOVA shown in Table 5.2. The test statistic for $\mathcal{H}_0 : \mu_1 = \mu_2 = \cdots = \mu_k$ is F $= 10.99/6.26 = 1.76$ which is not significant.

Such a model would be relevant, if we had selected seven specific bales, e.g., the bales with identification labels "AF37Q", "HK983", ..., and "BB837". Thus, $i = 1$ would refer to bale "AF37Q", and the probability distributions would refer to repeated sampling, but under such imaginative repeated sampling, $i = 1$ would always refer to this specific bale with label "AF37Q".

Random effects model

However, although there is not strong evidence against $\mathcal{H}_1$, we will not consider the bales to have the same purity. The idea behind the sampling was to *describe the variation* in the shipment, and the purity of the seven selected bales was not of interest in it self, but rather as representative for the variation in the shipment.

Therefore, instead of the *fixed effects* model in Chapter 3 on page 48, we shall introduce a *random effects* model.

Formulation of the model

As the individual groups (bales in Example 5.1) are selected at random from a large population, we shall use a *random model*:

DEFINITION 5.2 – ONE-WAY MODEL WITH RANDOM EFFECTS
Consider the random variables $Y_{ij}, i = 1, 2, \ldots, k;\ j = 1, 2, \ldots, n_i$
  A *one-way random effects* model for Y_{ij} is a model such that

$$Y_{ij} = \mu + U_i + \epsilon_{ij} \ , \tag{5.10}$$

with $\epsilon_{ij} \sim \mathrm{N}(0, \sigma^2)$ and $U_i \sim \mathrm{N}(0, \sigma_u^2)$, and where ϵ_{ij} are mutually independent, and also the U_i's are mutually independent, and finally the U_i's are independent of ϵ_{ij}.
  We shall put

$$N = \sum_{i=1}^{k} n_i \tag{5.11}$$

When all groups are of the same size, $n_i = n$, we shall say that the model is *balanced*.

▸ **Remark 5.2 – Parameters in the one-way random effects model**
Consider a one-way random effects model specified by (5.10). The (*fixed*) parameters of the model are $(\mu, \sigma^2, \sigma_u^2)$.
  Sometimes, the *signal to noise ratio*

$$\gamma = \frac{\sigma_u^2}{\sigma^2} \tag{5.12}$$

is used instead of σ_u^2. Thus, the parameter γ expresses the inhomogeneity between groups in relation to the internal variation in the groups. We shall use the term *signal/noise ratio* for the parameter γ. ◂

▸ **Remark 5.3 – The one-way model as a hierarchical model**
Putting $\mu_i = \mu + U_i$ we may formulate (5.10) as a *hierarchical model*, where we shall assume that

$$Y_{ij}|\mu_i \sim \mathrm{N}(\mu_i, \sigma^2) \ , \tag{5.13}$$

and in contrast to the *systematic/fixed model*, the bale level μ_i is modeled as a realization of a random variable,

$$\mu_i \sim N(\mu, \sigma_u^2), \tag{5.14}$$

where the μ_i's are assumed to be mutually independent, and Y_{ij} are *conditionally independent*, i.e., Y_{ij} are mutually independent in the conditional distribution of Y_{ij} for given μ_i.

Often we say that we have *hierarchical variation*, or that we have a hierarchical model, corresponding to a random variation between groups (nested within the constant factor), and the replication within groups is nested within the partitioning in groups.   ◄

▶ **Remark 5.4 – Interpretation of the one-way random effect model**
The random effects model will be a reasonable model in situations where the interest is not restricted alone to the experimental conditions at hand, but where the experimental conditions rather are considered as representative for a larger collection (population) of varying experimental conditions, in principle selected at random from that population.

As illustration of the difference between the random and the systematic models, we further note that the analysis of the systematic model puts emphasis on the assessment of the results in the individual groups, μ_i, and possible differences, $\mu_i - \mu_h$, between the results in specific groups, whereas the analysis of the random effects model aims at describing the variation between the groups, $\mathrm{Var}[\mu_i] = \sigma_u^2$.

Thus, in the example with the bales of wool, the probability distributions in the random effects model refer to imaginative repeated sampling from the shipment in such a way that the whole sampling process is repeated. Thus, e.g., $i = 1$ refers to the first bale selected, and *not* to a specific bale. In one realization of the sampling process, $i = 1$ might refer to the bale with identification label "AF37Q", but in another realization $i = 1$ could refer to the bale with identification label "BB837". Individual bales are simply considered as random representatives of the bales in the population.   ◄

Marginal and joint distributions

The hierarchical model structure of the mixed models leads to specific correlation structures as shown in the following.

THEOREM 5.1 – MARGINAL DISTRIBUTIONS IN THE RANDOM EFFECTS MODEL FOR ONE WAY ANALYSIS OF VARIANCE
The marginal distribution of Y_{ij} is a normal distribution with

$$E[Y_{ij}] = \mu$$

$$\mathrm{Cov}[Y_{ij}, Y_{hl}] = \begin{cases} \sigma_u^2 + \sigma^2 & \text{for } (i,j) = (h,l) \\ \sigma_u^2 & \text{for } i = h,\ j \neq l \\ 0 & \text{for } i \neq h \end{cases} \tag{5.15}$$

▶ **Remark 5.5 – Observations from the same group are correlated**
We note that there is a positive covariance between observations from the same group. This positive covariance expresses that observations within the same group will deviate in the same direction from the mean, μ, in the marginal distribution, viz. in the direction towards the group mean in question.

The coefficient of correlation,

$$\rho = \frac{\sigma_u^2}{\sigma_u^2 + \sigma^2} = \frac{\gamma}{1 + \gamma} \tag{5.16}$$

that describes the correlation within a group, is often termed the *intraclass correlation*.   ◀

▶ **Remark 5.6 – Distribution of individual group averages**
We finally note that the joint distribution of the group averages is characterized by

$$\mathrm{Cov}[\overline{Y}_{i\cdot}, \overline{Y}_{h\cdot}] = \begin{cases} \sigma_u^2 + \sigma^2/n_i & \text{for } i = h \\ 0 & \text{otherwise} \end{cases} \tag{5.17}$$

That is, that the k group averages $\overline{Y}_{i\cdot}$, $i = 1, 2, \ldots, k$ are mutually independent, and that the variance of the group average

$$\mathrm{Var}[\overline{Y}_{i\cdot}] = \sigma_u^2 + \sigma^2/n_i = \sigma^2(\gamma + 1/n_i)$$

includes the variance of the random component, $\sigma_u^2 = \sigma^2\gamma$, as well as the effect of the residual variance on the group average.

Thus, an increase of the sample size in the individual groups will improve the precision by the determination of the group mean α_i, but the variation between the individual group means is not reduced by this averaging.   ◀

▶ **Remark 5.7 – Observation vector for a group**
When we consider the set of observations corresponding to the i'th group as a n_i-dimensional column vector,

$$\boldsymbol{Y}_i = \begin{pmatrix} Y_{i1} \\ Y_{i2} \\ \vdots \\ Y_{in_i} \end{pmatrix} \tag{5.18}$$

we have, that the correlation matrix in the marginal distribution of $\boldsymbol{Y}_i$ is an *equicorrelation-matrix* of the form

$$\boldsymbol{E}_{n_i} = (1 - \rho)\boldsymbol{I}_{n_i} + \rho\boldsymbol{J}_{n_i} = \begin{pmatrix} 1 & \rho & \cdots & \rho \\ \rho & 1 & \cdots & \rho \\ \vdots & \vdots & \ddots & \vdots \\ \rho & \rho & \cdots & 1 \end{pmatrix} \tag{5.19}$$

where $\boldsymbol{J}_{n_i}$ is a $n_i \times n_i$-dimensional matrix consisting solely of 1's.

Thus, the set of observations $\boldsymbol{Y}_i$, $i = 1, 2, \ldots, k$ may be described as k *independent observations* of a n_i dimensional variable $\boldsymbol{Y}_i \sim \mathrm{N}_{n_i}(\boldsymbol{\mu}, \sigma^2 \boldsymbol{I}_{n_i} + \sigma_u^2 \boldsymbol{J}_{n_i})$, i.e., the dispersion matrix for $\boldsymbol{Y}_i$ is

$$
\begin{aligned}
\boldsymbol{V}_i &= \mathrm{D}[\boldsymbol{Y}_i] \\
&= \mathrm{E}[(\boldsymbol{Y}_i - \boldsymbol{\mu})(\boldsymbol{Y}_i - \boldsymbol{\mu})^T] \\
&= \begin{pmatrix}
\sigma_u^2 + \sigma^2 & \sigma_u^2 & \cdots & \sigma_u^2 \\
\sigma_u^2 & \sigma_u^2 + \sigma^2 & \cdots & \sigma_u^2 \\
\vdots & \vdots & \ddots & \vdots \\
\sigma_u^2 & \sigma_u^2 & \cdots & \sigma_u^2 + \sigma^2
\end{pmatrix}
\end{aligned}
\tag{5.20}
$$

Such a matrix is denoted a *compound symmetrical matrix*. ◄

▶ **Remark 5.8 – Covariance structure for the whole set of observations**
If we organize all observations in one column, organized according to groups, we observe that the $N \times N$-dimensional dispersion matrix $\mathrm{D}[\boldsymbol{Y}]$ is

$$
\boldsymbol{V} = \mathrm{D}[\boldsymbol{Y}] = \text{ Block diag}\{\boldsymbol{V}_i\}
\tag{5.21}
$$

where $\boldsymbol{V}_i$ is given by (5.20).

Analogously one finds that the correlation matrix for the whole set of observations is an $N \times N$-dimensional block matrix with the matrices $\boldsymbol{E}_{n_i}$ in the diagonal, and zeros outside, which illustrates that observations from *different groups are independent*, whereas observations from the *same group are correlated*. ◄

Test of hypothesis of homogeneity

Consider again the ANOVA shown in Table 5.2, and notice the notation (SSE, SSB, and SST) used in the table. As usual we have

$$
\mathrm{SSE} = \sum_{i=1}^{k} \sum_{j=1}^{n_i} (y_{ij} - \bar{y}_i)^2
$$

$$
\bar{\bar{y}} = \sum_i n_i \bar{y}_{i\cdot} \Big/ N
$$

$$
\mathrm{SSB} = \sum_{i=1}^{k} n_i (\bar{y}_i - \bar{\bar{y}})^2
$$

$$
\mathrm{SST} = \sum_{i=1}^{k} \sum_{j=1}^{n_i} (y_{ij} - \bar{\bar{y}})^2 = \mathrm{SSB} + \mathrm{SSE}
$$

Furthermore we introduce the *shrinkage factor*

$$w_i(\gamma) = \frac{1}{1 + n_i \gamma} \tag{5.22}$$

the importance of $w_i(\gamma)$ is seen as $\text{Var}[\overline{Y}_{i\cdot}] = \sigma^2 / w_i(\gamma)$.

Under the random effects model, the hypothesis that the varying experimental conditions do not have an effect on the observed values, is formulated as

$$\mathcal{H}_0 : \sigma_u^2 = 0. \tag{5.23}$$

The hypothesis is tested by comparing the variance ratio with the quantiles in a $F(k-1, N-k)$-distribution.

Theorem 5.2 – Test of the hypothesis of homogeneity in the random effects model
Under the model specified by (5.13) and (5.14) the likelihood ratio test for the hypothesis (5.23) has the test statistic

$$Z = \frac{\text{SSB}/(k-1)}{\text{SSE}/(N-k)} \tag{5.24}$$

Large values of z are considered as evidence against the hypothesis.

Under the hypothesis (5.23), Z will follow a $F(k-1, N-k)$-distribution. In the balanced case, $n_1 = n_2 = \ldots = n_k = n$, we can determine the distribution of Z also under the alternative hypothesis. In this case we have

$$Z \sim (1 + n\gamma)F(k-1, N-k). \tag{5.25}$$

Estimation of parameters

The (fixed) parameters are μ, σ^2 and γ or σ_u^2. In this section we will present various estimates for these parameters.

▸ **Remark 5.9 – Balanced design**
The *balanced design* or *balanced case* is defined as an experimental design in which the same number of observations is taken for each combination of the experimental factors, i.e., in the *balanced case* we have

$$n_1 = n_2 = \cdots = n_k = n . \tag{5.26}$$

◂

▸ **Remark 5.10 – Confidence interval for the variance ratio**
In the balanced case one may use (5.25) to construct a *confidence interval for the variance ratio* γ. Using (5.25), one finds that a $1 - \alpha$ confidence interval for γ, i.e., an interval (γ_L, γ_U), satisfying

$$P[\gamma_L < \gamma < \gamma_U] = 1 - \alpha$$

is obtained by using

$$\gamma_L = \frac{1}{n}\left(\frac{Z}{F(k-1,N-k)_{1-\alpha/2}} - 1\right)$$

$$\gamma_U = \frac{1}{n}\left(\frac{Z}{F(k-1,N-k)_{\alpha/2}} - 1\right)$$

(5.27)

where Z is given by (5.24).   ◄

THEOREM 5.3 – MOMENT ESTIMATES IN THE RANDOM EFFECTS MODEL
*Under the model given by (5.13) and (5.14) one finds the moment estimates
for the parameters μ, σ^2 and σ_u^2 by*

$$\widetilde{\mu} = \overline{\overline{Y}}_{..}$$
$$\widetilde{\sigma}^2 = \text{SSE}/(N-k)$$
$$\widetilde{\sigma_u}^2 = \frac{\text{SSB}/(k-1) - \text{SSE}/(N-k)}{n_0} = \frac{\text{SSB}/(k-1) - \widetilde{\sigma}^2}{n_0}$$

(5.28)

where the weighted average group size n_0 *is given by*

$$n_0 = \frac{\sum_1^k n_i - \left(\sum_1^k n_i^2 / \sum_1^k n_i\right)}{k-1} = \frac{N - \sum_i n_i^2/N}{k-1}$$

(5.29)

Proof It may be shown that

$$E[\text{SSE}/(N-k)] = \sigma^2$$
$$E[\text{SSB}/(k-1)] = \sigma^2 + n_0\sigma_u^2$$ ∎

▸ **Remark 5.11 – Distribution of "residual" sum of squares**
In the balanced case we have that

$$\text{SSE} \sim \sigma^2 \chi^2(k(n-1))$$
$$\text{SSB} \sim \{\sigma^2/w(\gamma)\}\chi^2(k-1)$$

and that SSE and SSB are independent.   ◄

▸ **Remark 5.12 – Unbiased estimates for variance ratio in the balanced case**
In the balanced case, $n_1 = n_2 = \cdots = n_k = n$, we can provide explicit unbiased
estimators for γ and $w(\gamma) = 1/(1+n\gamma)$. One has

$$\widetilde{w} = \frac{\text{SSE}/\{k(n-1)\}}{\text{SSB}/(k-3)}$$
$$\widetilde{\gamma} = \frac{1}{n}\left(\frac{\text{SSB}/(k-1)}{\text{SSE}/\{k(n-1)-2\}} - 1\right)$$

(5.30)

Table 5.3: *ANOVA table for the baled wool data.*

Variation	Sum of squares		f	$s^2 = SS/f$	$E[S^2]$
Between bales	SSB	65.9628	6	10.9938	$\sigma^2 + 4\sigma_u^2$
Within bales	SSE	131.4726	21	6.2606	σ^2

are *unbiased estimators* for $w(\gamma) = 1/(1+n\gamma)$ and for $\gamma = \sigma_u^2/\sigma^2$, respectively.

Proof Follows from the remark above together with the properties of the χ^2-distribution:

$$E\left[\frac{SSE}{k(n-1)}\right] = \sigma^2$$

$$E\left[\frac{k(n-1)-2}{SSE}\right] = \frac{1}{\sigma^2}$$

$$E\left[\frac{SSB}{k-1}\right] = \frac{\sigma^2}{w(\gamma)}$$

$$E\left[\frac{k-3}{SSB}\right] = \frac{w(\gamma)}{\sigma^2}$$

■

Example 5.2 – Balanced data, ANOVA-table
We shall now show the calculations for a one-way analysis of variance for Example 5.1 on page 160, and obtain the ANOVA shown in Table 5.3. The test statistic for the hypothesis $\mathcal{H}_0 : \sigma_u^2 = 0$, is

$$z = \frac{10.9938}{6.2606} = 1.76 < F_{0.95}(6, 21) = 2.57$$

The p-value is $P[F(6, 21) \geq 1.76] = 0.16$

Thus, the test fails to reject the hypothesis of no variation between the purity of the bales when testing at a 5% significance level. However, as the purpose is to describe the variation in the shipment, we will estimate the parameters in the random effects model, irrespective of the test result.

A 95% confidence interval for the ratio $\gamma = \sigma_u^2/\sigma^2$ is found using (5.27). As $F(6, 21)_{0.025} = 1/F(21, 6)_{0.975}$, one finds the interval

$$\gamma_L = \frac{1}{4}\left(\frac{1.76}{F(6, 21)_{0.975}} - 1\right) = 0.25 \times \left(\frac{1.76}{3.09} - 1\right) = -0.11$$

$$\gamma_U = \frac{1}{4}\left(\frac{1.76}{F(6, 21)_{0.025}} - 1\right) = 0.25 \times (1.76 \times 5.15 - 1) = 2.02$$

Maximum likelihood estimation

Theorem 5.4 – Maximum likelihood estimates for the parameters
under the random effects model
*Under the model given by (5.13) and (5.14) the maximum likelihood estimates
for μ, σ^2 and $\sigma_u^2 = \sigma^2\gamma$ are determined by*

a)   *For $\sum_i n_i^2(\bar{y}_{i.} - \bar{\bar{y}}_{..})^2 < \text{SSE} + \text{SSB}$ one obtains*

$$\widehat{\mu} = \bar{\bar{y}}_{..} = \frac{1}{N}\sum_i n_i\bar{y}_{i.}$$

$$\widehat{\sigma}^2 = \frac{1}{N}(\text{SSE} + \text{SSB})$$

$$\widehat{\gamma} = 0$$

b)   *For $\sum_i n_i^2(\bar{y}_{i.} - \bar{\bar{y}}_{..})^2 > \text{SSE} + \text{SSB}$ the estimates are determined as solution to*

$$\widehat{\mu} = \frac{1}{W(\widehat{\gamma})}\sum_{i=1}^{k} n_i w_i(\widehat{\gamma})\bar{y}_{i.}. \tag{5.31}$$

$$\widehat{\sigma}^2 = \frac{1}{N}\left\{\text{SSE} + \sum_{i=1}^{k} n_i w_i(\widehat{\gamma})(\bar{y}_{i.} - \widehat{\mu})^2\right\} \tag{5.32}$$

$$\frac{1}{W(\widehat{\gamma})}\sum_{i=1}^{k} n_i^2 w_i(\widehat{\gamma})^2(\bar{y}_{i.} - \widehat{\mu})^2 = \frac{1}{N}\left\{\text{SSE} + \sum_{i=1}^{k} n_i w_i(\widehat{\gamma})(\bar{y}_{i.} - \widehat{\mu})^2\right\} \tag{5.33}$$

where $w_i(\gamma)$ is given by (5.22) and

$$W(\gamma) = \sum_{i=1}^{k} n_i w_i(\gamma).$$

The solution to (5.31) and (5.33) has to be determined by iteration.

Proof The result follows by looking at the log-likelihood based upon the
marginal distribution of the set of observations. This log-likelihood is – except
for an additive constant

$$l(\mu, \sigma^2, \gamma) = -\frac{\text{SSE}}{2\sigma^2} - \frac{1}{2\sigma^2}\sum_{i=1}^{k} n_i w_i(\gamma)(\bar{y}_{i.} - \mu)^2$$

$$-\frac{N}{2}\log(\sigma^2) + \frac{1}{2}\sum_{i=1}^{k}\log(w_i(\gamma))$$

∎

▶ **Remark 5.13 – The maximum likelihood estimate $\widehat{\mu}$ is a weighted average of the group averages**
It follows from (5.31), that $\widehat{\mu}$ is a weighted average of the group averages, $\overline{y}_{i\cdot}$, with the estimates for the marginal precisions

$$\sigma^2 n_i w_i(\gamma) = \frac{\sigma^2}{\operatorname{Var}[\overline{Y}_{i\cdot}]}$$

as weights. We have the marginal variances

$$\operatorname{Var}[\overline{Y}_{i\cdot}] = \sigma_u^2 + \frac{\sigma^2}{n_i} = \frac{\sigma^2}{n_i}(1 + n_i\gamma) = \frac{\sigma^2}{n_i w_i(\gamma)}$$

When the experiment is *balanced*, i.e., when $n_1 = n_2 = \cdots = n_k$, then all weights are equal, and one obtains the simple result that $\widehat{\mu}$ is the crude average of the group averages.   ◀

▶ **Remark 5.14 – The estimate for σ^2 utilizes also the variation between groups**
We observe that the estimate for σ^2 is not only based upon the variation within groups, SSE, but the estimate does also utilize the knowledge of the variation between groups, as

$$\mathrm{E}[(\overline{Y}_{i\cdot} - \mu)^2] = \operatorname{Var}[\overline{Y}_{i\cdot}] = \frac{\sigma^2}{n_i w_i(\gamma)}$$

and therefore, the terms $(\overline{y}_{i\cdot} - \mu)^2$ contain information about σ^2 as well as γ.   ◀

▶ **Remark 5.15 – The estimate for σ^2 is not necessarily unbiased**
We observe further that – as usual with ML-estimates of variance – the estimate for σ^2 is not necessarily unbiased.

Instead of the maximum likelihood estimate above, it is common practice to adjust the estimate. In Section 5.4 on page 181 we shall introduce the so-called *residual maximum likelihood* (REML) estimates for σ^2 and σ_u^2, obtained by considering the distribution of the residuals.   ◀

▶ **Remark 5.16 – Maximum-likelihood-estimates in the balanced case**
In the balanced case, $n_1 = n_2 = \cdots = n_k$ the weights

$$w_i(\gamma) = \frac{1}{1 + n\gamma}$$

do not depend on i, and then (5.31) reduces to

$$\widehat{\mu} = \frac{1}{k}\sum_{i=1}^{k}\overline{y}_{i+} = \overline{\overline{y}}_{++} \,,$$

which is the same as the moment estimate.

When $(n-1)\,\mathrm{SSB} > \mathrm{SSE}$ then the maximum likelihood estimate corresponds to an inner point.

The equations (5.32) and (5.33) for determination of σ^2 and $\gamma = \sigma_u^2/\sigma^2$ become

$$N\sigma^2 = \mathrm{SSE} + \frac{\mathrm{SSB}}{1+n\gamma}$$

$$N\,\frac{n}{1+n\gamma}\,\frac{\mathrm{SSB}}{k} = \mathrm{SSE} + \frac{\mathrm{SSB}}{1+n\gamma}$$

with the solution

$$\widehat{\sigma}^2 = \frac{\mathrm{SSE}}{N-k}$$

$$\widehat{\gamma} = \frac{1}{n}\left[\frac{\mathrm{SSB}}{k\widehat{\sigma}^2} - 1\right]$$

$$\widehat{\sigma}_b^2 = \frac{\mathrm{SSB}/k - \widehat{\sigma}^2}{n}$$

◀

Estimation of random effects, BLUP-estimation

In a mixed effects model, it is not clear what fitted values and residuals are. Our best prediction for subject i is not given by the overall mean, μ nor is it given by the individual averages. Let us illustrate this, and, as previously, the discussion is linked to Example 5.1.

It may sometimes be of interest to estimate the random effects (which for the considered example are the mean values in the selected bales). We write the joint density for the observable random variables Y and the unobservable μ. This may be termed the *hierarchical likelihood* (see Lee and Nelder (2001) and Lee and Nelder (1996))

$$L(\mu, \sigma^2, \gamma, \mu) = f(y|\mu)f(\mu)$$

$$L = \frac{1}{\sigma^N}\exp\left\{-\frac{\mathrm{SSE} + \sum_i n_i(\overline{y}_i - \mu_i)^2}{2\sigma^2}\right\}\frac{1}{(\gamma\sigma^2)^{k/2}}\exp\left\{-\frac{\sum_i(\mu_i - \mu)^2}{2\gamma\sigma^2}\right\}$$

leading to the log-likelihood

$$\ell(\mu, \sigma^2, \gamma, \mu) = -(N/2)\log(\sigma^2) - \frac{\mathrm{SSE} + \sum_i n_i(\overline{y}_i - \mu_i)^2}{2\sigma^2}$$

$$- (k/2)\log(\gamma\sigma^2) - \frac{\sum_i(\mu_i - \mu)^2}{2\gamma\sigma^2}$$

Differentiating with respect to μ_i we find

$$\frac{\partial}{\partial\mu_i}\ell(\mu, \sigma^2, \gamma, \mu) = \frac{n_i(\overline{y}_i - \mu_i)}{\sigma^2} - \frac{\mu_i - \mu}{\gamma\sigma^2}\ ,$$

which is zero for

$$\mu_i \left(\frac{n_i}{\sigma^2} + \frac{1}{\gamma \sigma^2} \right) = \frac{n_i}{\sigma^2} \bar{y}_i + \frac{1}{\gamma \sigma^2} \mu$$

The equations shall be solved together with (5.31) to (5.33) by iteration.
We observe that the solution satisfies

$$\mu_i \frac{n_i \gamma + 1}{\gamma \sigma^2} = \frac{n_i \gamma \bar{y}_i + \mu}{\gamma \sigma^2}$$

i.e.,

$$\mu_i = \frac{n_i \gamma}{n_i \gamma + 1} \bar{y}_i + \frac{1}{n_i \gamma + 1} \mu \qquad (5.34)$$

and hence,

$$\mu_i = \Big(1 - w_i(\gamma) \Big) \bar{y}_i + w_i(\gamma) \mu \qquad (5.35)$$

Thus, the estimate for μ_i is a weighted average between the individual bale
averages, $\bar{y}_i$ and the overall average $\hat{\mu}$ with weights $(1 - w_i(\gamma))$ and $w_i(\gamma)$,
where the shrinkage factor is

$$w_i(\gamma) = \frac{1}{1 + n_i \gamma} \qquad (5.36)$$

The solution is called the *best linear unbiased predictor* (BLUP). In Section 5.4
on page 182 we shall return to the construction of BLUP estimates of the
random effects in general.

You may compare the estimate (5.35) with the expression for the (Bayesian)
posterior mean in a normal distribution (see, e.g., Wasserman (2004), Chapter
11). We shall return to the interpretation of the BLUP estimate as an empirical
Bayesian estimate in Section 5.5 on page 189.

Figure 5.1 illustrates the influence of the shrinkage factor, w on the individual estimates of purity. When $w = 0$ ($\tilde{\sigma}^2 = 0$), we may use the individual
estimates, and for $w = 1$ ($\gamma = 0$), one should use the overall average as
predictor also for the individual means.

Example 5.3 – Bales data analyzed in R
Let us illustrate how to analyze the bales data introduced in Example 5.1
using R.

You will observe that R uses

i) `lm` for (general) linear models

ii) `glm` for generalized linear models

iii) `lme` for (general) linear mixed-effects models

iv) `nlme` for nonlinear mixed-effect models

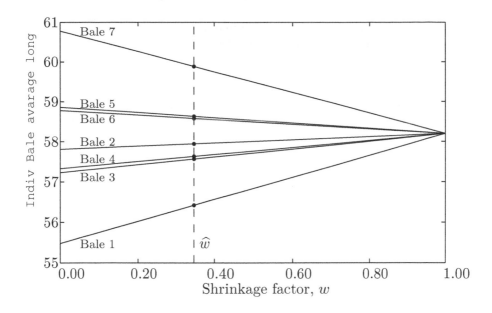

Figure 5.1: *Simultaneous estimation of bale means as a function of the shrinkage factor.*

These methods have a number of generic functions **anova**(), **coef**(), **fitted**(), fixed.**effects**(), **plot**(), **predict**(), **print**(), **residuals**(), random .**effects**(), and **summary**().

In a lme-model it is necessary to specify two *formulae*, one for the fixed effects, and another for the random effects. Further there is a possibility for specifying a large variety of covariance structures for the groups.

For the wool bales one would write in R

```
wool.lme1 <- lme(purity ~ 1, random=~1|Bale,data=wool)
```

```
summary(wool.lme1)
Linear mixed-effects model fit by REML
 Data: wool
       AIC      BIC    logLik
  138.8586 142.7461 -66.4293

Random effects:
 Formula:  ~ 1 | Bale
         (Intercept) Residual
StdDev:     1.087795 2.502115

Fixed effects: purity ~ 1
```

```
              Value Std.Error DF  t-value p-value
(Intercept) 58.03643 0.6266057 21 92.62033  <.0001

Standardized Within-Group Residuals:
     Min         Q1        Med         Q3       Max
 -2.5877 -0.2882734 0.01656582 0.5471141 2.368107

Number of Observations: 28
Number of Groups: 7
>

ranef(wool.lme1)
    (Intercept)
'1' -1.10170548
'2' -0.09856151
'3' -0.34719483
'4' -0.30521778
'5'  0.35457541
'6'  0.31798003
'7'  1.18012416

intervals(wool.lme1)
Approximate 95% confidence intervals

 Fixed effects:
                lower     est.   upper
(Intercept) 56.73333 58.03643 59.33953

 Random Effects:
  Level: Bale
                    lower     est.   upper
sd((Intercept)) 0.2939755 1.087795 4.025159

 Within-group standard error:
    lower     est.   upper
 1.851305 2.502115 3.381713
```

5.3   More examples of hierarchical variation

As already indicated, it often happens that the sampling procedure induces a hierarchical variation in data. Such an extra variation may act as a disturbing effect, or the purpose of the sampling might be to analyze this effect.

In the following we shall present examples of both situations.

Analysis of blocked experiments

Disturbing random effect

Now assume that we for the example in Table 5.1 on page 161 had analyzed the first two samples in each bale (sample 1 and 2) with one method (method A), and the last two samples (sample 3 and 4) with another method (method B), and that the whole experiment had the purpose of investigating the difference between the two methods.

If we had just one sample with each method from each bale, the situation had been the standard setup for paired data (the t-test for paired data). Then we would simply have determined the difference between the two test results from each bale and investigated a hypothesis of $\delta = 0$, where δ is the mean value in the distribution of the differences, $D_i \sim N(\delta, \sigma_d^2)$. Hereby we would have eliminated the variation between bales when assessing the difference. We have used each bale as its own *control*.

In the literature on Design of Experiments one would say that the experiment is a *block experiment* with each bale constituting a block.

Such block structure may take many forms, some examples are termed "split plot", "repeated measurements", just to mention a few. The help pages on the lme method and the literature on mixed effect models give guidance in the analysis of such data. For more information on design of experiments we refer to, e.g., Montgomery (2005).

Random coefficient regression lines

Random effect of prime interest

Sometimes the purpose of the sampling is to assess the random effect. Thus, the purpose of sampling the bales in Table 5.1 on page 161 was to assess the magnitude of the two levels of variation in order to achieve an efficient balance between number of bales selected, and number of samples from each bale.

A random effect description might also be relevant for other parameters than the level. The following motivating example illustrates that a random effect description is relevant for the slope of a regression line.

Example 5.4 – Ramus bone length
Table 5.4 on the next page (selected from Elston and Grizzle (1962)) shows the length in [mm] of the ramus bone for five randomly selected boys in the age 8-10 years. For each boy the bone length was measured four times, at age 8, 8.5, 9 and 9.5 years. The data are represented graphically in Figure 5.2. For the purpose of the example we shall model these data by a linear regression model (although a model that respects that the growth between two measurements can only be positive might be a more realistic model). If we had only been interested in these five specific boys, we would have

Table 5.4: *Length of ramus bone for five boys at four points in time. The reduced age is given by the actual age minus the average age of 8.75 years. Estimated parameters of the fixed effects model (5.37).*

Boy	Reduced age, x_i = age − 8.75				β_{i1}	β_{i2}
	−0.75	−0.25	0.25	0.75		
A	52.5	53.2	53.3	53.7	53.175	0.74
B	51.2	53.0	54.3	54.5	53.250	2.24
C	51.2	51.4	51.6	51.9	51.525	0.46
D	52.1	52.8	53.7	55.0	53.400	1.92
E	50.7	51.7	52.7	53.3	52.100	1.76
Average	51.54	52.42	53.12	53.68	52.690	1.424

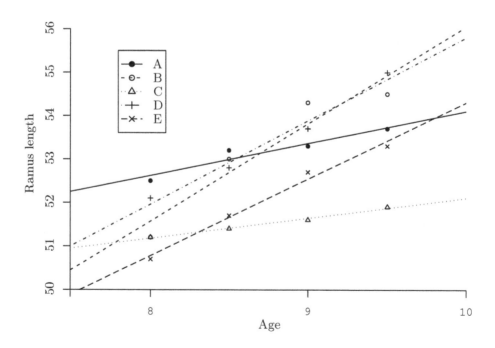

Figure 5.2: *Length of ramus bone for five boys versus boys age. Measurements and fixed effect linear regression model predictions.*

formulated a fixed effects model

$$Y_{ij} = \beta_{i1} + x_{ij}\beta_{i2} + \epsilon_{ij}, \ i = 1, 2, \ldots, 5; \ j = 1, 2, 3, 4, \qquad (5.37)$$

where ϵ_{ij} are assumed independent $N(0, \sigma^2)$-distributed.

For *each boy* we have the design matrix

$$X = \begin{pmatrix} 1.0 & -0.75 \\ 1.0 & -0.25 \\ 1.0 & 0.25 \\ 1.0 & 0.75 \end{pmatrix} \text{ with } X^T X = \begin{pmatrix} 4.00 & 0.00 \\ 0.00 & 1.25 \end{pmatrix}$$

and hence

$$(X^T X)^{-1} = \begin{pmatrix} 0.25 & 0.00 \\ 0.00 & 0.80 \end{pmatrix}$$

The estimates of the individual regression coefficients, $\widehat{\beta}_i$ determined as

$$\widehat{\beta}_i = (X^T X)^{-1} X^T y_i \qquad (5.38)$$

are given in Table 5.4.

In R one might have used a model formula like:

```
> formula = ramus ~ Boy+agered+Boy:agered
```

viz. a model allowing separate intercept and slope for each boy. The analysis under this model gives a strong indication that the boys are different (i.e., with separate parameters for each boy).

However, since we are not interested in the individual boys as such, but consider them as a sample of boys, so we will use a *random effects* model.

The observations from the i'th boy are modelled by

$$Y_i = X\beta + XU_i + \epsilon_i, \ i = 1, 2, \ldots, k$$

where the two-dimensional random effect contains the random deviations from the overall values of the intercept and slope, and where

$$U_i \sim N_2(0, \sigma^2\Psi), \ \ \epsilon_i \sim N_n(0, \sigma^2 I_n), \qquad (5.39)$$

and where U_i, U_j are mutually independent for $i \neq j$, and ϵ_i and ϵ_j are mutually independent for $i \neq j$, and further are U_i and ϵ_j independent.

The covariance matrix Ψ denotes the covariance matrix in the population distribution of intercepts and slopes with the measurement error σ^2 extracted as a factor.

The marginal distribution

The marginal distribution of Y_i under the model (5.39) is given by

$$Y_i \sim N_n(X\beta, \sigma^2[I_n + X\Psi X^T]),$$

and the marginal distribution of the crude estimates (5.38) is

$$\widehat{\beta}_i \sim N_2(\beta, \sigma^2[(X^T X)^{-1} + \Psi]). \tag{5.40}$$

It is noticed that the distribution is influenced as well by the design matrix X, as by the covariance matrix Ψ in the distribution of U. As X is known we note that the distribution may provide information about Ψ. In Section 5.4 on page 181 we shall return to the estimation of these parameters.

In R the estimation may be performed by the lme method, and in the following we will briefly outline the principles and comment on the results.

Assuming the data is found in ramus we write

```
ramus.lme1 <- lme(ramus ~1+ agered,
random=~1+agered|Boy,data=ramus)

ramus.lme1
Linear mixed-effects model fit by REML
  Data: ramus
  Log-restricted-likelihood: -16.48603
  Fixed: ramus ~ 1 + agered
 (Intercept) agered
       52.69  1.424

Random effects:
 Formula:  ~ 1 + agered | Boy
 Structure: General positive-definite
              StdDev    Corr
(Intercept) 0.8172897 (Intercept)
     agered 0.7323660 0.586
   Residual 0.2939388

Number of Observations: 20
Number of Groups: 5

ranef(ramus.lme1)
   (Intercept)      agered
A    0.4458929 -0.5334634
B    0.5561130  0.7300490
C   -1.1380612 -0.9038613
D    0.6912308  0.4741513
E   -0.5551755  0.2331244

coefficients(ramus.lme1)
   (Intercept)      agered
A     53.13589 0.8905366
B     53.24611 2.1540490
```

```
C    51.55194 0.5201387
D    53.38123 1.8981513
E    52.13482 1.6571244
```

The fixed effects part of the model leads to estimates of the overall values of the parameters. These overall estimates for the intercept and slope are $\hat{\beta}_1 = 52.69$ and $\hat{\beta}_2 = 1.424$, respectively. The random effects part of the model describes the deviations from the fixed effects parameters, and the predictions for each boy are listed as output both without and with considerations to the fixed effects part.

The parameters of the random effects part of the model can be specified as

$$\hat{\sigma}^2 \widehat{\boldsymbol{\Psi}} = 0.2939^2 \begin{pmatrix} 7.7312 & 4.0573 \\ 4.0573 & 6.2072 \end{pmatrix} = \begin{pmatrix} 0.8173^2 & 0.3506 \\ 0.3506 & 0.7323^2 \end{pmatrix} \tag{5.41}$$

where the estimated correlation coefficient (0.586) is used to state the off-diagonal value of the covariance matrix $\widehat{\boldsymbol{\Psi}}$.

In conclusion it is seen that the average length of the ramus bone for boys at age 8.75 is 52.69 [mm], and the average growth rate is 1.42 [mm/year]. Finally, the correlation coefficient shows a positive relation between the length of the ramus bone at age 8.75 and the growth rate.

5.4 General linear mixed effects models

Introduction

The hierarchical linear normal distribution models, we considered in the previous sections are special cases of a general class of models, the so-called *linear mixed effects models*. In the following we shall sketch the general theory for this model class.

For a more thorough treatment, see, e.g., Pawitan (2001) or Pinheiro (2000).

DEFINITION 5.3 – LINEAR MIXED EFFECTS MODEL
The model

$$\boldsymbol{Y} = \boldsymbol{X\beta} + \boldsymbol{ZU} + \boldsymbol{\epsilon} \tag{5.42}$$

with $\boldsymbol{X}$ and $\boldsymbol{Z}$ denoting known matrices, and where $\boldsymbol{\epsilon} \sim \mathrm{N}(\boldsymbol{0}, \boldsymbol{\Sigma})$ and $\boldsymbol{U} \sim \mathrm{N}(\boldsymbol{0}, \boldsymbol{\Psi})$ are independent is called a *mixed general linear model*. In the general case may the covariance matrices $\boldsymbol{\Sigma}$ and $\boldsymbol{\Psi}$ depend on some unknown parameters, $\boldsymbol{\psi}$, that also need to be estimated.

The parameters $\boldsymbol{\beta}$ are called *fixed effects* or *systematic effects*, while the quantities $\boldsymbol{U}$ are called *random effects*.

It follows from the independence of U and ϵ that

$$D\left[\begin{pmatrix}\epsilon\\U\end{pmatrix}\right]=\begin{pmatrix}\Sigma & 0\\0 & \Psi\end{pmatrix}$$

As described in Section 5.1 the model may also be interpreted as a *hierarchical model*

$$U \sim N(0, \Psi) \tag{5.43}$$

$$Y|U = u \sim N(X\beta + Zu, \Sigma) \tag{5.44}$$

with the probability density functions

$$f_U(u; \psi) = \frac{1}{(\sqrt{2\pi})^q \det(\Psi)}\, \exp\left[\frac{-1}{2}\, u^T \Psi^{-1} u\right] \quad \text{for } u \sim \mathbb{R}^q \tag{5.45}$$

and

$$f_{Y|u}(y; \beta) = \frac{1}{(\sqrt{2\pi})^N \det(\Sigma)} \times \exp\left[-\tfrac{1}{2}\,(y - X\beta - Zu)^T \Sigma^{-1}(y - X\beta - Zu)\right] \tag{5.46}$$

for $y \sim \mathbb{R}^N$.

It then follows that the *marginal distribution* of Y is a normal distribution with

$$E[Y] = X\beta \tag{5.47}$$

$$D[Y] = \Sigma + Z\Psi Z^T \tag{5.48}$$

We shall introduce the symbol V for the dispersion matrix in the marginal distribution of Y, i.e.,

$$V = \Sigma + Z\Psi Z^T \tag{5.49}$$

The matrix V may grow rather large and cumbersome to handle.

Example 5.5 – One-way model with random effects

The one-way model with random effects

$$Y_{ij} = \mu + U_i + e_{ij}$$

that was introduced in Definition 5.2 on page 162 may be formulated as

$$Y = X\beta + ZU + \epsilon \tag{5.50}$$

with

$$X = 1_N$$
$$\beta = \mu$$
$$U = (U_1, U_2, \ldots, U_k)^T$$
$$\Sigma = \sigma^2 I_N$$
$$\Psi = \sigma_u^2 I_k$$

where 1_N is a column of 1's. The i, j'th element in the $N \times k$ dimensional matrix Z is 1, if y_{ij} belongs to the i'th group, otherwise it is zero.

Estimation of fixed effects and variance parameters

The fixed effect parameters β and the variance parameters ψ are estimated from the marginal distribution of Y. The log-likelihood for all the parameters is

$$\ell(\beta, \psi; y) = -\tfrac{1}{2} \log |V| - \tfrac{1}{2}(y - X\beta)^T V^{-1}(y - X\beta) \qquad (5.51)$$

For a fixed value of ψ the score function with respect to β (cfn. (3.28) on page 49) is

$$\ell_\beta^T(\beta; \psi) = X^T \left[V^{-1}y - V^{-1}X\beta \right] \qquad (5.52)$$

and, hence, for fixed ψ the estimate of β is found as the solution of

$$(X^T V^{-1} X)\beta = X^T V^{-1} y \qquad (5.53)$$

This is the well-known *weighted least squares (WLS)* formula. In some software systems the solution is called the generalised least squares (GLS). Note, however, that the solution may depend on the unknown variance parameters ψ as we saw in the case of the unbalanced one-way random effect model.

The profile log-likelihood for the variance parameter ψ is therefore

$$\ell(\psi) = -\tfrac{1}{2} \log |V| - \tfrac{1}{2}(Y - X\widehat{\beta})^T V^{-1}(Y - X\widehat{\beta}) \qquad (5.54)$$

where $\widehat{\beta}$ is determined by putting (5.52) equal to zero.

The observed Fisher information for β is

$$I(\widehat{\beta}) = X^T V^{-1} X \qquad (5.55)$$

from which an estimate for the dispersion matrix for $\widehat{\beta}$ is determined as

$$\mathrm{Var}[\widehat{\beta}] = (X^T V^{-1} X)^{-1} \qquad (5.56)$$

In order to determine estimates for the variance parameters ψ we shall modify the profile likelihood (5.54) for ψ in order to compensate for the estimation of β (see, e.g., Pawitan (2001), Example 10.12).

The *modified profile log-likelihood* is

$$\begin{aligned} \ell_m(\psi) = &-\tfrac{1}{2} \log |V| - \tfrac{1}{2} \log |X^T V^{-1} X| \\ &-\tfrac{1}{2}(Y - X\widehat{\beta})^T V^{-1}(Y - X\widehat{\beta}) \end{aligned} \qquad (5.57)$$

When $\widehat{\beta}$ depends on ψ it is necessary to determine the solution by iteration, and it is in general rather complicated to determine an analytical expression for (5.57). Pawitan (2001) Section 17.5 provides a simpler approach that does not require determination of V.

▸ **Remark 5.17 – REML-method**

The modification (5.57) to the profile likelihood equals the so-called *residual maximum likelihood* (REML)-method derived by Patterson and Thompson (1971) using the marginal distribution of the *residual* $(\boldsymbol{y} - \boldsymbol{X}\widehat{\boldsymbol{\beta}}_\psi)$.

In REML the problem of biased variance components is solved by setting the fixed effects estimates equal to the WLS solution above in the likelihood function and then maximising it to find the variance component terms only. The reasoning is that the fixed effects cannot contribute with information on random effects leading to a justification of not estimating these parameters in the same likelihood.

The method is also termed[1] *restricted maximum likelihood* method because the model may be embedded in a more general model for the group observation vector $\boldsymbol{Y}_i$ (5.18) where the random effects model *restricts* the correlation coefficient in the general model.

Considering (5.51) it is observed that the REML-estimates are obtained by minimising

$$(\boldsymbol{Y} - \boldsymbol{X}\boldsymbol{\beta})^T \boldsymbol{V}^{-1}(\boldsymbol{\psi})(\boldsymbol{Y} - \boldsymbol{X}\boldsymbol{\beta}) + \log|\boldsymbol{V}^{-1}(\boldsymbol{\psi})| + \log|\boldsymbol{X}^T\boldsymbol{V}^{-1}(\boldsymbol{\psi})\boldsymbol{X}|$$

A comparison with the full likelihood function in (5.51) shows that it is the variance term $\log|\boldsymbol{X}^T\boldsymbol{V}^{-1}(\boldsymbol{\psi})\boldsymbol{X}|$ which is associated with the estimation of $\boldsymbol{\beta}$ that causes the REML estimated variance components to be unbiased.

If accuracy of estimates of the variance terms are of greater importance than bias then the full maximum likelihood should be considered instead. An optimal weighting between bias and variance of estimators is obtained by the estimators optimising the so-called *Godambe Information* – see, e.g., Heyde (1997).

In balanced designs REML gives the classical moment estimates of variance components (constrained to be non-negative). See Harville (1977) for a general review and discussion of variance component estimation.   ◂

Estimation of random effects

Formally, the random effects, $\boldsymbol{U}$ are not parameters in the model, and the usual likelihood approach does not make much sense for "estimating" these random quantities. It is, however, often of interest to assess these "latent", or "state" variables.

We formulate a so-called *hierarchical likelihood* (Lee and Nelder (1996)) by writing the joint density for observable as well as unobservable random quantities $(\boldsymbol{Y}, \boldsymbol{U})$,

$$f(\boldsymbol{y}, \boldsymbol{u}; \boldsymbol{\beta}, \boldsymbol{\psi}) = f_{Y|u}(\boldsymbol{y}; \boldsymbol{\beta})f_U(\boldsymbol{u}; \boldsymbol{\psi})$$

---

[1]In the US the method is most often referred to as *restricted maximum likelihood*.

where $f_{Y|u}(\boldsymbol{y}; \boldsymbol{\beta})$ and $f_U(\boldsymbol{u}; \boldsymbol{\psi})$ are given by (5.45) and (5.46). Therefore one has

$$\ell(\boldsymbol{\beta}, \boldsymbol{\psi}, \boldsymbol{u}) = -\tfrac{1}{2}\log|\boldsymbol{\Sigma}| - \tfrac{1}{2}(\boldsymbol{y} - \boldsymbol{X}\boldsymbol{\beta} - \boldsymbol{Z}\boldsymbol{u})^T\boldsymbol{\Sigma}^{-1}(\boldsymbol{y} - \boldsymbol{X}\boldsymbol{\beta} - \boldsymbol{Z}\boldsymbol{u})$$
$$-\tfrac{1}{2}\log|\boldsymbol{\Psi}| - \tfrac{1}{2}\boldsymbol{u}^T\boldsymbol{\Psi}^{-1}\boldsymbol{u} \tag{5.58}$$

and, hence

$$\frac{\partial}{\partial\boldsymbol{u}}\ell(\boldsymbol{\beta}, \boldsymbol{\psi}, \boldsymbol{u}) = \boldsymbol{Z}^T\boldsymbol{\Sigma}^{-1}(\boldsymbol{y} - \boldsymbol{X}\boldsymbol{\beta} - \boldsymbol{Z}\boldsymbol{u}) - \boldsymbol{\Psi}^{-1}\boldsymbol{u} \tag{5.59}$$

By putting the derivative equal to zero and solving with respect to $\boldsymbol{u}$ one finds that the estimate $\widehat{\boldsymbol{u}}$ is solution to

$$(\boldsymbol{Z}^T\boldsymbol{\Sigma}^{-1}\boldsymbol{Z} + \boldsymbol{\Psi}^{-1})\boldsymbol{u} = \boldsymbol{Z}^T\boldsymbol{\Sigma}^{-1}(\boldsymbol{y} - \boldsymbol{X}\boldsymbol{\beta}) \tag{5.60}$$

where the estimate $\widehat{\boldsymbol{\beta}}$ is inserted in place of $\boldsymbol{\beta}$.

The solution to (5.60) is termed the *best linear unbiased predictor* (BLUP-estimate) because it may be shown that the estimator minimizes

$$\mathrm{E}\left[(\widehat{U}_i - \mathrm{E}[U_i])^2\right]$$

among all linear (in the observations) estimators $\widehat{U}_i$, that are unbiased for $\mathrm{E}[U_i]$, i.e., they satisfy $\mathrm{E}[\widehat{U}_i] = \mathrm{E}[U_i]$.

BLUPs replace the random effects $\boldsymbol{U}$ by their conditional means $\widehat{\boldsymbol{u}}$ given the data, and then make predictions using these values:

$$\widehat{\boldsymbol{Y}} = \boldsymbol{X}\widehat{\boldsymbol{\beta}} + \boldsymbol{Z}\widehat{\boldsymbol{u}}.$$

The second derivative of the log-likelihood with respect to $\boldsymbol{u}$ is

$$\frac{\partial^2}{\partial\boldsymbol{u}\partial\boldsymbol{u}^T}\ell(\boldsymbol{\beta}, \boldsymbol{\psi}, \boldsymbol{u}) = -\boldsymbol{Z}^T\boldsymbol{\Sigma}^{-1}\boldsymbol{Z} - \boldsymbol{\Psi}^{-1}$$

which shows that the observed Fisher information wrt. $\boldsymbol{u}$ is

$$I(\widehat{\boldsymbol{u}}) = (\boldsymbol{Z}^T\boldsymbol{\Sigma}^{-1}\boldsymbol{Z} + \boldsymbol{\Psi}^{-1})$$

that may be used to assess the uncertainty on $\widehat{\boldsymbol{u}}$.

Simultaneous estimation of $\boldsymbol{\beta}$ and $\boldsymbol{u}$

The estimates for $\boldsymbol{\beta}$ and for $\boldsymbol{u}$ are those values that simultaneously maximize $\ell(\boldsymbol{\beta}, \boldsymbol{\psi}, \boldsymbol{u})$ (5.58) for a fixed value of $\boldsymbol{\psi}$.

From (5.58) we have

$$\frac{\partial}{\partial\boldsymbol{\beta}}\ell(\boldsymbol{\beta}, \boldsymbol{\psi}, \boldsymbol{u}) = \boldsymbol{X}^T\boldsymbol{\Sigma}^{-1}(\boldsymbol{y} - \boldsymbol{X}\boldsymbol{\beta} - \boldsymbol{Z}\boldsymbol{u})$$

Combining this with (5.59) and putting equal to zero, we obtain the so-called *mixed model equations*

$$\begin{pmatrix} X^T\Sigma^{-1}X & X^T\Sigma^{-1}Z \\ Z^T\Sigma^{-1}X & Z^T\Sigma^{-1}Z + \Psi^{-1} \end{pmatrix} \begin{pmatrix} \beta \\ u \end{pmatrix} = \begin{pmatrix} X^T\Sigma^{-1}y \\ Z^T\Sigma^{-1}y \end{pmatrix} \qquad (5.61)$$

The estimates from these simultaneous equations are precisely the estimates that result from (5.52) and (5.60). The equations facilitate the estimation of β and u without calculation of the marginal variance V, or its inverse.

The estimation may be performed by an *iterative back-fitting algorithm* as follows:

i) Start with an estimate for β, e.g., the usual least squares estimate

$$\widehat{\beta} = (X^TX)^{-1}X^Ty$$

and iterate between steps ii) and iii) below, until convergence has been achieved.

ii) Calculate an adjusted observation

$$y^{adj} = y - X\widehat{\beta}$$

and estimate u from a random effects model $y^{adj} = Zu + \epsilon$ by

$$(Z^T\Sigma^{-1}Z + \Psi^{-1})u = Z^T\Sigma^{-1}y^{adj}$$

iii) Recalculate the adjusted observation

$$y^{adj} = y - Zu$$

and estimate β from a fixed effects model $y^{adj} = X\beta + \epsilon$ by

$$(X^T\Sigma^{-1}X)\beta = X^T\Sigma^{-1}y^{adj}$$

It is, however, a problem that in most cases the variance parameters enter in the matrices Σ and Ψ. In these cases it will be necessary to determine estimates $\widehat{\Sigma}$ and $\widehat{\Psi}$ (e.g., by REML), and then solve the mixed model equations

$$\begin{pmatrix} X^T\widehat{\Sigma}^{-1}X & X^T\widehat{\Sigma}^{-1}Z \\ Z^T\widehat{\Sigma}^{-1}X & Z^T\widehat{\Sigma}^{-1}Z + \widehat{\Psi}^{-1} \end{pmatrix} \begin{pmatrix} \beta \\ b \end{pmatrix} = \begin{pmatrix} X^T\widehat{\Sigma}^{-1}y \\ Z^T\widehat{\Sigma}^{-1}y \end{pmatrix}$$

instead of (5.61). Thereafter updated estimates $\widehat{\Sigma}$ and $\widehat{\Psi}$ are determined. The process is continued until convergence has been achieved.

Interpretation as empirical Bayes estimate

It is seen from (5.60), restated here

$$(Z^T \Sigma^{-1} Z + \Psi^{-1}) u = Z^T \Sigma^{-1} (y - X \widehat{\beta})$$

that the BLUP-estimate $\widehat{u}$ for the random effects has been "shrunk" towards zero, as it is a weighted average of the direct estimate, $(y - X\widehat{\beta})$, and the prior mean, $E[U] = 0$, where the weights are the precision Ψ^{-1} in the distribution of U.

5.5 Bayesian interpretations

Conditional, simultaneous and marginal distributions

We recall the definition of a conditional probability density function, and the rules relating conditional and marginal moments (e.g., Madsen 2008):

$$E[Y] = E_X[\, E[\, Y|X\,] \,] \tag{5.62a}$$
$$\text{Var}[Y] = E_X[\, \text{Var}[\, Y|X\,] \,] + \text{Var}_X[\, E[\, Y|X\,] \,] \tag{5.62b}$$
$$\text{Cov}[Y, Z] = E_X[\, \text{Cov}[Y, Z]\, |X \,] + \text{Cov}_X[\, E[\, Y|X\,], E[\, Z|X\,] \,] \tag{5.62c}$$

In settings where $f_X(x)$ expresses a so-called "subjective probability distribution" (possibly degenerate), the expression

$$f_{X|Y=y}(x) = \frac{f_{Y|X=x}(y) f_X(x)}{\int f_{Y|X=x}(y) f_X(x) dx} \tag{5.63}$$

for the conditional distribution of X for given $Y = y$ is termed *Bayes' theorem*, named after the British Presbyterian minister and Fellow of the Royal Society, Thomas Bayes (1702-1761). In such settings, the distribution $f_X(\cdot)$ of X is called the *prior distribution*, and the conditional distribution with density function $f_{X|Y=y}(x)$ given by (5.63) is called the *posterior distribution* after observation of $Y = y$.

The theorem is useful in connection with hierarchical models where the variable X denotes a non-observable state (or parameter) that is associated with the individual experimental object, and Y denotes the observed quantities. In such situations one may often describe the conditional distribution of Y for given state $(X = x)$, and one will have observations of the marginal distribution of Y. However, in general it is not possible to observe the states (x), and therefore the distribution $f_X(x)$ is not observed directly. This situation arises in many contexts such as hidden Markov models (HMM) (see Cappé, Moulines, and Rydén 2005 or Ephrain and Merhav 2002), or state space models, where inference about the state (X) can be obtained using the so-called Kalman Filter, see Madsen (2008), Ljung (1987), Maybeck (1982), or Harvey (1996). In general the Kalman Filter provides estimates of the conditional mean and covariance only – not the entire distribution.

A Bayesian formulation

In this section we discuss the use of Bayes' theorem in situations where the "prior distribution", $f_X(x)$, has a frequency interpretation. As seen previously in Section 5.2 on page 160 the one-way random effects model may be formulated in a Bayesian framework.

We may identify the $N(\cdot, \sigma_u^2)$-distribution of $\mu_i = \mu + U_i$ as the *prior distribution*. The statistical model for the data is such that for given μ_i, are the Y_{ij}'s independent and distributed like $N(\mu_i, \sigma^2)$.

In a Bayesian framework, the *conditional distribution* of μ_i given $\overline{Y}_i = \overline{y}_i$ is termed the *posterior distribution* for μ_i.

THEOREM 5.5 – THE POSTERIOR DISTRIBUTION OF μ_i
Consider the one-way model with random effects in Definition 5.2 on page 162

$$Y_{ij}|\mu_i \sim N(\mu_i, \sigma^2) \tag{5.64a}$$

$$\mu_i \sim N(\mu, \sigma_u^2) \tag{5.64b}$$

where μ, σ^2 and σ_u^2 are known.

The posterior distribution of μ_i after observation of $y_{i1}, y_{i2}, \ldots, y_{in}$ is a normal distribution with mean and variance

$$E[\mu_i \mid \overline{Y}_i = \overline{y}_i] = \frac{\mu/\sigma_u^2 + n_i \overline{y}_i/\sigma^2}{1/\sigma_u^2 + n_i/\sigma^2} = w\mu + (1-w)\overline{y}_i \tag{5.65a}$$

$$Var[\mu_i|\overline{Y}_i = \overline{y}_i] = \frac{1}{\dfrac{1}{\sigma_u^2} + \dfrac{n}{\sigma^2}} \tag{5.65b}$$

where

$$w = \frac{\dfrac{1}{\sigma_u^2}}{\dfrac{n}{\sigma^2} + \dfrac{1}{\sigma_u^2}} = \frac{1}{1 + n\gamma} \tag{5.66}$$

with $\gamma = \sigma_u^2/\sigma^2$.

Proof The proof may be found in Example 11.2 in Wasserman (2004). ∎

▶ **Remark 5.18 – Interpretation of parameters**
We observe that the posterior mean is a weighted average of the prior mean μ, and sample result $\overline{y}_i$ with the corresponding *precisions* (reciprocal variances) as weights.

Note that the weights only depend on the signal/noise ratio γ, and not on the numerical values of σ^2 and σ_u^2; therefore we may express the posterior mean as

$$E[\mu_i \mid \overline{Y}_i = \overline{y}_i] = \frac{\mu/\gamma + n_i \overline{y}_i}{1/\gamma + n_i}$$

The expression (5.65b) for the posterior variance simplifies, if instead we consider the *precision*, i.e., the reciprocal variance

$$\frac{1}{\sigma_{post}^2} = \frac{1}{\sigma_u^2} + \frac{n_i}{\sigma^2} \tag{5.67}$$

Thus, we have that *the precision in the posterior distribution is the sum of the precision in the prior distribution and the sampling precision.* In terms of the signal/noise ratio, γ, with $\gamma_{prior} = \sigma_u^2/\sigma^2$ and $\gamma_{post} = \sigma_{post}^2/\sigma^2$ we have

$$\frac{1}{\gamma_{post}} = \frac{1}{\gamma_{prior}} + n_i$$

and

$$\mu_{post} = w\mu_{prior} + (1 - w)\bar{y}_i \tag{5.68}$$

with

$$w = \frac{1}{1 + n\gamma_{prior}}$$

in analogy with the BLUP-estimate (5.34).   ◀

Estimation under squared error loss

For comparison purposes we shall now introduce some concepts from statistical decision theory.

Assume that the discrepancy between a set of estimates $d_i(\boldsymbol{y})$ and the true parameter values μ_i, $i = 1, \ldots, k$ is measured by the *squared error loss function*

$$L(\boldsymbol{\mu}, \boldsymbol{d}(\boldsymbol{y})) = \left[\sum_{i=1}^{k} (d_i(\boldsymbol{y}) - \mu_i)^2\right]. \tag{5.69}$$

Averaging over the distribution of $\boldsymbol{Y}$ for given value of $\boldsymbol{\mu}$ we obtain the *risk* of using the estimator $\boldsymbol{d}(\boldsymbol{Y})$ when the true parameter is $\boldsymbol{\mu}$

$$R(\boldsymbol{\mu}, \boldsymbol{d}(.)) = \frac{1}{k} \operatorname{E}_{\boldsymbol{Y}|\boldsymbol{\mu}}\left[\sum_{i=1}^{k} (d_i(\boldsymbol{y}) - \mu_i)^2\right]. \tag{5.70}$$

THEOREM 5.6 – RISK OF THE ML-ESTIMATOR IN THE ONE-WAY MODEL
Let $\boldsymbol{d}^{ML}(\boldsymbol{Y})$ denote the maximum likelihood estimator for $\boldsymbol{\mu}$ in the one-way model with fixed effects (5.64a) with $\boldsymbol{\mu}$ arbitrary,

$$d_i^{ML}(\boldsymbol{Y}) = \overline{Y}_i = \frac{1}{n}\sum_{j=1}^{n} Y_{ij}. \tag{5.71}$$

The risk of this estimator is

$$R(\boldsymbol{\mu}, \boldsymbol{d}^{ML}) = \frac{\sigma^2}{n} \tag{5.72}$$

regardless of the value of $\boldsymbol{\mu}$.

Proof The result may be proved by considering the partition of the estimation error (risk) into *variance* and *bias*

$$
\begin{aligned}
R(\boldsymbol{\mu}, \boldsymbol{d}(.)) &= \frac{1}{k} \ \mathrm{E}\left[\sum \left(d_i(\boldsymbol{Y}) - \mu_i\right)^2\right] \\
&= \frac{1}{k} \ \mathrm{E}\left[\sum \left\{(d_i(\boldsymbol{Y}) - \mathrm{E}[d_i])^2 + (\mathrm{E}[d_i] - \mu_i)^2\right\}\right] \\
&= \frac{1}{k} \ \sum \left(\mathrm{Var}[d_i] + \delta_i^2\right)
\end{aligned}
\tag{5.73}
$$

where $\delta_i = \mathrm{E}[d_i(\boldsymbol{Y})] - \mu_i$ denotes the *bias* for the estimator $d_i(\boldsymbol{Y})$. As the maximum likelihood estimator is *unbiased*, $\mathrm{E}[d_i^{ML}(\boldsymbol{Y})] = \mu_i$ the risk is simply the variance, $\mathrm{Var}[\overline{Y}_i] = \sigma^2/n$. ∎

▸ **Remark 5.19 – Bayes risk for the ML-estimator**
Introducing the further assumption that $\boldsymbol{\mu}$ may be considered as a random variable with the (prior) distribution (5.64b) we may determine *Bayes risk* of $\boldsymbol{d}^{ML}(\cdot)$ under this distribution as

$$r((\boldsymbol{\mu}, \gamma), \boldsymbol{d}^{ML}) = \mathrm{E}_\mu(R(\boldsymbol{\mu}, \boldsymbol{d}^{ML})$$

with $R(\boldsymbol{\mu}, \boldsymbol{d}^{ML})$ given by (5.72). Clearly, as $R(\boldsymbol{\mu}, \boldsymbol{d}^{ML})$ does not depend on $\boldsymbol{\mu}$ we have that the Bayes risk is

$$r((\boldsymbol{\mu}, \gamma), \boldsymbol{d}^{ML}) = \frac{\sigma^2}{n}.$$ ◂

▸ **Remark 5.20 – The Bayes estimator in the one-way random effects model**
The *Bayes estimator* $\boldsymbol{d}^B(\boldsymbol{Y})$ is the estimator that minimizes the Bayes risk. We find the Bayes estimator

$$d_i^B(\boldsymbol{Y}) = \mathrm{E}[\mu_i \mid \overline{Y}_i] \tag{5.74}$$

with $\mathrm{E}[\mu_i \mid \overline{Y}_i]$ given by (5.65a).

It may be shown that the Bayes risk of this estimator is the *posterior variance*,

$$r((\boldsymbol{\mu}, \gamma), \boldsymbol{d}^B) = \frac{1}{\dfrac{1}{\sigma_u^2} + \dfrac{n}{\sigma^2}} = \frac{\sigma^2/n}{1 + 1/(n\gamma)} \tag{5.75}$$

where the posterior variance, $\text{Var}[\mu_i|\overline{Y}_i = \overline{y}_i]$ has been given by (5.65b). We observe that the Bayes risk of the Bayes estimator is less than that of the maximum likelihood estimator. ◀

Although the maximum-likelihood estimator, d_i^{ML}, possesses the properties that usually are required of a good estimator, the estimator is unbiased with minimum variance, but nevertheless (or maybe therefore) is the average (expected value) estimation error larger than for the Bayes estimator d_i^B.

The stochastic "dependence" between the individual μ_i that is specified in the prior distribution of $\boldsymbol{\mu}$, expresses that the parameter-values do not vary arbitrarily, but the means of the k samples will group around μ in accordance with a normal distribution with variance $\gamma\sigma^2$. The Bayes estimator

$$d_i^B(\boldsymbol{Y}) = w\mu + (1 - w)d_i^{ML}(\boldsymbol{Y})$$

utilizes this information by pulling the observed group averages $d_i^{ML}(\boldsymbol{y})$ towards the center of gravity μ. It may be shown that the *shrinkage factor*, $1 - w$, is determined as the coefficient of μ_i in the regression of μ_i on $\overline{Y}_i$ in the joint distribution of μ_i and $\overline{Y}_i$. We have

$$1 - w = \frac{\text{Cov}[\overline{Y}_i, \mu_i]}{\text{Var}[\overline{Y}_i]} = \frac{\text{Var}[\text{E}[\overline{Y}_i|\mu_i]]}{\text{E}[\text{Var}[\overline{Y}_i|\mu_i]] + \text{Var}[\text{E}[\overline{Y}_i|\mu_i]]} = \frac{n\gamma}{1 + n\gamma}$$

It may be shown that

$$r((\boldsymbol{\mu}, \gamma), d^B) = (1 - w)r((\boldsymbol{\mu}, \gamma), \boldsymbol{d}^{ML}) \tag{5.76}$$

which expresses that the reduction in the Bayes risk by using the Bayes estimator $\boldsymbol{d}^B$ instead of the maximum likelihood estimator $\boldsymbol{d}^{ML}$ is proportional to the shrinkage factor $(1 - w)$. The more the shrinkage, i.e., the smaller the value of $(1 - w)$, the larger the gain by using the Bayes estimator. This is not surprising as the shrinkage factor $(1 - w)$ is small when the variance between the parameter values is small as compared to the measurement uncertainty σ^2/n in the determination of the individual parameter values.

The empirical Bayes approach

When the parameters (μ, γ) in the prior distribution are unknown, one may utilize the whole set of observations $\boldsymbol{Y}$ for estimating μ, γ and σ^2, as it was described in Section 5.2 on page 169. We have

$$\widehat{\mu} = \overline{\overline{Y}}_{..} = \frac{1}{k}\sum_{i=1}^{k}\overline{Y}_{i.}, \quad \widehat{\sigma}^2 = \frac{\text{SSE}}{k(n-1)}$$

with $\text{SSE} \sim \sigma^2\chi^2(k(n-1))$ and $\text{SSB} \sim \sigma^2(1+n\gamma)\chi^2(k-1)$. As SSE and SSB are independent with

$$\text{E}\left[\frac{k-3}{\text{SSB}}\right] = \frac{1}{\sigma^2(1+n\gamma)}$$

we find that

$$\mathrm{E}\left[\frac{\widehat{\sigma}^2}{\mathrm{SSB}/(k-3)}\right] = \frac{1}{1+n\gamma} = w .$$

We shall, however, introduce the estimator

$$\widetilde{\sigma}^2 = \frac{\mathrm{SSE}}{k(n-1)+2} \tag{5.77}$$

and utilize that

$$\widehat{w} = \frac{\widetilde{\sigma}^2}{\mathrm{SSB}/(k-3)} \tag{5.78}$$

We observe that $\widehat{w}$ may be expressed by the usual F-test statistic as

$$\widehat{w} = \frac{k-3}{k-1}\frac{k(n-1)}{k(n-1)+2}\frac{1}{F} .$$

Substituting μ and w by the estimates $\widehat{\mu}$ and $\widehat{w}$ in the expression (5.65a) for the posterior mean

$$d_i^B(\boldsymbol{Y}) = \mathrm{E}[\mu_i|\overline{Y}_{i.}] = w\mu + (1-w)\overline{Y}_{i.}$$

we obtain the estimator

$$d_i^{EB}(\boldsymbol{Y}) = \widehat{w}\widehat{\mu} + (1-\widehat{w})\overline{Y}_{i.} \tag{5.79}$$

The estimator is called an *empirical Bayes estimator*, as it utilizes an estimate for the "prior distribution" based on the *empirical* (the observed) distribution of $\overline{Y}_{i.}$.

The following theorem shows the properties of the empirical Bayes estimator.

THEOREM 5.7 – BAYES RISK OF THE EMPIRICAL BAYES ESTIMATOR
Under the assumptions of Remark 5.19 on page 188 we have that

$$r((\boldsymbol{\mu},\gamma),\boldsymbol{d}^{EB}) = \frac{1}{n}\left(1 - \frac{2(k-3)}{\{k(n-1)+2\}(1+n\gamma)}\right)$$

Proof The proof follows by simple, but tedious integrations.   ∎

▶ **Remark 5.21 – The Bayes risk for the empirical Bayes estimator is less than the Bayes risk for the maximum likelihood estimator**
Comparing with Remark 5.19 we find that, when $k > 3$, then the prior risk for the empirical Bayes estimator $\boldsymbol{d}^{EB}$ is smaller than for the maximum likelihood estimator $\boldsymbol{d}^{ML}$. The smaller the value of the signal/noise ratio γ, the larger the difference in risk for the two estimators.   ◀

It may further be shown that also the *risk* is uniformly improved by the shrinkage introduced in the empirical Bayes estimator, viz.

$$R(\boldsymbol{\mu}, \boldsymbol{d}^{EB}) < R(\boldsymbol{\mu}, \boldsymbol{d}^{ML}) \tag{5.80}$$

holds for arbitrary vectors $\boldsymbol{\mu}$, see Stein (1955) and Efron and Morris (1972). In other words: *the mean squared error for the empirical Bayes estimator is less than that for the maximum-likelihood estimator.* This result is also known as Stein's paradox, see Wassserman p. 204.

▸ **Remark 5.22 – Intuitive rationale for shrinkage estimation**
Consider the decomposition (5.73) of the mean squared error into bias and variance. For $w < 1$ we have that $\mathrm{E}[w\overline{Y}_i] = w\mu_i \neq \mu_i$, but $\mathrm{Var}[w\overline{Y}_i] = w^2 \mathrm{Var}[\overline{Y}_i] < \sigma^2/n$. Thus, introducing the *shrinkage* by w we introduce a trade-off between bias and variance by reducing the variance at the expense of a small bias, $\delta_i = (1 - w)\mu_i$. ◂

In practical statistical applications with a large number of unknown parameters such a technique to reduce prediction error is often termed *regularization*. Notice the similarity with the regularization techniques previously introduced in Section 3.7 on page 64. Other methods for shrinkage estimation are considered in, e.g., Brown, Tauler, and Walczak (2009) and Öjelund, Madsen, and Thyregod (2001).

5.6 Posterior distributions for multivariate normal distributions

The results in the previous section can rather easily be generalized to the multivariate case as briefly outlined in the following.

THEOREM 5.8 – POSTERIOR DISTRIBUTION FOR MULTIVARIATE NORMAL DISTRIBUTIONS
Let $\boldsymbol{Y} \mid \boldsymbol{\mu} \sim \mathrm{N}_p(\boldsymbol{\mu}, \boldsymbol{\Sigma})$ and let $\boldsymbol{\mu} \sim \mathrm{N}_p(\boldsymbol{m}, \boldsymbol{\Sigma}_0)$, where $\boldsymbol{\Sigma}$ and $\boldsymbol{\Sigma}_0$ are of full rank, p, say.

Then the posterior distribution of $\boldsymbol{\mu}$ after observation of $\boldsymbol{Y} = \boldsymbol{y}$ is given by

$$\boldsymbol{\mu} \mid \boldsymbol{Y} = \boldsymbol{y} \sim \mathrm{N}_p(\boldsymbol{W}\boldsymbol{m} + (\boldsymbol{I} - \boldsymbol{W})\boldsymbol{y}, (\boldsymbol{I} - \boldsymbol{W})\boldsymbol{\Sigma}) \tag{5.81}$$

with $\boldsymbol{W} = \boldsymbol{\Sigma}(\boldsymbol{\Sigma}_0 + \boldsymbol{\Sigma})^{-1}$ and $\boldsymbol{I} - \boldsymbol{W} = \boldsymbol{\Sigma}_0(\boldsymbol{\Sigma}_0 + \boldsymbol{\Sigma})^{-1}$

▸ **Remark 5.23**
If we let $\boldsymbol{\Psi} = \boldsymbol{\Sigma}_0\boldsymbol{\Sigma}^{-1}$ denote the *generalized ratio* between the variation between groups, and the variation within groups, in analogy with the signal

to noise ratio introduced in (5.12), then we can express the weight matrices W and $I - W$ as

$$W = (I + \Psi)^{-1} \quad \text{and} \quad I - W = (I + \Psi)^{-1}\Psi \qquad \blacktriangleleft$$

THEOREM 5.9 – POSTERIOR DISTRIBUTION IN REGRESSION MODEL
Let Y denote a $n \times 1$ dimensional vector of observations, and let X denote a $n \times p$ dimensional matrix of known coefficients.

Assume that $Y \mid \beta \sim N_n(X\beta, \sigma^2 V)$ and that the prior distribution of β is

$$\beta \sim N_p(\beta_0, \sigma^2 \Lambda)$$

where Λ is of full rank. Then the posterior distribution of β after observation of $Y = y$ given by

$$\beta \mid Y = y \sim N_p(\beta_1, \sigma^2 \Lambda_1)$$

with

$$\beta_1 = W\beta_0 + W\Lambda X^T V^{-1} y \qquad (5.82)$$

where $W = (I + \Gamma)^{-1}$, $\Gamma = \Lambda X^T V^{-1} X$ and

$$\Lambda_1 = (I + \Gamma)^{-1}\Lambda = W\Lambda \qquad (5.83)$$

Proof The proof follows by using Theorem 5.8 on the preceding page. ∎

▶ **Remark 5.24 – The posterior mean expressed as a weighted average**
If X is of full rank, then $X^T V^{-1} X$ may be inverted, and we find the posterior mean

$$\beta_1 = W\beta_0 + (I - W)\widehat{\beta}$$

where $\widehat{\beta}$ denotes the usual least squares estimate for β,

$$\widehat{\beta} = (X^T V^{-1} X)^{-1} X^T V^{-1} y \qquad \blacktriangleleft$$

5.7 Random effects for multivariate measurements

Let us finally introduce the analysis of variance for multivariate measurements. Hence, let us now consider the situation where the individual observations are p-dimensional vectors, and consider the model

$$X_{ij} = \mu + \alpha_i + \epsilon_{ij}, \; i = 1, 2, \ldots, k; \; j = 1, 2, \ldots, n_{ij} \qquad (5.84)$$

where μ, α_i and ϵ_{ij} denotes p-dimensional vectors and where ϵ_{ij} are mutual independent and normally distributed, $\epsilon_{ij} \in N_p(0, \Sigma)$, and where Σ denotes the $p \times p$-dimensional covariance matrix. For simplicity we will assume that Σ has full rank.

For the fixed effects model we further assume

$$\sum_{i=1}^{k} n_i \boldsymbol{\alpha}_i = 0$$

Given these assumptions we find $\boldsymbol{Z}_i = \sum_j \boldsymbol{X}_{ij} \sim N_p(n_i(\boldsymbol{\mu} + \boldsymbol{\alpha}_i), n_i \Sigma)$.

In the case of multivariate observations the variation is described by $p \times p$-dimensional SS matrices.

Let us introduce the notation

$$\overline{\boldsymbol{X}}_{i+} = \sum_{j=1}^{n_i} \boldsymbol{X}_{ij}/n_i \tag{5.85}$$

$$\overline{\boldsymbol{X}}_{++} = \sum_{i=1}^{k} \sum_{j=1}^{n_i} \boldsymbol{X}_{ij}/N = \sum_{i=1}^{k} n_i \overline{\boldsymbol{X}}_{i+} / \sum_{i=1}^{k} n_i \tag{5.86}$$

as descriptions of the group averages and the total average, respectively.

Furthermore introduce

$$\textbf{SSE} = \sum_{i=1}^{k} \sum_{j=1}^{n_i} (\boldsymbol{X}_{ij} - \overline{\boldsymbol{X}}_{i+})(\boldsymbol{X}_{ij} - \overline{\boldsymbol{X}}_{i+})^T \tag{5.87}$$

$$\textbf{SSB} = \sum_{i=1}^{k} n_i (\overline{\boldsymbol{X}}_{i+} - \overline{\boldsymbol{X}}_{++})(\overline{\boldsymbol{X}}_{i+} - \overline{\boldsymbol{X}}_{++})^T \tag{5.88}$$

$$\textbf{SST} = \sum_{i=1}^{k} \sum_{j=1}^{n_i} (\boldsymbol{X}_{ij} - \overline{\boldsymbol{X}}_{++})(\boldsymbol{X}_{ij} - \overline{\boldsymbol{X}}_{++})^T \tag{5.89}$$

as a description of the variation between groups (**SSE**), between groups (**SSB**), and the total variation (**SST**).

As previously illustrated in Figure 3.1 on page 52 and discussed, e.g., in Section 3.4 we have the Pythagorean relation

$$\textbf{SST} = \textbf{SSE} + \textbf{SSB} \tag{5.90}$$

Random effects model

Let us consider the random effects model for the p-dimensional observations as introduced in the previous section.

Consider the following model:

$$\boldsymbol{X}_{ij} = \boldsymbol{\mu} + \boldsymbol{u}_i + \boldsymbol{\epsilon}_{ij}, \ i = 1, \ldots, k; \quad j = 1, 2, \ldots, n_i. \tag{5.91}$$

where $\boldsymbol{u}_i$ now are independent, $\boldsymbol{u}_i \sim N_p(\boldsymbol{0}, \Sigma_0)$, and where $\boldsymbol{\epsilon}_{ij}$ is independent, $\boldsymbol{\epsilon}_{ij} \sim N_p(\boldsymbol{0}, \Sigma)$. Finally, $\boldsymbol{u}$ and $\boldsymbol{\epsilon}$ are independent.

THEOREM 5.10 – THE MARGINAL DISTRIBUTION IN THE CASE OF MULTIVARI-
ATE p-DIMENSIONAL OBSERVATIONS
Consider the model introduced in (5.91). Then the marginal density *of* $\boldsymbol{Z}_i = \sum_j \boldsymbol{X}_{ij}$ *is*

$$N_p(n_i\boldsymbol{\mu}, n_i\boldsymbol{\Sigma} + n_i^2\boldsymbol{\Sigma}_0)\text{-}distribution$$

and the marginal density for $\overline{\boldsymbol{X}}_{i+}$ *is*

$$N_p(\boldsymbol{\mu}, \frac{1}{n_i}\boldsymbol{\Sigma} + \boldsymbol{\Sigma}_0)$$

Finally, we have that **SSE** *follows a Wishart distribution*

$$\mathbf{SSE} \in Wis_p(N - k, \boldsymbol{\Sigma})$$

and **SSE** *is independent of* $\overline{\boldsymbol{X}}_{i+}$, $i = 1, 2, \ldots, k$.

Proof Omitted, but follows from the reproducibility of the normal density
and the variation separation theorem in Madsen (2008).   ∎

DEFINITION 5.4 – GENERALIZED SIGNAL TO NOISE RATIO
Let us introduce the generalized signal to noise ratio as the p-dimensional
matrix $\boldsymbol{\Gamma}$ representing the ratio between the variation between groups and
the variation within groups $\boldsymbol{\Gamma} = \boldsymbol{\Sigma}_0 \boldsymbol{\Sigma}^{-1}$.

THEOREM 5.11 – MOMENT ESTIMATES FOR THE MULTIVARIATE RANDOM
EFFECTS MODEL
Given the assumptions from Theorem 5.10 we find the moment estimates for
$\boldsymbol{\mu}, \boldsymbol{\Sigma}$ *and* $\boldsymbol{\Sigma}_0$ *as*

$$\tilde{\boldsymbol{\mu}} = \overline{x}_{++}$$
$$\tilde{\boldsymbol{\Sigma}} = \frac{1}{N - k} \mathbf{SSE} \tag{5.92}$$
$$\tilde{\boldsymbol{\Sigma}}_0 = \frac{1}{n_0} \left(\frac{\mathbf{SSB}}{k - 1} - \tilde{\boldsymbol{\Sigma}} \right)$$

Proof Omitted, but follows by using that

$$\mathrm{E}[\overline{\boldsymbol{X}}_{++}] = \boldsymbol{\mu}, \quad \mathrm{E}[\mathbf{SSE}] = (N - k)\boldsymbol{\Sigma}, \quad \text{and} \quad \mathrm{E}[\mathbf{SSB}] = (k - 1)(\boldsymbol{\Sigma} + n_0\boldsymbol{\Sigma}_0)$$

where n_0 is given by (5.29).   ∎

The moment estimate for $\boldsymbol{\Sigma}_0$ is unbiased, but not necessarily non-negative
definite. Hence it might be more appropriate to consider the maximum
likelihood estimates which are obtained by numerical optimization of the
likelihood function given in the following theorem.

Table 5.5: *Results of three repeated calibrations of six flow-meters at two flows.*

| | Flow, $[\mathrm{m}^3/\mathrm{h}]$ | | | | | |
| | Calibration 1 | | Calibration 2 | | Calibration 3 | |
Meter	0.1	0.5	0.1	0.5	0.1	0.5
41	−2.0	1.0	2.0	3.0	2.0	2.0
42	5.0	3.0	1.0	1.0	2.0	2.0
43	2.0	1.0	−3.0	−1.0	1.0	0.0
44	4.0	4.0	−1.0	2.0	3.0	5.0
45	4.0	2.0	0.0	1.0	−1.0	0.0
46	5.0	9.0	4.0	8.0	6.0	10.0

THEOREM 5.12 – MLE FOR THE MULTIVARIATE RANDOM EFFECTS MODEL
Still under the assumptions from Theorem 5.10 we find the maximum likelihood estimates (MLEs) for $\boldsymbol{\mu}$, $\boldsymbol{\Sigma}$ and $\boldsymbol{\Sigma}_0$ by maximizing the log-likelihood

$$\ell(\boldsymbol{\mu}, \boldsymbol{\Sigma}, \boldsymbol{\Sigma}_0; \overline{x}_{1+}, \dots, \overline{x}_{k+})$$

$$= -\frac{N-k}{2}\log(\det(\boldsymbol{\Sigma})) - \frac{1}{2}\mathrm{tr}((\mathbf{SSE})\boldsymbol{\Sigma}^{-1}) - \sum_{i=1}^{k}\left[\log\left(\det\left(\frac{\boldsymbol{\Sigma}}{n_i} + \boldsymbol{\Sigma}_0\right)\right)\right.$$

$$\left. + \frac{1}{2}(\overline{x}_{i+} - \boldsymbol{\mu})^T\left(\frac{\boldsymbol{\Sigma}}{n_i} + \boldsymbol{\Sigma}_0\right)^{-1}(\overline{x}_{i+} - \boldsymbol{\mu})\right] \quad (5.93)$$

with respect to $\boldsymbol{\mu} \in \mathbb{R}^p$ and $\boldsymbol{\Sigma}$ and $\boldsymbol{\Sigma}_0$ in the space of non-negative definite $p \times p$ matrices.

Proof Omitted, but follows from the fact that $\mathbf{SSE}$ follows a $\mathrm{Wis}_p(N-k, \boldsymbol{\Sigma})$-distribution and that $\mathbf{SSE}$ and $\overline{\boldsymbol{X}}_{i+}$, $i = 1, 2, \dots, k$ are independent, and further that $\overline{\boldsymbol{X}}_{i+} \in \mathrm{N}_p(\boldsymbol{\mu}, \boldsymbol{\Sigma}/n_i + \boldsymbol{\Sigma}_0)$ are independent, $i = 1, 2, \dots, k$. ∎

Since no explicit solution exists the maximum likelihood estimates must be found using numerical procedures. In order to ensure that the covariance matrices are non-negative definite, the matrices are often parameterized by scalar variances and correlation coefficients.

Let us illustrate the results by an example where each observation is two-dimensional.

Example 5.6 – Variation between measurement errors for flow meters

Table 5.5 shows the results of three repeated calibrations of six flow meters selected at random from a population of flow-meters. Each of the six meters were calibrated at the same two flows, 0.1 $[\mathrm{m}^3/\mathrm{h}]$ and 0.5 $[\mathrm{m}^3/\mathrm{h}]$, respectively. The measurements are illustrated graphically in Figure 5.3.

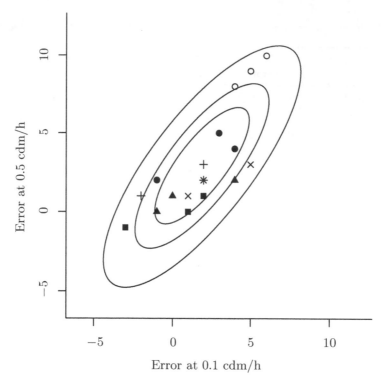

Figure 5.3: *Total variation. Results for calibration of each of six flow meters at two flows. The three repetitions for a meter are indicated using the same symbol.*

It is seen that there is a close connection between repeated calibrations of the same meter with a rather small variation between repeated calibrations – and a somewhat larger variation between calibrations on different meters.

In order to describe the variation of measurement errors we set up the following model:

$$X_{ij} = \mu + u_i + \epsilon_{ij}, \ i = 41, 42, \ldots, 46; \quad j = 1, 2, 3$$

where u_i are independent, $u_i \sim N_2(0, \Sigma_0)$, and where ϵ_{ij} are mutually independent, $\epsilon \sim N_2(0, \Sigma)$. The situation resembles the case with repeated sampling from bales of wool in Example 5.1 on page 160, only in this case each observation is two-dimensional.

By using (5.92) we find the moment estimates

$$\tilde{\mu} = \begin{pmatrix} 1.61 \\ 2.89 \end{pmatrix}$$

and the so-called uncertainty of the calibration

$$\tilde{\Sigma} = \mathbf{SSE}/12 = \begin{pmatrix} 5.56 & 2.42 \\ 2.42 & 1.28 \end{pmatrix}$$

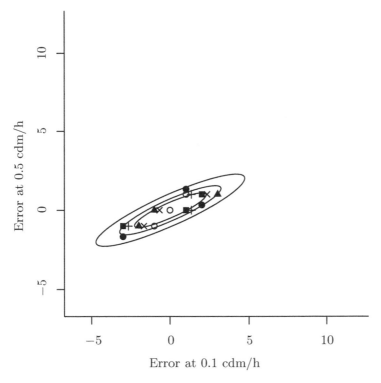

Figure 5.4: *Within group variation. The variation of the variation with groups (Repeatability variation).*

and covariance matrix for the population of flow meters

$$\widetilde{\Sigma}_0 = \frac{1}{3}[\mathbf{SSB}/5 - \mathbf{SSE}/12] = \begin{pmatrix} 2.65 & 5.01 \\ 5.01 & 10.00 \end{pmatrix}$$

The decomposition of the total variation in "within groups" and "between groups" is illustrated in Figures 5.4 and 5.5.

5.8   Hierarchical models in metrology

The simple hierarchical models outlined in this section are important within metrology, where the concepts of *repeatability* and *reproducibility* conditions are important and often used instead of the more general term *precision*.

The following descriptions follow closely to the definitions in the International Vocabulary of Basic and General Terms in Metrology (VIM):

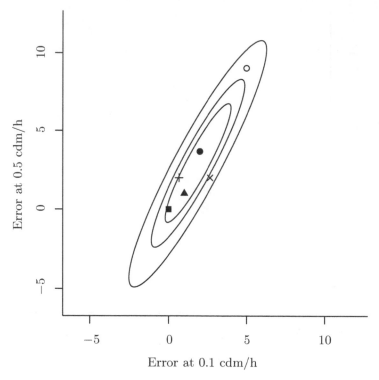

Figure 5.5: *Between group variation. An illustration of the variation between groups; the deviation from the average for three repetitions for each of six flow meters calibrated at two flows.*

Repeatability conditions   Conditions where independent test results are obtained with the same method under identical conditions on identical test items

    i) in the same laboratory

    ii) by the same operator

    iii) using the same equipment

    iv) within short intervals of time

Reproducibility conditions   Conditions where independent test results are obtained under different conditions on identical test items, e.g.,

    i) in different laboratories

    ii) with different operators

    iii) using different equipment

5.9   General mixed effects models

The preceding part of this chapter has focused on the important class of linear normal mixed effects models. This section will present a collection of methods to deal with nonlinear and non-normal mixed effects models. In general it will be impossible to obtain closed form solutions and hence numerical methods must be used, and therefore some guidance about software and some numerical details will be provided.

As previously estimation and inference will be based on likelihood principles. The general mixed effects model can be represented by its likelihood function:

$$L_M(\boldsymbol{\theta}; \boldsymbol{y}) = \int_{\mathbb{R}^q} L(\boldsymbol{\theta}; \boldsymbol{u}, \boldsymbol{y}) d\boldsymbol{u} \qquad (5.94)$$

where $\boldsymbol{y}$ is the observed random variables, $\boldsymbol{\theta}$ is the model parameters to be estimated, and $\boldsymbol{U}$ is the q unobserved random variables or effects. The likelihood function L is the joint likelihood of both the observed and the unobserved random variables. The likelihood function for estimating $\boldsymbol{\theta}$ is the marginal likelihood L_M obtained by integrating out the unobserved random variables.

The integral in (5.94) is generally difficult to solve if the number of un-observed random variables is more than a few, i.e., for large values of q. A large value of q significantly increases the computational demands due to the product rule which states that if an integral is sampled in m points per dimension to evaluate it, the total number of samples needed is m^q, which rapidly becomes infeasible even for a limited number of random effects.

The likelihood function in (5.94) gives a very broad definition of mixed models: the only requirement for using mixed modeling is to define a joint likelihood function for the model of interest. In this way mixed modeling can be applied to any likelihood based statistical modeling. Examples of applications are linear mixed models (LMM) and nonlinear mixed models (NLMM), see, e.g., Bates and Watts (1988), generalized linear mixed models, see, e.g., McCulloch and Searle (2001), but also models based on Markov chains, ODEs or SDEs are considered in the literature; see e.g., Beal and Sheiner (1980), Tornøe, Jacobsen, and Madsen (2004), Mortensen (2009), Mortensen et al. (2007), and Klim et al. (2009). The presentation in this section is inspired mostly by Mortensen (2009).

Hierarchical models

As for the Gaussian linear mixed models in Section 5.4 it is useful to formulate the model as a *hierarchical model* containing a *first stage model*

$$f_{Y|u}(\boldsymbol{y}; \boldsymbol{u}, \boldsymbol{\beta}) \qquad (5.95)$$

which is a model for the data given the random effects, and a *second stage model*

$$f_U(\boldsymbol{u}; \boldsymbol{\Psi}) \tag{5.96}$$

which is a model for the random effects. The total set of parameters is $\boldsymbol{\theta} = (\boldsymbol{\beta}, \boldsymbol{\Psi})$. Hence the joint likelihood is given as

$$L(\boldsymbol{\beta}, \boldsymbol{\Psi}; \boldsymbol{u}, \boldsymbol{y}) = f_{Y|u}(\boldsymbol{y}; \boldsymbol{u}, \boldsymbol{\beta}) f_U(\boldsymbol{u}; \boldsymbol{\Psi}) \tag{5.97}$$

To obtain the likelihood for the model parameters $(\boldsymbol{\beta}, \boldsymbol{\Psi})$ the unobserved random effects are again integrated out. The likelihood function for estimating $(\boldsymbol{\beta}, \boldsymbol{\Psi})$ is as before the marginal likelihood

$$L_M(\boldsymbol{\beta}, \boldsymbol{\Psi}; \boldsymbol{y}) = \int_{\mathbb{R}^q} L(\boldsymbol{\beta}, \boldsymbol{\Psi}; \boldsymbol{u}, \boldsymbol{y}) d\boldsymbol{u} \tag{5.98}$$

where q is the number of random effects, and $\boldsymbol{\beta}$ and $\boldsymbol{\Psi}$ are the parameters to be estimated.

Grouping structures and nested effects

For nonlinear mixed models where no closed form solution to (5.94) is available it is necessary to invoke some form of numerical approximation to be able to estimate the model parameters. The complexity of this problem is mainly dependent on the dimensionality of the integration problem which in turn is dependent on the dimension of $\boldsymbol{U}$ and in particular the *grouping structure* in the data for the random effects. These structures include a *single grouping*, *nested grouping*, *partially crossed* and *crossed random effects*.

For problems with only one level of grouping the marginal likelihood can be simplified as

$$L_M(\boldsymbol{\beta}, \boldsymbol{\Psi}; \boldsymbol{y}) = \prod_{i=1}^{M} \int_{\mathbb{R}^{q_i}} f_{Y|u_i}(\boldsymbol{y}; \boldsymbol{u}_i, \boldsymbol{\beta}) f_{U_i}(\boldsymbol{u}_i; \boldsymbol{\Psi}) d\boldsymbol{u}_i \tag{5.99}$$

where q_i is the number of random effects for group i and M is the number of groups. Instead of having to solve an integral of dimension q it is only necessary to solve M smaller integrals of dimension q_i. In typical applications there is often just one or only a few random effects for each group, and this thus greatly reduces the complexity of the integration problem. If the data has a nested grouping structure a reduction of the dimensionality of the integral similar to that shown in (5.99) can be performed.

An example of a nested grouping structure is data collected from a number of schools, a number of classes within each school and a number of students from each class. However, if some students change school during the study, the random effects structure is suddenly partially crossed and the simplification in (5.99) no longer applies.

If the nonlinear mixed model is extended to include any structure of random effects such as crossed or partially crossed random effects it is required to evaluate the full multi-dimensional integral in (5.98) which may be computational infeasible even for rather low numbers of random effects. For this reason estimation in models with crossed random effects is not supported by any of the standard software packages for fitting NLMMs such as nlme (Pinheiro et al. 2008), SAS NLMIXED (SAS Institute Inc. 2004), and NONMEM (Beal and Sheiner 2004). However, estimation in these models can efficiently be handled using the multivariate Laplace approximation, which only samples the integrand in one point common to all dimensions, as described in the next section.

5.10 Laplace approximation

The Laplace approximation will be outlined in the following. A thorough description of the Laplace approximation in nonlinear mixed models is found in Wolfinger and Lin (1997).

For a given set of model parameters $\boldsymbol{\theta}$ the joint log-likelihood $\ell(\boldsymbol{\theta}, \boldsymbol{u}, \boldsymbol{y}) = \log(L(\boldsymbol{\theta}, \boldsymbol{u}, \boldsymbol{y}))$ is approximated by a second order Taylor approximation around the optimum $\tilde{\boldsymbol{u}} = \widehat{\boldsymbol{u}}_\theta$ of the log-likelihood function w.r.t. the unobserved random variables $\boldsymbol{u}$, i.e.,

$$\ell(\boldsymbol{\theta}, \boldsymbol{u}, \boldsymbol{y}) \approx \ell(\boldsymbol{\theta}, \tilde{\boldsymbol{u}}, \boldsymbol{y}) - \frac{1}{2}(\boldsymbol{u} - \tilde{\boldsymbol{u}})^T \boldsymbol{H}(\tilde{\boldsymbol{u}})(\boldsymbol{u} - \tilde{\boldsymbol{u}}) \qquad (5.100)$$

where the first-order term of the Taylor expansion disappears since the expansion is done around the optimum $\tilde{\boldsymbol{u}}$ and $\boldsymbol{H}(\tilde{\boldsymbol{u}}) = -\ell''_{uu}(\boldsymbol{\theta}, \boldsymbol{u}, \boldsymbol{y})|_{u=\tilde{u}}$ is the *negative* Hessian of the joint log-likelihood evaluated at $\tilde{\boldsymbol{u}}$ which will simply be referred to as "the Hessian."

Using the approximation in (5.100) in (5.98) the Laplace approximation of the marginal log-likelihood becomes

$$\ell_{M,LA}(\boldsymbol{\theta}, \boldsymbol{y}) = \log \int_{\mathbb{R}^q} \exp\left(\ell(\boldsymbol{\theta}, \tilde{\boldsymbol{u}}, \boldsymbol{y}) - \tfrac{1}{2}(\boldsymbol{u} - \tilde{\boldsymbol{u}})^T \boldsymbol{H}(\tilde{\boldsymbol{u}})(\boldsymbol{u} - \tilde{\boldsymbol{u}})\right) d\boldsymbol{u}$$

$$= \ell(\cdot) + \log \int_{\mathbb{R}^q} \exp\left(-\tfrac{1}{2}(\boldsymbol{u} - \tilde{\boldsymbol{u}})^T \boldsymbol{H}(\tilde{\boldsymbol{u}})(\boldsymbol{u} - \tilde{\boldsymbol{u}})\right) d\boldsymbol{u}$$

$$= \ell(\cdot) + \log \left|\frac{2\pi}{\boldsymbol{H}(\tilde{\boldsymbol{u}})}\right|^{\frac{1}{2}}$$

$$\times \int_{\mathbb{R}^q} \frac{\exp\left(-\tfrac{1}{2}(\boldsymbol{u} - \tilde{\boldsymbol{u}})^T \boldsymbol{H}(\tilde{\boldsymbol{u}})(\boldsymbol{u} - \tilde{\boldsymbol{u}})\right)}{(2\pi)^{\frac{q}{2}}|\boldsymbol{H}^{-1}(\tilde{\boldsymbol{u}})|^{\frac{1}{2}}} d\boldsymbol{u}$$

$$= \ell(\cdot) + \log \left|\frac{2\pi}{\boldsymbol{H}(\tilde{\boldsymbol{u}})}\right|^{\frac{1}{2}}$$

$$= \ell(\boldsymbol{\theta}, \tilde{\boldsymbol{u}}, \boldsymbol{y}) - \tfrac{1}{2}\log\left|\frac{\boldsymbol{H}(\tilde{\boldsymbol{u}})}{2\pi}\right| \qquad (5.101)$$

where the integral is eliminated by transforming it to an integration of a multivariate Gaussian density with mean $\tilde{u}$ and covariance $H^{-1}(\tilde{u})$. In the step in (5.101) the fraction in the determinant is inverted to avoid a matrix inversion of the Hessian. So far the parameterization is very flexible, however, for the important hierarchical model in (5.98) we have that $\theta = (\beta, \Psi)$.

▶ **Remark 5.25**
Notice that the Laplace approximation requires the log-likelihood to be optimized with respect to the unobserved random variables u for each set of model parameter values θ where the marginal log-likelihood is to be evaluated. Searching for the maximum likelihood estimate $\widehat{\theta}$ in large nonlinear models can require many function evaluations, so an efficient optimizer is required. ◀

The Laplace likelihood only approximates the marginal likelihood for mixed models with nonlinear random effects and thus maximizing the Laplace likelihood will result in some amount of error in the resulting estimates. However, in Vonesh (1996) it is shown that joint log-likelihood converges to a quadratic function of the random effect for increasing number of observations per random effect and thus that the Laplace approximation is asymptotically exact. In practical applications the accuracy of the Laplace approximation may still be of concern, but often improved numerical approximation of the marginal likelihood (such as Gaussian quadrature) may easily be computationally infeasible to perform. Another option for improving the accuracy is Importance sampling, which will be described later in this chapter.

Two-level hierarchical model

For the two-level or hierarchical model specified in Section 5.9 it is readily seen that the joint log-likelihood is two-level hierarchical mode

$$\ell(\theta, u, y) = \ell(\beta, \Psi, u, y) = \log f_{Y|u}(y; u, \beta) + \log f_U(u; \Psi) \qquad (5.102)$$

which implies that the Laplace approximation becomes

$$\ell_{M,LA}(\theta, y) = \log f_{Y|u}(y; \tilde{u}, \beta) + \log f_U(\tilde{u}; \Psi) - \frac{1}{2} \log \left| \frac{H(\tilde{u})}{2\pi} \right| \qquad (5.103)$$

It is clear that as long as a likelihood function of the random effects and model parameters can be defined it is possible to use the Laplace likelihood for estimation in a mixed model framework.

Gaussian second stage model

Let us now assume that the second stage model is zero mean Gaussian, i.e.,

$$u \sim N(0, \Psi) \qquad (5.104)$$

which means that the random effect distribution is completely described by its covariance matrix $\boldsymbol{\Psi}$. In this case the Laplace likelihood in (5.103) becomes

$$
\begin{aligned}
\ell_{M,LA}(\boldsymbol{\theta}, \boldsymbol{y}) = {}& \log f_{Y|u}(\boldsymbol{y}; \tilde{\boldsymbol{u}}, \boldsymbol{\beta}) - \frac{1}{2} \log |\boldsymbol{\Psi}| \\
& - \frac{1}{2} \tilde{\boldsymbol{u}}^T \boldsymbol{\Psi}^{-1} \tilde{\boldsymbol{u}} - \frac{1}{2} \log |\boldsymbol{H}(\tilde{\boldsymbol{u}})|
\end{aligned}
\tag{5.105}
$$

where it is seen that we still have no assumptions on the first stage model $f_{Y|u}(\boldsymbol{y}; \boldsymbol{u}, \boldsymbol{\beta})$.

If we furthermore assume that the first stage model is Gaussian

$$
\boldsymbol{Y}|\boldsymbol{U} = \boldsymbol{u} \sim N(\boldsymbol{\mu}(\boldsymbol{\beta}, \boldsymbol{u}), \boldsymbol{\Sigma})
\tag{5.106}
$$

then the Laplace likelihood can be further specified.

For the hierarchical Gaussian model it is rather easy to obtain a numerical approximation of the Hessian $\boldsymbol{H}$ at the optimum, $\tilde{\boldsymbol{u}}$

$$
\boldsymbol{H}(\tilde{\boldsymbol{u}}) \approx \boldsymbol{\mu}_u' \boldsymbol{\Sigma}^{-1} \boldsymbol{\mu}_u'^{\,T} + \boldsymbol{\Psi}^{-1}
\tag{5.107}
$$

where $\boldsymbol{\mu}_u'$ is the partial derivative with respect to $\boldsymbol{u}$. The approximation in (5.107) is called Gauss-Newton approximation. In some contexts estimation using this approximation is also called the First Order Conditional Estimation (FOCE) method.

Applying the Laplace approximation in R, the orange tree data

The following example is taken from Mortensen (2009) where it is shown how nonlinear mixed effects models (NLMM) can be estimated using the Laplace approximation in R. In the article in Mortensen (2009) the full nonlinear Gaussian case is considered, where the conditional distribution of the response given the random effects as well as the random effects are Gaussian. The considered model class is

$$
\boldsymbol{Y}|\boldsymbol{U} = \boldsymbol{u} \sim N(\boldsymbol{f}(\boldsymbol{\beta}, \boldsymbol{u}), \boldsymbol{\Sigma}(\boldsymbol{\lambda}))
\tag{5.108}
$$

$$
\boldsymbol{U} \sim N(\boldsymbol{0}, \boldsymbol{\Psi}(\boldsymbol{\psi}))
\tag{5.109}
$$

where $\boldsymbol{\beta}$ are fixed regression parameters, $\boldsymbol{u}$ is a q-vector of random effects, $\boldsymbol{\lambda}$ and $\boldsymbol{\psi}$ are variance parameters parameterizing the covariance matrices $\boldsymbol{\Sigma}$ and $\boldsymbol{\Psi}$ and $\boldsymbol{f}$ is the model function.

To illustrate the approach, a study of the growth of orange trees reported by Draper and Smith (1981) is used where the circumference of five trees is measured at seven time points. This dataset has been used for example by Lindstrom and Bates (1990) and Pinheiro and Bates (1995). In both cases a logistic growth model is fitted with a single random component, u_{1i}, allowing for a tree specific asymptotic circumference. The model can be written as

$$
y_{ij} = \frac{\beta_1 + u_{1i}}{1 + \exp[-(t_j - \beta_2)/\beta_3]} + \epsilon_{ij} , \quad i = 1, \dots, 5, \quad j = 1, \dots, 7,
\tag{5.110}
$$

with $\epsilon_{ij} \sim \mathrm{N}(0, \sigma^2)$ and $u_{1i} \sim \mathrm{N}(0, \sigma_{u1}^2)$ and mutually independent. For the model (5.110) the matrices $\boldsymbol{\Sigma}$ and $\boldsymbol{\Psi}$, as in (5.108) and (5.109), are both diagonal. Here, β_1 determines the asymptotic circumference, β_2 is the age at half this value, β_3 is related to the growth rate and t_j is the time in days since the beginning of the study. The maximum likelihood estimates (MLEs) of the fixed parameters along with standard errors for model (5.110) are given in Table 5.6. A plot of the data and model (5.110) is shown in Figure 5.6a.

A plot of residuals versus time (sampling occasion) shown in Figure 5.6b reveals an unmodelled variation with time. This is also noted in Millar (2004) where it is proposed to include a second random component, u_{2j}, for the sampling occasion, that is, crossed with the random component for trees leading to the following model

$$y_{ij} = \frac{\beta_1 + u_{1i} + u_{2j}}{1 + \exp[-(t_j - \beta_2)/\beta_3]} + \epsilon_{ij} \qquad (5.111)$$

with $u_{2j} \sim \mathrm{N}(0, \sigma_{u2}^2)$ and independent of u_{1i} and ϵ_{ij}. This successfully removes the most significant structure in the residuals. In this model, the effect of the sampling occasion, u_{2j}, is proportional to the model prediction. This is reasonable during the initial growth period, but unreasonable when the trees reach their asymptotic circumference. Another option would be to include u_{2j} additively in the exp-term in the denominator in (5.110) to make the random effects additive on the logit-scale. This makes the effect of the sampling occasion vanish as the trees approach their asymptotic circumference.

A closer look at the sampling scheme reveals, however, that the apparently random effect of the sampling occasion is caused by a seasonal effect and an irregular sampling pattern. In the residual plot in Figure 5.6b, it is seen that all samples are taken either in the spring (April or May) or in the fall (October) and that two periods are missing. A categorical seasonal effect, β_4 is therefore added to the model resulting in the following model

$$y_{ij} = \frac{\beta_1 + u_{1i}}{1 + \exp[-((t_j - \beta_2)/\beta_3 + s_j\beta_4)]} + \epsilon_{ij} \qquad (5.112)$$

where s_j is $-1/2$ and $1/2$ for samples taken in the spring and fall, respectively. The models (5.111) and (5.112) still show significant unmodelled serial correlation in the residuals within trees. This may be modelled with a continuous auto-regressive (CAR) process (see, e.g., Pinheiro (2000)) for the residuals by assuming

$$\mathrm{cov}(\epsilon_{ij}, \epsilon_{ij'}) = \sigma^2 \exp(-\phi|t_{j'} - t_j|/(365/2)), \quad \phi \geq 0 \qquad (5.113)$$

so the full covariance matrix is block diagonal with $\boldsymbol{\Sigma}(\phi, \sigma) = \boldsymbol{I}_5 \otimes \mathrm{cov}(\boldsymbol{\epsilon}_i)$ where $\otimes$ denotes the Kronecker product. The time is scaled so that $\rho = \exp(-\phi)$ can be interpreted as the correlation over half a year and therefore roughly between sampling occasions.

Table 5.6: *Parameter estimates (and standard errors) and log-likelihoods for models estimated for the orange tree data.*

Model	β_1	β_2	β_3	β_4	σ	σ_{u1}	σ_{u2}	ρ	$\log(L)$
(5.110)	192.1	727.9	348.1		7.84	31.6			-131.57
	(15.7)	(35.3)	(27.1)						
(5.111)	196.2	748.4	352.9		5.30	32.6	10.5		-125.45
	(19.4)	(62.3)	(33.3)						
(5.112)	217.1	857.5	436.8	0.322	4.79	36.0			-116.79
	(18.1)	(42.0)	(24.5)	(0.038)					
(5.111) + (5.113)	192.4	730.1	348.1		6.12	32.7	12.0	0.773	-118.44
	(19.6)	(63.8)	(34.2)						
(5.112) + (5.113)	216.2	859.1	437.8	0.330	5.76	36.7		0.811	-106.18
	(17.6)	(30.5)	(21.6)	(0.022)					

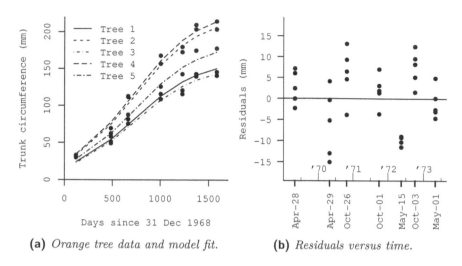

(a) *Orange tree data and model fit.* **(b)** *Residuals versus time.*

Figure 5.6: *Plots for model (5.110) for orange tree data.*

Model (5.111) with crossed random effects cannot easily, if at all, be fitted with standard software for NLMMs. The models can however all be estimated by means of the Laplace approximation implemented on a case-by-case basis.

The computational approach is based on estimating the parameters of the Laplace likelihood (5.101) by a general purpose quasi-Newton optimizer. To evaluate the Laplace likelihood for a set of parameters, $\boldsymbol{\theta}$, two quantities $\tilde{\boldsymbol{u}}$ and $\boldsymbol{H}(\boldsymbol{\theta}, \tilde{\boldsymbol{u}})$ as specified in (5.107) must be available. This leads to a nested optimization, since for every evaluation of the Laplace likelihood with a set of parameters, $\boldsymbol{\theta}$ in the outer optimization, the joint log-likelihood, ℓ has to be optimized over $\boldsymbol{u}$ in the inner optimization. A general purpose quasi-Newton optimizer is also used for the latter task. The only unknown quantity needed to evaluate $\boldsymbol{H}(\boldsymbol{\theta}, \tilde{\boldsymbol{u}})$ given $\boldsymbol{\theta}$ and $\tilde{\boldsymbol{u}}$ is the Jacobian, $\boldsymbol{f}'_u$ for which a finite difference approximation is used. Implementation of any NLMM consists of three functions: The model function, $\boldsymbol{f}$, the joint log-likelihood, ℓ in (5.102) and the Laplace likelihood in (5.101).

The starting values in the inner optimization are simply zero; the expectation of the random effects. Starting values for the regression parameters, $\boldsymbol{\beta}$ are based on plots of the data or previous fits of other models, potentially fixed effect versions. Starting values for variance and correlation parameters are qualified guesses based on plots of the data.

At convergence of the outer optimization, a finite difference approximation to the Hessian is used to obtain the variance-covariance matrix of the parameters.

Any inaccuracies in the estimation of $\tilde{\boldsymbol{u}}$ and $\boldsymbol{f}'_u$ are directly reflected as noise in the Laplace likelihood. For the gradient based estimation of the model parameters to converge smoothly, it is therefore important to obtain sufficiently good estimates of these quantities.

The variance parameters are optimized on the scale of the logarithm of the standard deviation to make the estimation unbounded and to make the log-likelihood surface more quadratic facilitating faster convergence. Because all terms in the Laplace likelihood (5.101) are evaluated on the log-scale, it can be evaluated for any variance parameter arbitrarily close to the boundary at zero (in finite computer arithmetic). This ensures that the optimization will proceed smoothly even if the MLE is zero. Further, it allows the likelihood to be profiled with respect to the variance parameters arbitrarily close to zero.

The optimizer nlminb in the base package in R is chosen for the inner and outer optimizations. The Jacobian is estimated using the numerical approximation implemented in jacobian in the numDeriv package Gilbert (2009). The hessian function, also from the numDeriv package, is used to obtain a finite difference estimation of the Hessian at the convergence of the outer optimization.

The model function, f for model (5.110) is defined as

```
> f <- function(beta, u) {
    (beta[1] + rep(u[1:5], each = 7))/
    (1 + exp((beta[2] - time)/beta[3])) }
```

The function returns a vector of the same length as the data with model predictions based on the 3 fixed effects in **beta**, the 5 random effects in u and the 7 time points in **time**. The joint negative log-likelihood based on (5.102) is defined as

```
> l <- function(u, beta, sigma, sigma.u) {
    -sum(dnorm(x = circumference, mean = f(beta, u),
            sd = sigma, log = TRUE)) -
    sum(dnorm(x = u[1:5], sd = sigma.u, log = TRUE)) }
```

using two vectorized calls to the univariate normal density function **dnorm**, because the conditional distribution of the observations and the distribution of the random effects are mutually independent normal. This is the *negative* joint log-likelihood, because standard optimization algorithms by default minimize rather than maximize.

Based on the implementations of the model function and the joint log-likelihood, the Laplace approximation to the marginal log-likelihood $\ell_{LA}(\boldsymbol{\theta})$ is implemented as

```
> l.LA <- function(theta) {
    beta <- theta[1:3]
    sigma <- exp(theta[4])
    sigma.u <- exp(theta[5])
    est <- nlminb(start = rep(0,5), objective = l, beta = beta,
            sigma = sigma, sigma.u = sigma.u)
    u <- est$par
```

```
l.u <- est$objective
Jac.f <- jacobian(func = f, x = u, beta = beta)
H <- crossprod(Jac.f)/sigma^2 + diag(1/sigma.u^2, 5)
l.u + 1/2 * log(det(H/(2 * pi))) }.
```

where the parameters to be estimated are $\theta = (\beta, \log \sigma, \log \sigma_{u1})$. The call to nlminb in l.LA performs the inner optimization and computes $\tilde{u}$, and the Hessian, H is computed as in (5.107) based on the Jacobian, f'_u.

The maximum likelihood fit of model (5.110) is obtained by performing the outer optimization with the call

```
> fit <- nlminb(theta0, l.LA)
```

where the starting values, θ^0 are inferred from Figure 5.6a. The resulting estimates are given in Table 5.6.

The code can be changed to estimate model (5.111) with crossed random effects. The model has two crossed random components u_{1i} and u_{2j} for tree and time and the full vector of random effects is thus $u = [u_{11}, \ldots, u_{15}, u_{21}, \ldots, u_{27}]^T$. The model function f is modified to include the 7 new random effects for sampling occasion by adding the term

rep(u[6:12],5)

to

beta[1] + **rep**(u[1:5], each = 7).

To accommodate the additional random effects with standard deviation σ_{u2} in the joint log-likelihood, the function l is updated by adding

-**sum**(**dnorm**(x=u[6:12], **sd**=sigma.u2, **log**=TRUE))

to the existing code.

The only change to the Laplace likelihood, l.LA is in the adaption of the change in the covariance matrix for the random effects, Ψ to the Hessian, H in (5.107); the term

diag(1/sigma.u^2,5)

is replaced by

diag(**c**(**rep**(1/sigma.u^2,5), **rep**(1/sigma.u2^2, 7))).

The results are shown in Table 5.6.

To fit model (5.112), the only change to the previously defined functions f, l and l.LA for model (5.110) is the addition of the term

beta[4] * season

in the **exp**-term in f. The estimate of model (5.112) is also shown in Table 5.6. The likelihood of model (5.112) is considerably higher than that of model (5.111) at the same expense of parameters.

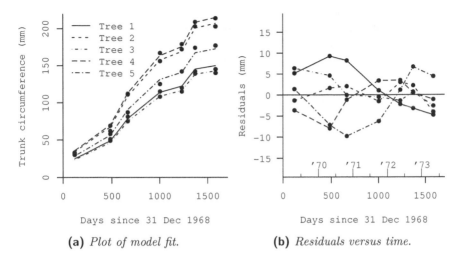

(a) *Plot of model fit.*   **(b)** *Residuals versus time.*

Figure 5.7: *Plots for model* (5.112) *for orange tree data.*

The model fit for model (5.112) is shown in Figure 5.7a. By comparing this to the fit of model (5.110) in Figure 5.6a, it appears that model (5.112) seems to capture the variation between sampling occasions. This is also verified by a plot of residuals versus time in Figure 5.7b, where the residuals within sample occasions are now centered around zero and smaller than in Figure 5.6b. The plots for model (5.111) are very similar to those in Figure 5.7 for model (5.112) and therefore omitted.

In Figure 5.7b, the residuals for each tree have been connected by lines to illustrate that a positive auto-correlation is present. Only small changes to the estimation scheme are required to accommodate any correlation or covariance structure in the residuals. For implementing a CAR process as in (5.113) for the within-tree residuals in the estimation of the models (5.111) and (5.112), we implement the covariance matrix, Σ in (5.113) as

```
> Sigma.CAR <- function(phi, sigma) {
    diff <- (time[1:7] - rep(time[1:7], each=7))
    delta.t <- matrix(diff / (365 / 2), nrow = 7, ncol = 7)
    P <- sigma^2 * exp( - phi * abs(delta.t))
    kronecker( diag(5), P) }
```

where `delta.t` is a matrix of time differences and `P` is $\text{cov}(\epsilon_i)$. To accommodate the CAR process in the residuals in models (5.111) and (5.112), the model functions remain as previously described and the joint log-likelihood is defined as

```
> l <- function(u, beta, sigma, sigma.u, sigma.u2, phi) {
    Sigma <- Sigma.CAR(phi, sigma)
```

```
    resid <- circumference - f(beta, u)
    0.5 * (log(det(2*pi*Sigma)) + crossprod(resid, solve(Sigma,
        resid))) - sum(dnorm(x = u[1:5], sd = sigma.u, log=TRUE)) -
        sum(dnorm(x = u[6:12], sd = sigma.u2, log=TRUE)) }
```

where the notable difference from previously is that the first part of l is now written as the logarithm of a multivariate normal density using the full residual covariance matrix \small Sigma (in model (5.112) the last call to **dnorm** concerning σ_{u2} is excluded). The term $(y - \mu)^T \Sigma^{-1}(y - \mu)$ in the normal density function is computed using

```
crossprod(resid, solve(Sigma, resid))
```

since this is numerically more stable and more efficient than computing the term directly as defined. The only change to l.LA to accommodate (5.113) is in the computation of the Hessian, D, where

```
crossprod(Jac.f)/sigma^2
```

is changed to

```
crossprod(Jac.f, solve(Sigma, Jac.f)).
```

To make the estimation of the correlation parameter, ϕ in the CAR process (5.113) unbounded, it is optimized on the log-scale. The estimates of models (5.111) and (5.112) with the CAR process (5.113) are shown in Table 5.6. For both models, the CAR process is a significant improvement with p-values <0.001 based on likelihood ratio tests. For model (5.112), the correlation over half a year, and therefore roughly between sampling occasions, is $\hat{\rho} = 0.81$, which is equivalent to the correlation coefficient in a discrete AR(1) model, where account is taken of missing sampling occasions. This corresponds to the strong auto-correlation seen in Figure 5.7b between successive sampling occasions.

The presented estimation scheme using the multivariate Laplace approximation offers a very large flexibility in the specification of NLMMs at the cost of only a rather limited amount of coding. It provides an option to fit models when standard software falls short. Especially models with crossed random effects are not (at least easily) handled by any currently available software package such as NONMEM, SAS NLMIXED, or nlme for R/S-Plus. In this way the approach presented here fills a gap left by standard software for NLMMs.

Validation

The profile likelihood is an inferential tool in its own right, and it can be used to make likelihood based confidence intervals instead of having to rely on the Wald approximation. For a scalar parameter τ, the profile likelihood is defined as $L_P(\tau; y) = \sup_\zeta L((\tau, \zeta); y)$, where ζ are nuisance parameters. The approach contains a single loop over the parameter of interest with repeated

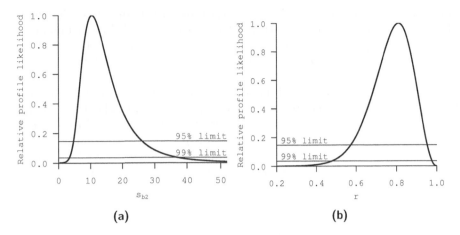

Figure 5.8: *Relative profile likelihoods for (a) the variance parameter for the random effects of sampling occasion in model (5.111) and (b) the correlation over half a year in model (5.112) + (5.113). The horizontal lines indicate 95% and 99% confidence intervals.*

optimization with respect to the remaining nuisance parameters. The profile likelihood can be interpolated by a spline (e.g., **spline** in R) to reduce the number of values of τ for which the likelihood has to be optimized to produce a smooth curve. Figure 5.8a shows the relative profile likelihood for the variance parameter for the random effects of sampling occasion in model (5.111), and Figure 5.8b shows the relative profile likelihood for the half-year correlation ρ in model (5.112) with the seasonal effect and the CAR residual structure. The horizontal lines at 0.1465 and 0.03625 define 95% and 99% confidence intervals based on the usual χ_1^2-asymptotics of the likelihood ratio statistic. The profile likelihood confidence bounds can be found by numerically solving for the intersection of the spline function with these threshold (e.g., using **uniroot** in R). The figures show which values of the parameters are supported by the data and which are of negligible likelihood relative to the MLE. The figures also illustrate the effect of the arguably arbitrary choice of confidence level.

Automatic differentiation

Another simpler, more efficient, and more direct way to use the Laplace transformation technique outlined here is via the open source software package AD Model Builder, which takes advantages of *automatic differentiation*. Some techniques used in AD Model Builder will be presented in the following and introduced more formally in Section 5.10.

Any calculation done via a computer program can be broken down to a long chain of simple operations like '+', '−', '.', '/', 'exp', 'log', 'sin', 'cos', 'tan', '$\sqrt{}$', and so on. It is simple to write down the analytical derivative of each of these operations by themselves. If our log-likelihood function ℓ consisted of only a few of these simple operations, then it would be tractable to use the chain rule $(f \circ g)'(x) = f'(g(x))g'(x)$ to find the analytical gradient $\nabla \ell$ of the log-likelihood function ℓ.

In practical applications this chain of simple operations often becomes so long that it is intractable to write down the analytical gradient for our log-likelihood function. In those cases most statistical software use finite differences (e.g. forward difference $\partial \ell / \partial \theta_i \approx (\ell(\theta + \Delta \theta_i e_i) - \ell(\theta)) / \Delta \theta_i$) to approximate the gradient, but that can be inaccurate and slow. The computational cost of evaluating the finite difference approximation of the gradient $\nabla \ell$ in a model with n model parameters is equivalent to the computational cost of evaluating the log-likelihood $n + 1$ times.

Automatic differentiation is a technique where the chain rule is used by the computer program itself. When the program evaluates the log-likelihood it keeps track of all the operations used along the way, and then runs the program backwards (reverse mode automatic differentiation) and uses the chain rule to update the derivatives one simple operation at a time.

Automatic differentiation is accurate, and the computational cost of evaluating the gradient is surprisingly low:

THEOREM 5.13
The computational cost of evaluating the gradient of the log-likelihood $\nabla \ell$ with reverse mode automatic differentiation is less than four times the computational cost of evaluating the log-likelihood function ℓ itself. This holds no matter how many parameters the model contain.

Proof See Griewank (2000).   ∎

The surprising part is that computational cost does not depend on how many parameters the model contain. There is however a practical concern. The computational cost mentioned above is measured in the number of operations, but reverse mode automatic differentiation requires all the intermediate variables in the calculation of the negative log-likelihood to be stored in the computer's memory, so if the calculation is lengthy, for instance consisting of a long iterative procedure, then the memory requirements can be enormous.

Automatic differentiation combined with the Laplace approximation

Finding the gradient of the Laplace approximation of the marginal log-likelihood is challenging, because the approximation itself includes the result of a function minimization, and not just a straightforward sequence of simple operations. It is however possible, but requires up to third order derivatives to be computed internally by clever successive application of automatic differentiation. The procedure is described in Skaug and Fournier (2006).

Importance sampling

Importance sampling (Ripley 1987) is a re-weighting technique for approximating integrals w.r.t. a density f by simulation in cases where it is not feasible to simulate from the distribution with density f. Instead it uses samples from a different distribution with density g, where the support of g includes the support of f.

For general mixed effects models it is possible to simulate from the distribution with density proportional to the second order Taylor approximation

$$\widetilde{L}(\boldsymbol{\theta}, \hat{\boldsymbol{u}}_\theta, \boldsymbol{Y}) = \exp\left\{\ell(\boldsymbol{\theta}, \hat{\boldsymbol{u}}_\theta, \boldsymbol{Y}) - \tfrac{1}{2}(\boldsymbol{u} - \hat{\boldsymbol{u}}_\theta)^T(-\ell''_{uu}(\boldsymbol{\theta}, \boldsymbol{u}, \boldsymbol{Y})|_{\boldsymbol{u}=\hat{\boldsymbol{u}}_\theta})(\boldsymbol{u} - \hat{\boldsymbol{u}}_\theta)\right\}$$

as, apart from a normalization constant, it is the density $\phi_{\hat{u}_\theta, \hat{V}_\theta}(\boldsymbol{u})$ of a multivariate normal with mean $\hat{\boldsymbol{u}}_\theta$ and covariance

$$\hat{\boldsymbol{V}}_\theta = \boldsymbol{H}^{-1}(\hat{\boldsymbol{u}}_\theta) = (-\ell''_{uu}(\boldsymbol{\theta}, \boldsymbol{u}, \boldsymbol{Y})|_{\boldsymbol{u}=\hat{\boldsymbol{u}}_\theta})^{-1}. \tag{5.114}$$

The integral to be approximated is given by (5.94) and can be rewritten as:

$$L_M(\boldsymbol{\theta}, \boldsymbol{Y}) = \int L(\boldsymbol{\theta}, \boldsymbol{u}, \boldsymbol{Y})d\boldsymbol{u} = \int \frac{L(\boldsymbol{\theta}, \boldsymbol{u}, \boldsymbol{Y})}{\phi_{\hat{u}_\theta, \hat{V}_\theta}(u)}\phi_{\hat{u}_\theta, \hat{V}_\theta}(\boldsymbol{u})d\boldsymbol{u}.$$

So if $u^{(i)}$, $i = 1, \dots, N$ is simulated from the multivariate normal distribution with mean $\hat{\boldsymbol{u}}_\theta$ and covariance $\hat{\boldsymbol{V}}_\theta$, then the integral can be approximated by the mean of the importance weights

$$L_M(\boldsymbol{\theta}, \boldsymbol{Y}) = \frac{1}{N}\sum \frac{L(\boldsymbol{\theta}, \boldsymbol{u}^{(i)}, \boldsymbol{Y})}{\phi_{\hat{u}_\theta, \hat{V}_\theta}(\boldsymbol{u}^{(i)})} \tag{5.115}$$

Notice that the importance sampling approximation of the marginal likelihood (5.115) can be used to check if the Laplace approximation is accurate, and to improve it in cases where it is not. If the importance sampling approximation is used in combination with automatic differentiation, then the same random number seed must be used in all function evaluations (for all θ values), as the function must be differentiable.

AD Model Builder

AD Model Builder is a programming language that builds on C++. It includes helper functions for reading in data, defining model parameters, and implementing and optimizing the negative log-likelihood function. The central feature is automatic differentiation (AD), which is implemented in such a way that the user rarely has to think about it at all. AD Model Builder can be used for fixed effects models, but in addition it includes Laplace approximation and importance sampling for dealing with general mixed effects models, which is why it is mentioned here and in the following example.

AD Model Builder is developed by Dr. Dave Fournier and was a commercial product for many years. Recently AD Model Builder has been placed in the public domain (see http://www.admb-project.org).

Table 5.7: *Germination of seeds data.*

Plate	Extract	Seed	r	n
1	0	0	10	39
2	0	0	23	62
3	0	0	23	81
4	0	0	26	51
5	0	0	17	39
6	0	1	5	6
7	0	1	53	74
8	0	1	55	72
9	0	1	32	51
10	0	1	46	79
11	0	1	10	13
12	1	0	8	16
13	1	0	10	30
14	1	0	8	28
15	1	0	23	45
16	1	0	0	4
17	1	1	3	12
18	1	1	22	41
19	1	1	15	30
20	1	1	32	51
21	1	1	3	7

Example 5.7 – Logistic regression with over-dispersion in AD Model Builder
The data for this example is taken from Crowder (1978). The proportion of seeds that germinated on each of $N = 21$ plates are of interest. The 21 plates are classified by seed (0=bean or 1=cucumber), and by type of root extract (0 or 1). Let r_i be the number germinated and n_i be the total number seeds on plate i (see Table 5.7).

The model assumed is a logistic regression, but with an unobserved random variable corresponding to each plate:

$$r_i \sim \text{Bin}(n_i, p_i), \quad \text{where}$$

$$p_i = \text{logit}^{-1}(\mathbf{X}\boldsymbol{\beta} + B_i), \quad \text{and}$$

$$B_i \sim \text{N}(0, \sigma^2) \quad \text{independent.}$$

The design matrix $\mathbf{X}$ above corresponds to the full two-way model with interaction between extract and seed. The model is implemented in AD Model Builder by:

```
DATA_SECTION
  init_int N
  init_vector r(1,N)
  init_vector n(1,N)
  init_matrix X(1,N,1,4)

PARAMETER_SECTION
  init_vector beta(1,4)
  init_number logSigma
  random_effects_vector B(1,N)
  sdreport_number sigma
  vector logitp(1,N)
  vector p(1,N)
  objective_function_value jnll

PROCEDURE_SECTION
  jnll=0.0;
  sigma=exp(logSigma);
  logitp=X*beta+B;
  p=elem_div(exp(logitp),(1.0+exp(logitp)));
  for(int i=1; i<=N; ++i){
    jnll+=-log_comb(n(i),r(i))-log(p(i))*r(i)-log(1.0-p(i))*(n(i)-r(
        i));
    jnll+=0.5*(log(2.0*M_PI*square(sigma))+square(B(i)/sigma));
  }
```

Notice that the program naturally partitions the coding of the negative log-likelihood into three parts. The DATA_SECTION where data are read in. The PARAMETER_SECTION where the model parameters are defined. All quantities with a type starting with init_ are considered as fixed effects parameters to be optimized. All quantities starting random_effects are considered as unobserved random variables. The PARAMETER_SECTION is also where variables for intermediate calculations can be defined, and where the objective function is declared. Finally the PROCEDURE_SECTION is where the joint negative log-likelihood $-\ell(\theta, u, Y)$ is calculated.

When the program is compiled (first to a C++ program, then to a binary) and executed, it estimates, in the default mode, the model parameters θ by minimizing the marginal negative log-likelihood $-\ell_M(\theta, Y)$, where the marginal negative log-likelihood is evaluated in each step by the Laplace approximation. If desired, importance sampling can be switched on by a command line flag to the binary without altering the program.

Let us now consider an example which indicates that the approach can be used for time series data as well.

Example 5.8 – Discrete valued time series

The data for this example originates from Zeger (1988). It consists of monthly counts of US polio cases y_i, $i = 1, \ldots, n$, in a period of $n = 168$ months starting in 1970. The model proposed by Zeger (1988) and extended by Chan and Ledolter (1995) to analyze the data is a state space model, where the count observations y_i are assumed to follow a Poisson distribution with intensities λ_i, where

$$\log(\lambda_i) = \boldsymbol{X}_i^T \boldsymbol{\beta} + u_i. \tag{5.116}$$

Here $\boldsymbol{X}_i$ is a 6-dimensional vector of covariates and $\boldsymbol{\beta}$ is the corresponding fixed effect model parameters. The random effects u_i are assumed to follow a first order autoregressive process.

$$u_i = a u_{i-1} + \epsilon_i, \tag{5.117}$$

where $\epsilon_i \sim N(0, \sigma^2)$, $i > 1$ is a sequence of independent identically distributed random variables (sometimes called a *white noise process*), and a and σ are model parameters. The first random effect is assumed to follow the stationary distribution of the first order auto regressive process $u_1 \sim N(0, \sigma^2/(1 - a^2))$ (see Madsen (2008)).

The joint likelihood becomes:

$$L(a, \boldsymbol{\beta}, \sigma; \boldsymbol{y}) = \varphi_{0, \frac{\sigma^2}{1 - a^2}}(u_1) \prod_{i=2}^{n} \left(\varphi_{0, \sigma^2}(u_i - a u_{i-1}) \right) \prod_{i=1}^{n} \left(p_{\lambda_i}(y_i) \right), \tag{5.118}$$

where φ_{μ, σ^2} is the pdf of the normal distribution with mean μ and variance σ^2, and p_λ is the pdf of the Poisson distribution with mean λ. The following shows how this model can be implemented in AD Model Builder. The implementation shown here is a slightly simplified version of the implementation in the AD Model Builder example catalog (http://www.admb-project.org), but otherwise identical:

```
DATA_SECTION
  init_int n
  init_vector y(1,n)
  init_int p
  init_matrix X(1,n,1,p)

PARAMETER_SECTION
  init_vector b(1,p,1)
  init_bounded_number a(-1,1,2)
  init_number log_sigma(2)
  random_effects_vector u(1,n,2)
  objective_function_value jnll
```

```
PROCEDURE_SECTION
  int i;

  sf1(log_sigma,a,u(1));

  for (i=2;i<=n;i++){
    sf2(log_sigma,a,u(i),u(i-1),i);
  }

  for (i=1;i<=n;i++){
    sf3(u(i),b,i);
  }

SEPARABLE_FUNCTION void sf1(const dvariable& ls,
                           const dvariable& aa,
                           const dvariable& u_1)
  jnll += ls - 0.5*log(1-square(aa))
          +0.5*square(u_1/exp(ls))*(1-square(aa));

SEPARABLE_FUNCTION void sf2(const dvariable& ls,
                           const dvariable& aa,
                           const dvariable& u_i,
                           const dvariable& u_i1, int i)
  jnll += ls +.5*square((u_i-aa*u_i1)/exp(ls));

SEPARABLE_FUNCTION void sf3(const dvariable& u_i ,
                           const dvar_vector& bb, int i)
  dvariable eta = X(i)*bb + u_i;
  dvariable lambda = exp(eta);
  jnll -= y(i)*eta - lambda;
```

Notice that the likelihood function (5.118) separates naturally into a product where each factor only depends on one or two of the random effects. This fact can be utilized to implement the Laplace approximation more efficiently, as the covariance matrix of the random effects in (5.101) becomes sparse, and hence all the elements known to be zero can be skipped in the calculations. The tool for utilizing this in AD Model Builder is SEPARABLE _FUNCTION. The joint negative log likelihood must be written as a sum of separable functions, where each separable function call uses only a few of the random effects. Besides in time series models (as illustrated here) exploiting the separable structure can give substantially more efficient computations with nested and crossed random effects and for models with a certain spatial covariance structures.

The estimates from the AD Model Builder implementation of this model are almost identical to the estimates reported by Kuk and Cheng (1999) using a Monte Carlo method. The interesting part is that a standard tool like AD Model Builder is able to handle such a nonlinear and non-Gaussian random effects model efficiently with little more custom coding than simply specifying the joint likelihood function.

The germination case and the polio case are just two simple examples of models that can easily be handled with the combination of Laplace approximation and automatic differentiation. The AD Model Builder example catalog (see http://admb-project.org/) includes: Generalized additive models, discrete valued time series, ordered categorical responses, Poisson regression with spatially correlated random effects, stochastic volatility models, gamma distributed random effects, and many more.

5.11   Mixed effects models in R

Background

The nlme library, Pinheiro (2000), includes functions for fitting mixed-effects models in R, lme for linear models and nlme for nonlinear models. The package is not included in the default installation of R but can be installed by calling

```
> install.packages("nlme")
```

Once the package is installed it can be used in a session by calling

```
> library(nlme)
```

Douglas Bates, one of the authors of nlme, and Martin Maechler have written a new library for fitting mixed-effects models, lme4. An important feature of this new package is that it can handle fully and partially crossed random effects gracefully. There are functions for fitting linear mixed models, lmer, generalized linear mixed models, glmer and nonlinear mixed models, nlmer. However, this section will focus on the lme function.

Model formulae

In a lme-model it is necessary to specify two *formulae*, one for the fixed effects, and another for the random effects. Further there is a possibility for specifying a large variety of covariance structures for the groups. The fixed effects part is a two-sided linear formula specifying the dependent variable and the fixed effects in the model (as in lm). The random effects part is typically a one-sided linear formula specifying the random effects and the grouping structure in the model. A random-effects term consists of two expressions separated by a vertical bar, |, which can be read as given or by. The expression on the right of the | symbol is evaluated as a factor, the grouping factor for the term.

For simple scalar random effects terms, the expression on the left should be 1 resulting in one random effect for each level of the grouping factor.

Model fitting

A typical call to the lme function is:

```
fit <- lme(y ~ a, data, ~ 1|b)
```

where y is the response variable, a is a fixed factor, b is a random factor and *data* is a dataframe. After a model has been fitted using lme, some of the same *generic* functions that are used for the general linear model can by used. A list of some useful functions is given in Section 3.12 on page 81. In addition, there exists some useful functions such as fixef and ranef returning the fixed and random effects respectively.

5.12   Problems

Exercise 5.1
To investigate the effect of drying of beech wood on the humidity percentage, the following experiment was conducted. Each of 20 planks was dried in a certain period of time. Then the humidity percentage was measured in 5 depths and 3 widths for each plank:

depth 1:	close to the top	width 1:	close to the side
depth 5:	in the center	width 3:	in the center
depth 9:	close to the bottom	width 2:	between 1 and 3
depth 3:	between 1 and 5		
depth 7:	between 5 and 9		

There are therefore $3 \cdot 5 = 15$ measurements for each plank and all together 300 observations. The data can be found in the file **planks.csv**. The first row of the file contains the name of the variables. The data can be imported into R with the following command:

```
dat<-read.table("planks.csv",sep=',',head=T)
```

Question 1 Perform some initial explorative analysis. The following R lines will produce some profile plots that can be useful.

```
with(planks, interaction.plot(width,plank,humidity,legend=F))
with(planks, interaction.plot(depth,plank,humidity,legend=F))
with(planks, interaction.plot(width,depth,humidity,legend=F))
with(planks, interaction.plot(depth,width,humidity,legend=F))
```

Question 2 Write out a natural model for the data.

Question 3 Fit the model to the data and reduce it if possible.

Question 4 Give estimates of the parameters in the model.

Exercise 5.2

Spinach heated to 90 or 100 degrees Celsius was vacuum packed and stored for 0, 1 or 2 weeks before the packs were opened and chill stored in normal atmosphere for 0, 1 or 2 days. Then the color was measured on a Hunter Lab. Two of the color coordinates, a and b (measuring respectively something like red and yellow color), were recorded. The variable batch is a blocking variable referring to two batches of spinach. The data can be found in the file spinach.csv. The first row of the file contains the name of the variables. The data can be imported into R with the following command:

```
dat<-read.table("spinach.csv",sep=',',head=T)
```

Question 1 Write down all the factors relevant for the analysis, and their levels and mutual structure. Are they crossed or nested, for example?

Question 2 Analyze the effect of the different factors on the two color measurements and summarize the significant effects.

Exercise 5.3

In an experiment with 4 sorts of blueberries the fructification (the number of berries as percent of the number of flowers earlier in the season) was determined for 5 twigs on each of 6 bushes for every sort. The data can be found in the file berries.csv. The first row of the file contains the name of the variables. The data can be imported into R with the following command:

```
dat<-read.table("berries.csv",sep=',',head=T)
```

Question 1 Write down the factors relevant for the analysis, and their levels and mutual structure.

Question 2 Write down a natural model for the data.

Question 3 Fit the model to the data and reduce it if possible.

Question 4 Give estimates for the parameters in the model.

Exercise 5.4

Table 5.8 shows the tensile strength of some rubber (measured as pounds per square inch).

Assume first that A, B, C, and D denote different methods for the production of the rubber.

Question 1 Suggest a reasonable model for the tensile strength data.

Question 2 Estimate the model parameters.

Assume now that A, B, C and D denote four different boxes from a large stock of rubber.

Question 3 Suggest a model given the new assumption.

Table 5.8: *Tensile strength of some rubber (measured as pounds per square inch).*

A	B	C	D
3210	3225	3220	3545
3000	3320	3410	3600
3315	3165	3320	3580
	3145	3370	3485

Question 4 Estimate the model parameters.

Question 5 Discuss the differences between the two type of models and the associated model parameters.

Exercise 5.5
A number of sediment samples from the geological formations FF, PF and GF were analyzed with respect to the content of Cu. The measured concentrations are assumed to be log-normally distributed.

Let X_{ij} denote the concentration in sample No. j and formation i, and consider $Y_{ij} = \log X_{ij}$. Based on the measured concentration we have the results shown in Table 5.9.

Table 5.9: *Data for Exercise 5.5.*

Geology	Number of obs	$\overline{Y}_i$	$\sum_j (Y_{ij} - \overline{Y}_i)^2$
FF	58	2.450	38.120
PF	33	2.141	6.769
GF	25	2.108	12.105

Question 1 Suggest a model and use this model to perform a test to see if there a significant (use 5 pct. level) difference between the concentration of Cu in the three geological formations.

Question 2 Estimate the model parameters.

Based on geological information it is expected that the geological formations FF and PF are different.

Question 3 Use another statistical test to evaluate whether it can be assumed that these two formations are different.

Exercise 5.6
This problem considers the tensile strength of some cord. In order to make an analysis of the cord a number of samples have been taken on five different days. The results are given in Table 5.10.

It is assumed that the observations of the tensile strength under the same experimental conditions are Gaussian distributed.

Table 5.10: *Tensile strength of some cord.*

Number of obs	Average [kg]	SSE [kg^2]
4	512.32	875.44
8	493.88	1808.72
7	517.04	1948.57
8	488.82	1340.68
10	497.23	1226.80

Question 1 Conduct a test for evaluating whether there is a difference from day to day.

Question 2 Estimate the parameters of a reasonable model for the variations of the tensile strength of the cord.

Exercise 5.7

An automatic machine for filling of soap powder has three funnels, A, B, and C, for filling the soap powder in cartons. For each of the funnels it is assumed that the observations of weight of powder in the cartons are independent and Gaussian distributed with a mean value for each of the funnels and variance 4 g^2. In a given period of time the characteristics of the production is given in Table 5.11.

Table 5.11: *Data for Exercise 5.7.*

Funnel	Fraction of production	Mean weight [g]	Variance [g^2]
A	40 pct	497	4
B	30 pct	500	4
C	30 pct	504	4

After being filled with soap powder the cartons are collected in packs with 10 cartons in each package.

Question 1 Calculate the mean and variance for the mean weight of the cartons in each package assuming that the cartons from the three funnels are packed randomly.

Question 2 Calculate the mean and variance for the mean weight of the cartons in each package assuming the package consists of cartons from only one of the funnels (and hence that 40% of the packages contains cartons from funnel A, 30% of packages contains cartons only from funnel B, etc.).

Question 3 Calculate the covariance between the weight of two cartons from the same package corresponding to each of the situations described in the questions above.

Exercise 5.8

The orange tree data was in this chapter analyzed in R via the Laplace approximation. The code for the Laplace approximation was written for this specific problem. A general Laplace approximation is built into AD Model Builder, and it uses Automatic differentiation (up to third order) to be more efficient.

Analyze the orange tree data via AD Model Builder.

Exercise 5.9

Linear Gaussian state-space model can be analyzed efficiently via the Kalman filter (Harvey (1996)). The AD aided Laplace approximation in AD Model Builder is able to handle more general models, but the cost in terms of efficiency, when it is applied in the linear Gaussian case, will be considered in this exercise.

Consider the simple random walk with observation noise model. Here u is unobserved and follows a random walk $u_{i+1} = u_i + \eta_i$ with $u_0 = 0$ and $\eta_i \sim N(0, \sigma_u^2)$, $i = 1, \ldots, n$, but the observations are $Y_i = u_i + \epsilon_i$ with $\epsilon_i \sim N(0, \sigma^2)$.

Question 1 Simulate three or more datasets from the model with different numbers of observations (e.g., $n=20$, $n=200$, $n=2000$).

Question 2 Estimate model parameters σ_u and σ, and the unobserved random walk u_i with confidence intervals using an ordinary Kalman filter (see, e.g., Madsen (2008)).

Question 3 Estimate the same parameter and the unobserved random walk via the Laplace approximation in AD Model builder.

Question 4 Compare and comment on run times and estimated quantities for the two solutions for each of the simulated cases.

Exercise 5.10

The strength of ready mixed concrete depends on the cement which is used in the production of concrete at the production plants. The strength is increasing from the time of the production until it reaches a steady state value after a rather long period of time, maybe six months. The strength of the concrete is measured after 28 days for control.

The cement is delivered in batches, and it is observed that the strength varies from batch to batch. Given the same batch of cement the strength is assumed to be Gaussian distributed.

In order to obtain a more efficient control for the strength of concrete an early value of the strength is measured after one week. The purpose of this exercise is to establish a model for the joint variation of the 7-days and 28-days strength.

Let us introduce $\boldsymbol{Y} = (Y_7, Y_{28})^T$, i.e., a vector containing both the early and the 28-day strength of concrete.

In the file concrete.csv (can be downloaded from the home-page of the book) both early (7-days) and 28-days observations of concrete strength are listed. The batch number for the cement used in the production is shown and the date of the production is also shown.

Question 1 Plot the observations of the 7-days and 28-days strength as a function of time.

Question 2 Estimate the mean concrete strength within batches.

Question 3 Formulate a multivariate mixed effect model for the simultaneous variation of strength.

Question 4 Estimate the correlation between the levels of early and 28-day concrete.

CHAPTER 6

Hierarchical models

6.1 Introduction, approaches to modeling of overdispersion

A characteristic property of the generalized linear models (Chapter 4 on page 87) is that the variance, $\text{Var}[Y]$ in the distribution of the response is a known function, $V(\mu)$, that only depends on the mean value μ, viz.

$$\text{Var}[Y_i] = \lambda_i V(\mu) = \frac{\sigma^2}{w_i} \, V(\mu)$$

where w_i denotes a known *weight*, associated with the i'th observation, and where σ^2 denotes a *dispersion parameter* common to all observations, irrespective of their mean.

As described in Section 4.5 on page 113, the dispersion parameter σ^2 does serve to express *overdispersion* in situations where the residual deviance is larger than what can be attributed to the variance function $V(\mu)$ and known weights w_i.

In this chapter we shall describe an alternative method for modeling overdispersion, viz. by *hierarchical models* analogous to the mixed effects models for the normally distributed observations in Chapter 5 on page 157. See also Lee and Nelder (2000) for a discussion of these two approaches.

A starting point in a hierarchical modeling is an assumption that the distribution of the random "noise" may be modeled by an exponential dispersion family (Binomial, Poisson, etc.), and then it is a matter of choosing a suitable (prior) distribution of the mean-value parameter μ. It seems natural to choose a distribution with a support that coincides with the mean value space $\mathcal{M}$ rather than using a normal distribution (with a support constituting all of the real axis $\mathbb{R}$).

In some applications an approach with a normal distribution of the canonical parameter is used. Such an approach is sometimes called *generalized linear mixed models* (GLMMs). Although such an approach is consistent with a formal requirement of equivalence between mean values space and support for the distribution of μ in the binomial and the Poisson distribution case, the resulting marginal distribution of the observation is seldom tractable, and the likelihood of such a model will involve an integral which cannot in general be computed explicitly. Also, the canonical parameter does not have a simple physical interpretation and, therefore, an additive "true value" + error, with a

Table 6.1: *The distribution of days with 0, 1, 2 or more episodes of thunderstorm at Cape Kennedy. (From Williford, Carter, and Hsieh (1974).)*

Number of episodes, z_i	Number of days, $\# i$	Poisson expected
0	803	791.85
1	100	118.78
2	14	8.91
3+	3	0.46

normally distributed "error" on the canonical parameter to describe variation between subgroups, is not very transparent. Instead, we shall describe an approach based on the so-called *standard conjugated distribution* for the mean parameter of the within group distribution for exponential families. These distributions combine with the exponential families in a simple way, and lead to marginal distributions that may be expressed in a closed form suited for likelihood calculations. In R these models are handled by the routine `lmer` (in the lme4 package).

Before the general introduction to *hierarchical generalized linear models* in Section 6.5 we will introduce the concept of conjugated distributions with some examples that correspond to a one-way model with random effects for various exponential family distributions.

6.2   Hierarchical Poisson Gamma model

We start with an illustrative example.

Example 6.1 – Variation between episodes of thunderstorm at Cape Kennedy
Table 6.1 (after Williford, Carter, and Hsieh (1974)) shows the distribution of the number of daily episodes of thunderstorms at Cape Kennedy, Florida, during the months of June, July and August for the 10-year period 1957–1966, total 920 days. All observational periods are $n_i = 1$ day. The data represents *counts* of events (episodes of thunderstorms) distributed in time, see Figure 6.1. A completely random distribution of the events would result in a Poisson distribution of the number of daily events.

The variance function for the Poisson distribution is $V(\mu) = \mu$; therefore, a Poisson distribution of the daily number of events would result in the variance in the distribution of the daily number of events being equal to the mean, $\hat{\mu} = \overline{y}_+ = 0.15$ thunderstorms per day. The empirical variance is $s^2 = 0.1769$, which is somewhat larger than the average. We further note that the observed distribution has *heavier tails* than the Poisson distribution. Thus, one might be suspicious of overdispersion.

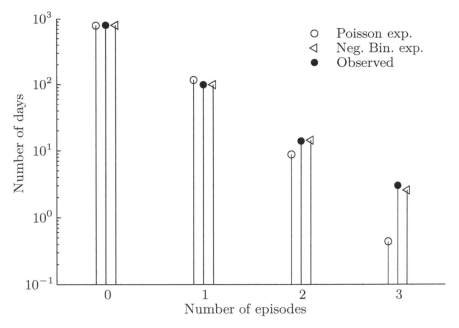

Figure 6.1: *Distribution of daily number of thunderstorm episodes at Cape Kennedy.*

Formulation of hierarchical model

THEOREM 6.1 – COMPOUND POISSON GAMMA MODEL
Consider a hierarchical model for Y specified by

$$Y|\mu \sim \text{Pois}(\mu), \tag{6.1a}$$

$$\mu \sim \text{G}(\alpha, \beta), \tag{6.1b}$$

i.e., a two stage model. In the first stage a random mean value μ is selected according to a Gamma distribution. The Y is generated according to a Poisson distribution with that value as mean value. Then the the marginal distribution of Y is a negative binomial distribution, $Y \sim \text{NB}(\alpha, 1/(1+\beta))$ *The probability function for Y is*

$$
\begin{aligned}
\text{P}[Y = y] &= g_Y(y; \alpha, \beta) \\
&= \frac{\Gamma(y+\alpha)}{y!\Gamma(\alpha)} \frac{\beta^y}{(\beta+1)^{y+\alpha}} \\
&= \binom{y+\alpha-1}{y} \frac{1}{(\beta+1)^\alpha} \left(\frac{\beta}{\beta+1}\right)^y, \quad \text{for } y = 0, 1, 2, \ldots
\end{aligned}
\tag{6.2}
$$

where we have used the convention

$$\binom{z}{y} = \frac{\Gamma(z+1)}{\Gamma(z+1-y)\, y!}$$

for z real and y integer values.

Proof As the method of proof is general for the derivation of marginal distributions in these situations, we shall present the proof: The probability function for the conditional distribution of Y for given μ

$$f_{Y|\mu}(y; \mu) = \frac{\mu^y}{y!} \exp(-\mu) \qquad (6.3)$$

and the probability density function for the distribution of μ is

$$f_\mu(\mu; \alpha, \beta) = \frac{1}{\beta\Gamma(\alpha)} \left(\frac{\mu}{\beta}\right)^{\alpha-1} \exp(-\mu/\beta) \qquad (6.4)$$

Therefore, the probability function for the *marginal distribution* of Y is determined from

$$
\begin{aligned}
g_Y(y; \alpha, \beta) &= \int_{\mu=0}^{\infty} f_{Y|\mu}(y; \mu) f_\mu(\mu; \alpha, \beta) d\mu \\
&= \int_{\mu=0}^{\infty} \frac{\mu^y}{y!} \exp(-\mu) \frac{1}{\beta\Gamma(\alpha)} \left(\frac{\mu}{\beta}\right)^{\alpha-1} \exp(-\mu/\beta) d\mu \qquad (6.5) \\
&= \frac{1}{\beta^\alpha \, y!\Gamma(\alpha)} \int_{\mu=0}^{\infty} \mu^{y+\alpha-1} \exp(-\mu(\beta+1)/\beta) \, d\mu
\end{aligned}
$$

We note that the integrand is the *kernel* in the probability density function for a gamma distribution, $G(y + \alpha, \beta/\{\beta+1\})$. As the integral of the density function shall equal one, we find by adjusting the norming constant that

$$\int_{\mu=0}^{\infty} \mu^{y+\alpha-1} \exp(-\mu/[\beta/(\beta+1)]) \, d\mu = \frac{\beta^{y+\alpha}\Gamma(y+\alpha)}{(\beta+1)^{y+\alpha}}$$

and then (6.2) follows. ∎

For integer values of α the negative binomial distribution is known as the distribution of the number of "failures" until the α'th success in a sequence of independent Bernoulli trials where the probability of success in each trial is $p = 1/(1 + \beta)$. For $\alpha = 1$ the distribution is known as the *geometric distribution*.

▶ **Remark 6.1 – Decomposition of the marginal variance, signal/noise ratio**

If $\mu \sim G(\alpha, \beta)$ then $E[\mu] = \alpha\beta$ and $Var[\mu] = \alpha\beta^2$ (see, e.g., Wasserman Sec. 3.4). Then, using (5.62a) we have the decomposition

$$Var[Y] = E[Var[Y|\mu]] + Var[E[Y|\mu]] = E[\mu] + Var[\mu] = \alpha\beta + \alpha\beta^2 \qquad (6.6)$$

of the total variation in variation within groups and between groups, respectively. In analogy with (5.12) we may introduce a signal/noise ratio as

$$\gamma = \frac{Var[E[Y|\mu]]}{E[Var[Y|\mu]]} = \frac{\alpha\beta^2}{\alpha\beta} = \beta \qquad (6.7)$$

◀

▶ **Remark 6.2 – Why use a Gamma distribution to describe variation between days?**
We might wonder why a Gamma distribution was chosen to represent the variation of μ between days. First of all, the support of the Gamma distribution, $0 < \mu < \infty$ conforms to the mean-value space, $\mathcal{M}$ for the Poisson distribution. Secondly, the two-parameter family of Gamma distributions is a rather flexible class of unimodal distributions, ranging from an exponential distribution ($\alpha = 1$) to fairly symmetrical distributions on the positive real line (large values of α). A third reason may be observed in the derivation of the marginal distribution of Y above. The fact that the kernel $\mu^{\alpha-1} \exp(-\mu/\beta)$ of the mixing distribution had the same structure as the kernel $\mu^y \exp(-\mu)$ of the likelihood function corresponding to the sampling distribution of Y. This feature had the consequence that the integral (6.5) has a closed form representation in terms of known functions.   ◀

Inference on individual group means

THEOREM 6.2 – CONDITIONAL DISTRIBUTION OF μ
Consider the hierarchical Poisson-Gamma model in (6.1), and assume that a value $Y = y$ has been observed.

Then the conditional distribution of μ for given $Y = y$ is a Gamma distribution,

$$\mu| \, Y = y \ \sim G(\alpha + y, \beta/(\beta + 1)) \tag{6.8}$$

with mean

$$E[\mu| \, Y = y] = \frac{\alpha + y}{(1/\beta + 1)} \tag{6.9}$$

Proof The conditional distribution is found using Bayes Theorem

$$g_\mu(\mu| \, Y = y) = \frac{f_{y,\mu}(y, \mu)}{g_Y(y; \alpha, \beta)}$$

$$= \frac{f_{y|\mu}(y; \mu)g_\mu(\mu)}{g_Y(y; \alpha, \beta)}$$

$$= \frac{1}{g_Y(y; \alpha, \beta)} \left[\frac{(\mu^y)}{y!} \exp(-\mu) \frac{1}{\beta\Gamma(\alpha)} \left(\frac{\mu}{\beta}\right)^{\alpha-1} \exp(-\mu/\beta) \right]$$

$$\propto \mu^{y+\alpha-1} \exp\left(-\mu(1 + 1/\beta)\right)$$

We identify the *kernel* of the probability density function

$$\mu^{y+\alpha-1} \exp(-\mu(1 + 1/\beta))$$

as the kernel of a Gamma distribution, $G(\alpha + y, \beta/(\beta + 1))$ ∎

▶ **Remark 6.3 – The Gamma distribution is conjugate to the Poisson**
In a Bayesian framework, we would identify the distribution of μ as the *prior distribution*, the distribution of $Y \mid \mu$ as the sampling distribution, and the conditional distribution of μ for given $Y = y$ as the *posterior* distribution.

When the posterior distribution belongs to the same distribution family as the prior one, we say that the prior distribution is *conjugate* with respect to that sampling distribution.

Using conjugate priors simplifies the modeling. To derive the posterior distribution, it is not necessary to perform the integration, as the posterior distribution is simply obtained by updating the parameters of the prior one. ◀

▶ **Remark 6.4 – The posterior mean as a weighted average**
Rewriting the posterior mean (6.9) in light of the decomposition (6.6) we find

$$E[\mu \mid Y = y] = \frac{\alpha + y}{(1/\beta + 1)} = \frac{\alpha\beta/\beta + y}{1/\beta + 1} = w\mu + (1 - w)y \qquad (6.10)$$

with the weight

$$
\begin{aligned}
w &= \frac{1/\beta}{1/\beta + 1} \\
&= \frac{E[\mathrm{Var}[Y \mid \mu]]}{\mathrm{Var}[E[Y \mid \mu]] + E[\mathrm{Var}[Y \mid \mu]]} \\
&= \frac{\dfrac{1}{\mathrm{Var}[E[Y \mid \mu]]}}{\dfrac{1}{\mathrm{Var}[E[Y \mid \mu]]} + \dfrac{1}{E[\mathrm{Var}[Y \mid \mu]]}}
\end{aligned}
$$

reflecting the decomposition of the total variation in variation within groups and between groups.   ◀

▶ **Remark 6.5 – Reparameterization of the Gamma distribution**
Instead of the usual parameterization of the gamma distribution of μ by its shape parameter α and scale parameter β, we may choose a parameterization by the mean value, $m = \alpha\beta$, and the signal/noise ratio $\gamma = \beta$

$$\gamma = \beta \qquad (6.11a)$$
$$m = \alpha\beta \qquad (6.11b)$$

The parameterization by m and γ implies that the degenerate one-point distribution of μ in a value m_0 may be obtained as limiting distribution for Gamma distributions with mean m_0 and signal/noise ratios $\gamma \to 0$. Moreover, under that limiting process the corresponding marginal distribution of Y (negative binomial) will converge towards a Poisson distribution with mean m_0. Thus, this parameterization allows us to construct a test for the hypothesis

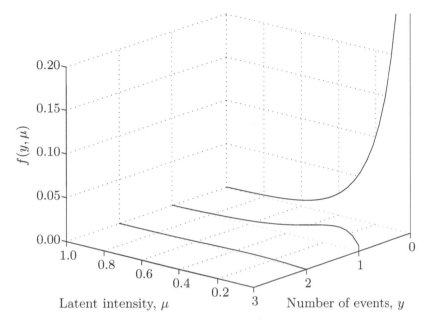

Figure 6.2: *Simultaneous distribution of daily number of thunderstorm episodes, y, and latent intensity, μ.*

$\mathcal{H}_0: \gamma = 0$ with the alternative $\mathcal{H}_a: \gamma > 0$ where $\mathcal{H}_0$ has the interpretation $Y \sim \text{Pois}(m)$ for some m and $\mathcal{H}_a$ corresponds to overdispersion with Y distributed like a negative binomial distribution. ◄

Example 6.2 – Thunderstorms (continued)
In order to estimate the parameters (m, γ) in the hierarchical model (6.1) on page 227 for the thunderstorm activity in Example 6.1 on page 226, the likelihood function is established by introducing the parameters (m, γ) in the probability function (6.2) and multiplying for all observations y. The corresponding log-likelihood is illustrated in Figure 6.3 on the next page. The log-likelihood is maximized with respect to m and γ by a general maximum-seeking algorithm resulting in the estimates $\widehat{m} = 0.1489$ [episodes/day], and $\widehat{\gamma} = 0.1939$. The estimated distribution has been illustrated in Figure 6.1 on page 227 and in Table 6.2 on the next page. It is seen that the negative binomial distribution provides a satisfactory fit.

By numerical maximization of the likelihood function we find $\widehat{m} = 0.1489$ [episodes/day] and $\widehat{\gamma} = 0.1939$. We may have determined a moment estimate by using that

$$\text{E}[Y] = m \approx \overline{y}_+ = 0.1489$$
$$\text{Var}[Y] = m(1 + \gamma) \approx s^2 = 0.1769$$

Table 6.2: *The distribution of days with 0, 1, 2 or more episodes of thunderstorm at Cape Kennedy compared with a Negative binomial distribution.*

Number of episodes, z_i	Number of days, $\# i$	Neg. bin. expected
0	803	802.92
1	100	100.15
2	14	14.38
3+	3	2.55

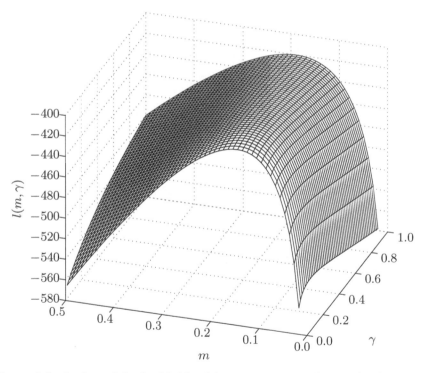

Figure 6.3: *Surface of the log-likelihood function corresponding to the data for the thunderstorm activity in Table 6.1 on page 226.*

leading to $\tilde{\gamma} = 0.1880$. It is not surprising that we obtain a better fit. It is well-known that meteorological phenomena are not randomly spread over the time axis, but there is a kind of slow dynamics in these phenomena that results in thunderstorms often occurring in clusters. We note, however, that modeling with the negative binomial distribution does not take the autocorrelation between daily observations into account. The negative binomial distribution provides a satisfactory description of the frequency of thunderstorms on randomly selected days, but if a description of the frequency of thunderstorms on successive days is desired, it is necessary to explicitly include a possible autocorrelation in the modeling.

6.3 Conjugate prior distributions

We will now introduce the general notion of a *standard conjugate* distribution for the mean parameter of the within group distribution to model the between group variation in standard situations where the between group distribution is considered to be unimodal. In the framework of generalized linear models, such models are termed *hierarchical generalized linear models*.

DEFINITION 6.1 – STANDARD CONJUGATE DISTRIBUTION FOR AN EXPO-NENTIAL DISPERSION FAMILY
Consider an exponential dispersion family $\mathrm{ED}(\mu, V(\mu)/\lambda)$ with density (4.2) for $\theta \in \Omega$. Let $\mathcal{M} = \tau(\Omega)$ denote the mean value space for this family. Let $m \in \mathcal{M}$ and consider

$$g_\theta(\theta; m, \gamma) = \frac{1}{C(m,\gamma)} \exp\left(\frac{\theta m - \kappa(\theta)}{\gamma}\right) \qquad (6.12)$$

with

$$C(m,\gamma) = \int_\Omega \exp\left(\frac{\theta m - \kappa(\theta)}{\gamma}\right) d\theta$$

for all (positive) values of γ for which the integral converges.
   Then (6.12) defines the density function of a probability distribution for θ. This distribution is called the *standard conjugate distribution* for θ corresponding to (4.2). The concept has its roots in the context of Bayesian parametric inference to describe a family of distributions whose densities have the structure of the likelihood kernel (see, e.g., Consonni and Veronese (1992)).

When the variance function, $V(\mu)$ for (4.2) is at most quadratic, the parameters m and γ have a simple interpretation in terms of the mean value parameter, $\mu = \tau(\theta)$, viz.

$$m = \mathrm{E}[\mu] \qquad (6.13)$$

$$\gamma = \frac{\mathrm{Var}[\mu]}{\mathrm{E}[\mathrm{Var}(\mu)]} \qquad (6.14)$$

with $\mu = E[Y|\theta]$, and with $\text{Var}(\mu)$ denoting the variance function (4.9), see Müller-Funk and Pukelsheim (1987). The use of the symbol γ in (6.14) is in agreement with our introduction of γ as *signal to noise ratio* for normally distributed observations (5.12) and for the Poisson-Gamma hierarchical model (6.7).

When the variance function for the exponential dispersion family (4.2) is at most quadratic, the standard conjugate distribution for μ coincides with the standard conjugate distribution (6.12) for θ. However, for the Inverse Gaussian distribution, the standard conjugate distribution for μ is improper (see Gutierez-Pena and Smith (1995)). In that case (6.13) and (6.14) are undefined.

The parameterization of the natural conjugate distribution for μ by the parameters m and γ has the advantage that location and spread are described by separate parameters. Thus, letting $\gamma \to 0$, the distribution of μ will converge towards a degenerate distribution with all its mass in m.

DEFINITION 6.2 – THE GENERALIZED ONE-WAY RANDOM EFFECTS MODEL
Now consider a hierarchical model with k randomly selected *groups*, $i = 1, 2, \ldots k$, and measurements Y_{ij}, $j = 1, \ldots, n_i$ from subgroup i.

i) Conditional on the group mean, μ_i, the measurements, Y_{ij} are independent and distributed according to an exponential dispersion model with mean μ_i, variance function, $V(\mu)$, and precision parameter λ (the sampling distribution).

ii) The group means, μ_i are independent random variables distributed according to the natural conjugate distribution of the sampling distribution.

The model may be thought of as a two-stage process: In the first stage, a group is selected, the group mean value μ_i is selected from the specified distribution. Then, with that value of μ_i a set of n_{ij} independent observations Y_{ij}, $j = 1, \ldots, n_i$ are generated according to the exponential dispersion model with that value, μ_i.

Marginal and simultaneous distributions

In analogy with Section 5.2 on page 163 we shall now describe the variance/covariance structure for the generalized one-way random effects model.

THEOREM 6.3 – MARGINAL DISTRIBUTIONS IN THE GENERALIZED ONE-WAY RANDOM EFFECTS MODEL
Consider a generalized one-way random effects model as specified in Definition 6.2.

The moments in the marginal distribution of Y_{ij} are

$$\mathrm{E}[Y_{ij}] = \mathrm{E}[\mathrm{E}[Y_i|\mu]] = \mathrm{E}[\mu] = m$$

$$\mathrm{Cov}[Y_{ij}, Y_{hl}] = \begin{cases} \mathrm{E}[V(\mu)]\{\gamma + 1/\lambda\} & \text{for } (i,j) = (h,l) \\ \mathrm{Var}[\mu] & \text{for } i = h,\ j \neq l \\ 0 & \text{for } i \neq h \end{cases} \tag{6.15}$$

Thus, the parameter, γ, in the distribution of the group means reflects the usual decomposition of total variation into variation within the groups and variation between the groups. As the variation, $V(\mu)$, within a specific group depends on the group mean, μ, the within-group variation is understood as an "average," $\mathrm{E}[V(\mu)]$.

The marginal distribution of the average in a group, $\overline{Y}_i$, has mean and variance

$$\mathrm{E}[\overline{Y}_i] = \mathrm{E}[\mathrm{E}[\overline{Y}_i|\mu]] = \mathrm{E}[\mu] = m \tag{6.16}$$

$$\mathrm{Var}[\overline{Y}_i] = \mathrm{E}[V(\mu)]\left(\gamma + \frac{1}{\lambda n_i}\right) \tag{6.17}$$

Thus, whenever $\mathrm{Var}[\mu] > 0]$, there is *overdispersion* in the sense that the variance in the marginal distribution exceeds the average variance in the within group distributions.

Measurements, Y_{ij} and Y_{ik} in the same group are correlated with *intraclass correlation*

$$\rho = \frac{\mathrm{Cov}[Y_{ij}, Y_{ik}]}{\mathrm{Var}[Y_{ij}]} = \frac{\mathrm{Var}[\mu]}{\mathrm{E}[V(\mu)]/\lambda + \mathrm{Var}[\mu]} = \frac{\gamma}{1/\lambda + \gamma} \tag{6.18}$$

and hence,

$$\gamma = \frac{\rho}{1 - \rho} \frac{1}{\lambda} \tag{6.19}$$

Estimation of fixed parameters

Maximum likelihood estimation

The likelihood function corresponding to a set of observations, $\overline{Y}_1, \overline{Y}_2, \ldots, \overline{Y}_k$ is constructed from the marginal probabilities for the group means, or group totals as a function of the parameters m and γ.

Estimation by the methods of moments

Now, consider the between-group sum of squares,

$$\mathrm{SSB} = \sum_{i=1}^{N} n_i \left(\overline{Y}_i - \overline{\overline{Y}}\right)^2 \tag{6.20}$$

with $\overline{\overline{Y}}$ denoting the usual (weighted) average of the subgroup averages,

$$\overline{\overline{Y}} = \frac{1}{\sum_i n_i} \sum_i n_i \overline{Y}_i$$

It may be shown that

$$E[\text{SSB}] = (k-1) \, E[V(\mu)] \left(\frac{1}{\lambda} + n_0 \gamma \right), \tag{6.21}$$

with n_0 denoting the *weighted average sample size* (5.29). Thus, a natural candidate for a method of moments estimate of the between-group dispersion, γ, is

$$S_2^2 = \frac{1}{N-1} \sum_{i=1}^{N} n_i (Y_i - \overline{Y})^2 \tag{6.22}$$

with expected value

$$E[S_2^2] = E[V(\mu)] \left(\frac{1}{\lambda} + n_0 \gamma \right). \tag{6.23}$$

Inference on individual group means, μ_i

THEOREM 6.4 – CONDITIONAL DISTRIBUTION OF GROUP MEAN μ_i AFTER OBSERVATION OF $\overline{y}_i$
Consider a generalized one-way random effects model as specified in Definition 6.2 on page 234 and assume that the sample from the i'th group resulted in the values $(y_{i1}, y_{i2}, \ldots, y_{in_i})$. Then the conditional distribution of the canonical parameter θ (the posterior distribution) given this sample result depends only on n and $\overline{y}_i$.

The probability density function in the conditional distribution of θ is

$$w\left(\theta \mid \overline{Y}_i = \overline{y}_i\right) = \frac{1}{C(m_1, \gamma_1)} \exp\left(\frac{\theta m_1 - \kappa(\theta)}{\gamma_1} \right) \tag{6.24}$$

with

$$m_1 = m_{post} = \frac{m/\gamma + n\lambda\overline{y}}{1/\gamma + n\lambda} \tag{6.25a}$$

$$\gamma_1 = \gamma_{post} = \frac{1}{1/\gamma + n\lambda}. \tag{6.25b}$$

Thus, the posterior distribution for μ is also of the same form as the prior distribution. We have just updated the parameters m and γ. The updating may be expressed

$$\frac{1}{\gamma_{post}} = \frac{1}{\gamma_{prior}} + n\lambda \tag{6.26a}$$

$$\frac{m_{post}}{\gamma_{post}} = \frac{m_{priori}}{\gamma_{priori}} + n\lambda\overline{y} \tag{6.26b}$$

Recall that $1/\gamma = E[V(\mu)]/\text{Var}[\mu]$ is a measure of the relative precision in the distribution of μ relative to the precision in the sampling distribution of the Y's, i.e.,

$$\gamma_{\text{post}} = \frac{\text{Var}[\mu|\bar{y}]}{E[V(\mu)|\bar{y}]} \,. \tag{6.27}$$

Hence, (6.26a) may be formulated as: the relative precision in the posterior distribution of μ is the sum of the relative prior precision and the relative sampling precision.

▸ **Remark 6.6 – Limit of the posterior distribution for $n_i \to \infty$**
It follows from (6.26a) above that the relative precision in the posterior distribution, $\gamma_{post} \to 0$ for $n_i \to \infty$.   ◂

6.4 Examples of generalized one-way random effects models

Binomial Beta model

Example 6.3 – Defective items in a batch production
Table 6.3 on the following page shows the distribution of the number of defective lids found in samples of 770 lids from each of 229 lots. (Source, J. H. Ford (1951).)

In total we find $z_+ = \sum_{i=1}^{229} z_i = 254$ defective lids in the 229 samples, i.e., $\bar{z}_+ = 254/229 = 1.11$ defective per sample, so we have $\bar{h}_+ = 1.11/770 = 0.0014$ defective per lid.

As the lids in the sample were selected from the lot by simple random sampling, and the production was considered to be stable, one would consider the number of defective lids in the samples to be distributed like a binomial distribution with $n = 770$ and the same underlying fraction defective, p, estimated to be $\hat{p} = 0.0014$.

In Table 6.3 on the next page we have shown the corresponding frequencies assuming a binomial distribution of defective lids, and Figure 6.4 on the following page shows the corresponding histograms. It is seen that the observed distribution has *heavier tails* than the binomial distribution, indicating overdispersion. There is no reason to doubt the binomial sampling distribution assumption as the samples were selected at random from the batches. A possible explanation of the overdispersion is that the process fraction of defective items is varying from batch to batch. Therefore, we will choose to model the variation between batches as random and described by the natural conjugate distribution.

Table 6.3: *Distribution of number of defective lids in samples of 770 lids from each of 229 lots.*

# lids defective, z_i	# samples	B(770, 0.0014), expected	Pl(770, 0.460, 319.1), expected
0	131	75.47	130.64
1	38	83.83	42.52
2	28	46.50	21.96
3	11	17.17	12.74
4	4	4.75	7.79
5	5	1.05	4.91
6	5	0.19	3.16
7	2	0.03	2.06
8	3	–	1.36
9	2	–	0.90
Total	229		

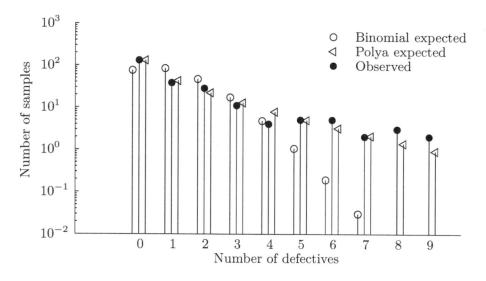

Figure 6.4: *Observed distribution of number of defective lids in samples of 770 lids from each of 229 productions compared with binomial and Polya distributions.*

Hierarchical Binomial-Beta distribution model

See also Example 11.1 in Wasserman (2004). The natural conjugate distribution to the binomial is a Beta-distribution of p (see Wassermann Section 2.4 and 3.4).

We have:

THEOREM 6.5
Consider the generalized one-way random effects model for $Z_1, Z_2, \ldots, Z_k$ given by

$$Z_i|p_i \sim \mathrm{B}(n, p_i) \tag{6.28a}$$

$$p_i \sim \mathrm{Beta}(\alpha, \beta) \tag{6.28b}$$

i.e., the conditional distribution of Z_i given p_i is a Binomial distribution, and the distribution of the mean value p_i is a Beta distribution. Then the marginal distribution of Z_i is a Polya distribution *with probability function*

$$\mathrm{P}[Z = z] = g_Z(z) = \binom{n}{z} \frac{\Gamma(\alpha + x)}{\Gamma(\alpha)} \frac{\Gamma(\beta + n - z)}{\Gamma(\beta)} \frac{\Gamma(\alpha + \beta)}{\Gamma(\alpha + \beta + n)} \tag{6.29}$$

for $z = 0, 1, 2, \ldots, n$.

The Polya distribution is named after the Hungarian mathematician G. Polya, who first described this distribution – although in another context. Instead of m, we shall use the symbol π to denote $\mathrm{E}[p]$. Thus, we shall use a parameterization given by

$$\pi = \frac{\alpha}{\alpha + \beta}, \qquad \gamma = \frac{1}{\alpha + \beta} \tag{6.30}$$

as

$$\mathrm{E}[p] = \pi = \frac{\alpha}{\alpha + \beta} \tag{6.31a}$$

$$\mathrm{Var}[p] = \frac{\alpha\beta}{(\alpha + \beta)^2(\alpha + \beta + 1)} = \frac{\pi(1 - \pi)}{\alpha + \beta + 1} \tag{6.31b}$$

$$\mathrm{E}[V(p)] = \mathrm{E}[p(1 - p)] = \frac{\pi(1 - \pi)}{1 + \gamma} \tag{6.31c}$$

The moments in the marginal distribution of Z are

$$\mathrm{E}[Z] = n\pi \tag{6.32a}$$

$$\mathrm{Var}[Z] = n\frac{\pi(1 - \pi)}{1 + \gamma}[1 + n\gamma] \tag{6.32b}$$

and the moments for the fraction $Y = Z/n$ are

$$\mathrm{E}[Y] = \pi \tag{6.33a}$$

$$\mathrm{Var}[Y] = \frac{\pi(1 - \pi)}{1 + \gamma}\left(\gamma + \frac{1}{n}\right). \tag{6.33b}$$

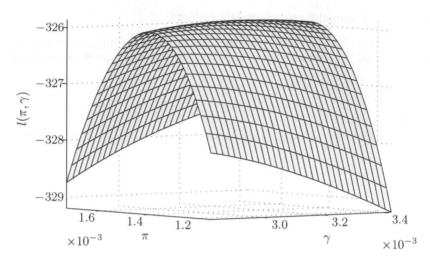

Figure 6.5: *Surface of log-likelihood for π and γ in beta-binomial model for number of defective lids corresponding to the data in Table 6.3 on page 238.*

Example 6.4 – Defective items in a batch production (continued)
Let $\overline{Y}_i = Z_i/770$. Then we find

$$s_2^2 = \sum_i = 1^{229}770(\overline{y}_i - \overline{\overline{y}})^2/228 = 0.00449.$$

Using the method of moments to estimate π and γ we find $\widetilde{\pi} = \overline{\overline{y}} = 0.0014$ and

$$\frac{\pi(1-\pi)}{1+\gamma}(1+770\gamma)$$

leading to $\widetilde{\gamma} = 0.00275$.

The log-likelihood surface is shown in Figure 6.5. The maximum likelihood estimate is found to be $\widehat{\pi} = 0.0014$ and $\widehat{\gamma} = 0.00313$.

Example 6.5 – Defective items in a batch production (continued)
Figure 6.6 shows the simultaneous distribution of number of defective lids in the sample and process fraction defective p. The conditional distribution of p for a given number of defectives in the sample may, by normalizing, be derived from the vertical walls.

Estimation of individual group means

THEOREM 6.6 – CONDITIONAL DISTRIBUTION OF p FOR GIVEN z
Consider the hierarchical binomial beta model in (6.28a) and assume that a value $Z = z$ has been observed. Then the conditional distribution of p is a Beta distribution

$$p|Z = z \sim \text{Beta}(\alpha + z, \beta + n - z) \tag{6.34}$$

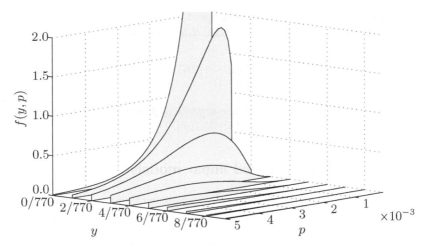

Figure 6.6: *Joint distribution of fraction of non-forming items in sample, y, and process fraction defective, p, in beta-binomial model for number of defective lids corresponding to the data in Table 6.3.*

with mean

$$E[p|Z = z] = \frac{\alpha + z}{\alpha + \beta + n}.$$ (6.35)

▸ **Remark 6.7 – The conditional mean as a weighted average**
We find the updating formulae

$$\frac{1}{\gamma_{post}} = \frac{1}{\gamma_{prior}} + n$$ (6.36a)

$$\pi_{post} = \frac{\alpha + z}{\alpha + \beta + n} = w\pi_{prior} + (1 - w)\frac{z}{n}$$ (6.36b)

with

$$w = \frac{\alpha + \beta}{\alpha + \beta + n} = \frac{1/\gamma_{prior}}{1/\gamma_{prior} + n}.$$ ◂

Normal distributions with random variance

As a non-trivial example of a hierarchical distribution we consider the hierarchical normal distribution model with random variance:

THEOREM 6.7
Consider a generalized one-way random effects model specified by

$$Y_i|\sigma_i^2 \sim N(\mu, \sigma_i^2)$$ (6.37a)

$$1/\sigma_i^2 \sim G(\alpha, 1/\beta)$$ (6.37b)

where σ_i^2 are mutually independent for $i = 1, \ldots, k$.

The marginal distribution of Y_i under this model is

$$\frac{Y_i - \mu}{\sqrt{\beta/\alpha}} \sim \mathrm{t}(2\alpha)$$

where $\mathrm{t}(2\alpha)$ is a t-distribution with 2α degrees of freedom, i.e., a distribution with heavier tails than the normal distribution.

6.5   Hierarchical generalized linear models

The model class was formulated by Lee and Nelder (1996) as a natural generalization of the generalized linear models to also incorporate random effects.

DEFINITION 6.3 – HIERARCHICAL GENERALIZED LINEAR MODEL
Consider a set of observations $\boldsymbol{Y} = (Y_1, Y_2, \ldots, Y_k)^T$ such that for a given value of a parameter θ the distribution of Y_i is given by an exponential dispersion model with density (4.2) and with canonical parameter space Ω (for θ), mean value $\mu = \kappa'(\theta)$, mean value space $\mathcal{M}$ (for μ) and canonical link $\theta = g(\mu)$.

Let the *conjugate distribution* of θ be given by (6.12), and the corresponding conjugate distribution of μ, (e.g., $f_Y(y)$ Poisson-distribution; $g_\mu(\mu)$ Gamma-distribution; link= log, i.e., $g(\mu) = \log(\mu)$).

The variables in a hierarchical generalized linear model are

  i) the observed responses $y_1, y_2, \ldots, y_k$ $(\in \mathcal{M})$

 ii) the unobserved state variables $u_1, u_2, \ldots, u_q$ $(\in \mathcal{M})$

iii) and the corresponding unobserved canonical variables $v_i = g(u_i)$. $(\in \Omega)$

The *linear predictor* is of the form

$$\boldsymbol{\theta} = g(\boldsymbol{\mu}|\boldsymbol{v}) = \boldsymbol{X}\boldsymbol{\beta} + \boldsymbol{Z}\boldsymbol{v} \tag{6.38}$$

in analogy with (5.42).

The distribution of $\boldsymbol{V} \in \Omega$ is a conjugated distribution to the canonical parameter θ. The derived distribution of $\boldsymbol{U} \in \mathcal{M}$ is the corresponding conjugated distribution to the mean value parameter μ such that $\mathrm{E}[\boldsymbol{U}] = \psi$

Example 6.6 – Poisson-gamma hierarchical generalized linear model
When the conditional distribution of $Y|\mu$ is a Poisson distribution, the distribution of V is constructed in such a way that the distribution of $U = \log(V)$ is a gamma distribution with mean value $\mathrm{E}[U] = \psi = 1$. It then follows that the distribution of Y is a negative binomial distribution with parameters determined by $\boldsymbol{X}\boldsymbol{\beta}$ and $\boldsymbol{Z}\boldsymbol{v}$.

▸ **Remark 6.8 – Estimation in a hierarchical generalized linear model**
Lee and Nelder (2001) have shown how the estimation can be performed by
an extended generalized linear model for y and ψ.

The mean value parameters $\boldsymbol{\beta}$ and $\boldsymbol{v}$ may be estimated by an iterative
procedure solving

$$\begin{pmatrix} X'\Sigma_0^{-1}X & X'\Sigma_0^{-1}Z \\ Z'\Sigma_0^{-1}X & Z'\Sigma_0^{-1}Z + \Sigma_1^{-1} \end{pmatrix} \begin{pmatrix} \widehat{\boldsymbol{\beta}} \\ \widehat{\boldsymbol{v}} \end{pmatrix} = \begin{pmatrix} X'\Sigma_0^{-1}z_0 \\ Z'\Sigma_0^{-1}z_0 + \Sigma_1^{-1}z_1 \end{pmatrix}$$

where $z_0 = \eta_0 + (\boldsymbol{y} - \boldsymbol{\mu}_0)(\partial\eta_0/\partial\mu_0)$ and $z_1 = \boldsymbol{v}_1 + (\psi_1) - u)(dv_1/du)$ are
the adjusted dependent variables for the distribution of y given v and of v,
respectively, in analogy with the estimation in generalized linear models.

In the one-way random effect models we have considered in the examples
the hierarchical likelihood estimates of group means are the empirical Bayes
estimates derived in the examples. ◂

The so-called *generalized linear mixed models* are also based on the linear
predictor (6.38), but instead of using the conjugate distribution, it is assumed
that V is normally distributed.

6.6 Problems

Exercise 6.1
Table 6.4 shows the number of times the air condition system failed in 10
airplanes during 1000 flight hours. The data is taken from Proschan (1963).

Table 6.4: *Number of times the air condition system failed in 10 airplanes during
1000 flight hours.*

Number of failures									
1	2	3	4	5	6	7	8	9	10
8	16	9	6	10	13	16	4	9	12

Question 1 Calculate $\bar{y}_+$ and s^2.

Question 2 Assume that $\mu \sim G(m/\gamma, \gamma)$. Find the moment estimates for m
and γ.

CHAPTER 7

Real life inspired problems

The following problems are considered in the chapter:

1) Dioxin emission Large numbers of experiments have been conducted at a number of Danish municipal solid waste (MSW) incinerator plants. During these experiments gas samples were collected and the dioxin emission estimated in the samples. Likewise a large number of possible explanatory variables were measured. The goal of the analysis is to build a model for the variations observed in the reported dioxin emission.

2) Used cars The prices of two car makes, VW and Ford Escort, listed for sale in a Danish newspaper during two weeks in 1999 have been recorded. Production year and mileage were also recorded. The goal of the analysis is to make a model to predict the depreciation of the value of VW and Ford Escort cars as they get older.

3) Young fish in the North Sea Fish of the tobis species were caught in September 1997 and 1998 at three specific locations in the North Sea and number of one-year-old fish was counted in the three samples. The goal of the analysis is to model the odds for the proportion of one-year-old fish for the sampled locations and years.

4) Traffic accidents In order to monitor road safety, police records of traffic accidents involving motor vehicles have been collected by Statistics Denmark for the years 1987–1991. In 1987 the Danish parliament passed a law making the use of headlights on cars in daylight mandatory from October 1st 1990. The purpose of the law was to reduce the number of head-on traffic accidents. The goal of the analysis is to describe the development of number of accidents during the given period and to test whether the law led to significant reduction in accidents.

5) Snail mortality An experiment has been conducted for studying the survival of snails. Groups of snails of two types, A and B, were held for periods of 1, 2, 3, or 4 weeks under controlled conditions. Temperature and humidity were kept at predefined levels. The main goal is to find out whether exposure, humidity, temperature or interactions between these have any effect on the survival probability for snails.

7.1 Dioxin emission

Background

Dioxin is the name generally given to a family of 75 different chemical compounds with a similar chemical structure and a set of biological effects, though their potency varies widely. Dioxins are organochlorines or compounds formed when a chlorine molecule binds with a carbon molecule during combustion or in some type of industrial production process. Dioxins cause cancer in humans and animals. Animal tests also show interference with reproduction, development and immune system function at low doses. This raises substantial concern about current human exposure levels.

Municipal solid waste and medical waste incinerators are the most significant sources of dioxin in the environment collectively accounting for roughly 85–87% of all known dioxin emission. It is, therefore, of interest to build a model for the variation of measured dioxin emission at a number of MSW incinerator plants. It is of particular interest to investigate whether the operating conditions influence the dioxin emission.

In order to create the data needed for setting up such a model, a large number of experiments have been conducted at a number of Danish MSW incinerator plants. During these experiments gas samples have been collected and the dioxin emission estimated in the samples. Likewise, a large number of possible explanatory variables have been measured. In planning the experiments, care must be taken to ensure that useful models and reliable conclusions can be formulated. Furthermore, dioxin measurements are difficult to obtain and the analyses are expensive. In order to ensure reliable conclusions and to obtain maximum information from the data, statistical designed experiments have been used. The experiments have been conducted at three Danish MSW plants. For one of the plants the experiment was repeated at a later time. The layout of an experiment is shown in Figure 7.1.

Data

The data can be found in the file `dioxin.csv`. The first row of the file contains the name of the variables referred to in the text. The data can be imported into R with the following command:

```
dat<-read.table("dioxin.csv",sep=',',head=T)
```

Dependent variable

The dependent variable is the concentration of dioxin in the combustion air. More precisely the concentration is measured as the total amount of dioxin in ng/m^3. In the data file the dependent variable is called DIOX. In the analysis, you should consider transforming the dependent variable.

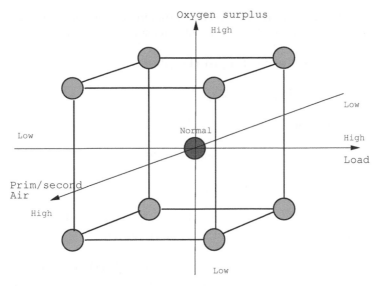

Figure 7.1: *Experimental setup corresponding to a single visit at a MSW plant.*

Explanatory variables

The total set of explanatory variables is most conveniently divided into block effects, active variables and passive variables. The block effects represent some conditions which we need to take into account in the modeling, the active variables are those varied according to the experimental plan and the passive variables are other measured variables which might influence the dependent variable.

In order to obtain the optimal amount of information, each measurement corresponds to one of the corners in a cubus, as shown in Figure 7.1. The order of experiments within the cubus is randomized. Even though the experiment is planned as described above, it is often rather difficult in practice to obtain the desired values of the active variables. Therefore, it is more reasonable to use the actual measured values related to the design. The active variables are, therefore, given both as according to the plan and the actual measured values.

The **block effects** are:

- MSW Plant (**PLANT**). The experiments have been conducted at three Danish MSW plants named **RENO_N**, **KARA**, and **RENO_S**.

- Time (**TIME**). For the plant **RENO_N**, the experiment was repeated at a later time point (1,2).

- Laboratory (**LAB**). Two laboratories have been used for the analysis, one in Denmark (KK) and one in the US (USA). It is very difficult (and expensive) to measure the amount of dioxin. Hence the data is assumed to be encumbered with considerable measurement noise.

The **active variables according to the plan** are:

- Oxygen surplus in gas (OXYGEN)

- Plant load (LOAD)

- Air distribution (primary/secondary) (PRSEK)

The **actual measured active variables** are:

- Measured oxygen surplus is called (O2). In the data file a value of the oxygen corrected by the mean is given by (O2COR).

- The normalized measured plant load is called (NEFFEKT). It is defined as the difference between the actual effect and the mean effect divided by the mean effect. A value of NEFFEKT $= -0.15$ means that the load is 15% less than the design load.

- The ratio between the primary air and the secondary air is called (QRAT).

The **passive variables** are:

- Gas flow (QROEG) (m^3/h)

- Combustion chamber temperature (TOVN) (°C)

- Gas temperature (TROEG) (°C)

- Pressure in the chamber (POVN)

- CO_2 (CO2) (ppm)

- CO (CO) (ppm)

- SO_2 (SO2) (mg/m^3)

- HCl (HCL) (mg/m^3)

- H_2O (H2O) (%)

CO_2, CO, SO_2, HCl and H_2O are measured in the gas.

The observations are numbered by a unique number as represented by OBSERV.

Analysis

The goal of this analysis is to find a model for the variations observed in the reported dioxin emission. For the analysis use $\alpha = 5\%$. Particularly, the following questions/tasks should be answered/solved:

Question 1 Start with some preliminary/explorative analysis of the data by making some plots.

Question 2 Set up a simple additive model only with the active variables according to the plan and the block variables. Reduce the model, if possible.

Question 3 Set up a similar model as before, but now use the measured values of the active variables along with the block variables. Reduce the model, if possible. Does the dioxin emission depend on the operating conditions (O2COR, NEFFEKT or QRAT)? If "yes," how can the dioxin emission be reduced by changing these conditions? Do you see any differences between the considered MSW plants? What about the two laboratories?

Question 4 Using the model with the measured active variables, predict the dioxin emission in the first visit to the RENO_N plant, analyzed in the KK laboratory with O2COR = 0.5, NEFFEKT = -0.01 and QRAT = 0.5 (you should disregard values of variables that you have removed from the model). Give a 95% prediction interval as well.

Question 5 Set up a final model, this time with the passive variables as well. You may want to consider including higher order terms in the model in order to find the model that gives the most complete description of the variation in the dioxin emission. Use residual analysis to validate the model and check if some observations are particularly influential. Give estimates of the parameters in the model and their uncertainties.

7.2 Depreciation of used cars

Background

For a car buyer it is of interest to know the depreciation of the value of the car as it gets older. To throw some light on this question, a prospective car buyer records the prices of two car makes, VW Golf and Ford Escort, listed for sale in a Danish newspaper during two weeks in 1999.

Data

The dataset usedcars.cvs shows for each car the car maker (VW or Ford), the model (Golf or Escort), the type, production year, mileage and price (in DKK). Clearly, there are too few observations to reasonably assess the influence of the "type," hence, the variation due to the car type is included in the random variation. The data can be imported into R with the following command:

```
dat<-read.table("usedcars.txt",header=T)
```

Analysis

The goal of the analysis is to make a model to predict the depreciation of VW Golf and Ford Escort cars as they get older. In particular, you should answer/solve the following questions/tasks:

Question 3 Plot price vs age and fit a linear model for each car make. Consider the sign variation of the residuals. Does the point scatter look like a random scatter around the best fitting line?

Question 4 Plot price vs log-age and fit a linear model for each car make. Which of the models do you prefer: price vs age, or price vs log-age?

Question 5 Does it make sense to compare the R^2-values for the fit of the two models. If not, why not?

Question 6 Formulate a comprehensive model for the data, and try to reduce this model. Keep the following questions in mind:

- Does the mileage provide a reasonable improvement of the precision of the prediction?

- Consider the linear relation between price and log-age. Are the two lines parallel? Is there only one line?

- Give an interpretation of the yearly depreciation of the price as expressed by the final model with a linear relation between price and log-age. Does the depreciation depend on age? May it be assumed that the yearly depreciation is the same for the two car makes?

7.3   Young fish in the North Sea

Background

50 fish of the tobis species were caught in September 1997 and 1998 at three specific locations in the North sea and the number of one-year-old fish counted in the three samples. The results are shown in Table 7.1.

Analysis

It is of interest to model the odds for the proportion of one-year-old fish for the sampled locations and years. In particular you should answer/solve the following questions/tasks:

Question 1 Consider the following model:

$$\log\left(p_{ij}/(1 - p_{ij})\right) = \rho + \alpha_i + \beta_j$$

Table 7.1: *Number of, and fraction of one-year-old fish in the North Sea.*

Year	Location			Year	Location		
	A	B	C		A	B	C
1997	9	9	2	1997	0.18	0.18	0.04
1998	38	30	15	1998	0.76	0.60	0.30
Number of one-year-old fish				*Proportion of one-year-old fish, y_{ij}*			

where index i, $i = 1, 2$ represents the years 1997 and 1998, and index j, $j = 1, 2, 3$ represents the three locations A, B and C.

Can it be assumed that the fraction of one-year-old fish at the three locations is the same? What about the year?

Question 2 Estimate $\widehat{p}_{11}$, the proportion of one-year-old fish in location A in 1997 using the model.

Question 3 Calculate the deviance contribution, $w_{22}d(y_{22}; \widehat{p}_{22})$, for location B in 1998 using the model.

Question 4 It is given that the two locations A and B are in the same part of the North Sea but location C is in another part. Can it be assumed that the relationship between odds for the proportion of one-year-old fish is the same for the two locations, A and B, in 1997 and 1998?

7.4 Traffic accidents

Background

In order to monitor road safety, police records of traffic accidents involving motor vehicles are collected by Statistics Denmark. Each accident is classified into one of 10 major categories. Moreover, the time of the accident is recorded. In each quarter the recordings are summarized. The dataset accidents_1.txt is an example of these recordings for the category of accidents occurring in the years 1987–1990 and involving vehicles on the same road going in opposite directions. Only accidents during daylight hours are included.

In 1987 the Danish parliament passed a law making the use of headlights on cars during daylight mandatory from October 1st 1990. The purpose of the law was to reduce the number of head-on traffic accidents. Opponents of the law were concerned that this intervention would not have the intended effect and, therefore, it was agreed that the possible effect of this intervention should be assessed by the end of 1991 and the law renegotiated. Clearly, the legislation would only influence accidents occurring during daylight, and only the category of accidents that involves vehicles on the same road driving in opposite directions. The dataset accidents_2.csv contains similar data as `accidents_1.csv`, but the period has been extended to cover up to and including the last quarter of 1991. A new variable has also been included to indicate whether daytime headlights were mandatory, or not.

Data

The two datasets can be inported into R with the following commands:

```
dat1<-read.table("accidents_1.txt",header=T)
dat2<-read.table("accidents_2.txt",header=T)
```

Analysis

The goal of the analysis is to describe the development of number of accidents during the given period and to test whether the law led to significant reduction in accidents. In particular, the following questions/tasks should be answered/solved.

Question 1 Fit a generalized linear model (Poisson, log) for the number of accidents vs, e.g., traffic index (offset), year and quartal using the data in `accidents_1.csv`. Reduce the model, if possible. There might be some multicollinearity between quartal and traffic index.

Question 2 Try to fit a model using the number of injured instead of number of accidents using the data in `accidents_1.csv`. Why is the Poisson distribution not appropriate in this case?

Question 3 Fit a suitable model to the data given in `accidents_2.csv`. Observe that the coefficient to the "time-variable" is negative, indicating a negative trend in the number of accidents, although not significant. Do you want to delete the trend variable from the model, or do you want to keep it in the model?

Question 4 Test whether the effect of the intervention is significant.

Question 5 Estimate the reduction in the number of accidents in the year 1987 if the intervention had been in effect in 1987, and give a confidence interval for that number. (Use the law of error propagation to obtain an approximation.)

7.5   Mortality of snails

Background

A designed experiment has been conducted by the Zoology Department, The University of Adelaide for studying the survival of snails. Groups of snails of two types, A and B, were held for periods of 1, 2, 3, or 4 weeks under controlled conditions. Temperature and humidity were kept at predefined levels.

Data

The data can be found in the file snails.csv. The data can be imported into R using the following command:

```
dat<-read.table('snails.txt', header=T)
```

The variables are as follows:

species	Snail species, A or B
exposure	Exposure in weeks (1, 2, 3, or 4)
humidity	Relative humidity (four levels)
temp	Temperature in degrees Celsius (three levels)
deaths	Number of deaths
n	Number of snails exposed

Analysis

The main goal is to find out whether exposure, humidity, temperature or interactions between these have any effect on the survival probability for snails.

Question 1 Decide whether to include exposure, humidity and temp as factor variables or numeric variables.

Question 2 Which distribution is appropriate to use for modeling the survival probability?

Question 3 Fit a model to the data and reduce it if possible. When a suitable model has been found, then perform a residual diagnostic on the model.

APPENDIX A

Supplement on the law of error propagation

The law of error propagation, also known as the *Delta method*, is often used to derive approximate expressions for the moments of a random variable, Y which is a known function of one or more random variables.

A.1 Function of one random variable

Assume that the random variable Y is a known function, $Y = h(X)$ of the random variable X. Assume further that the variation of X around its mean value is so narrow that a Taylor expansion is applicable. A Taylor expansion around $\mu = \mathrm{E}[X]$ yields

$$Y = h(X) \approx h(\mu) + (X - \mu)h'(\mu)$$

and, hence,

$$\mathrm{E}[Y] = \mathrm{E}[h(X)] \approx h(\mu)$$

$$\mathrm{Var}[Y] = \mathrm{Var}[h(X)] \approx \left(h'(\mu)\right)^2 \mathrm{Var}[X]$$

A.2 Function of several random variables

Assume that the random variable Y is a known function of the random variables $(X_1, X_2, \ldots, X_k)$ with expectations $(\mu_1, \mu_2, \ldots, \mu_k)$, i.e.,

$$Y = h(X_1, X_2, \ldots, X_k)$$

The Taylor approximation around the expected values $(\mu_1, \mu_2, \ldots, \mu_k)$ is

$$Y = h(X_1, X_2, \ldots, X_k)$$
$$\approx h(\mu_1, \mu_2, \ldots, \mu_k) + (X_1 - \mu_1)\frac{\partial h}{\partial x_1} + \cdots + (X_k - \mu_k)\frac{\partial h}{\partial x_k}$$

where the partial derivatives are evaluated at $(\mu_1, \mu_2, \ldots, \mu_k)$.

Assuming that the variation of the X_i's around their mean value is so narrow that the Taylor approximation is applicable, we find

$$E[Y] \approx h(\mu_1, \mu_2, \ldots, \mu_k)$$

$$\mathrm{Var}[Y] \approx \left(\frac{\partial h}{\partial x_1}\right)^2 + \cdots + \left(\frac{\partial h}{\partial x_k}\right)^2 + 2\frac{\partial h}{\partial x_1}\frac{\partial h}{\partial x_2}\mathrm{Cov}[X_1, X_2]$$

$$+ \cdots + 2\frac{\partial h}{\partial x_{k-1}}\frac{\partial h}{\partial x_k}\mathrm{Cov}[X_{k-1}, X_k]$$

When the X_i's are independent (or just uncorrelated) we have

$$\mathrm{Var}[Y] \approx \left(\frac{\partial h}{\partial x_1}\right)^2 + \cdots + \left(\frac{\partial h}{\partial x_k}\right)^2$$

Some probability distributions

This section mainly serves as a reference, specifying notation, properties of moments, etc., related to the various distributions.

A main purpose is also to discuss the exponential family of distributions and explain how some of the individual densities can be written either as a member of the *natural exponential family* of distributions, i.e., of the form

$$f_Y(y; \theta) = c(y) \exp(\theta y - \kappa(\theta)), \quad \theta \in \Omega \tag{B.1}$$

or as a member of the *exponential dispersion family* of distributions, i.e., of the form

$$f_Y(y; \theta) = c(y, \lambda) \exp(\lambda\{\theta y - \kappa(\theta)\}) \tag{B.2}$$

where λ is the *index parameter*. For a further description of the above families we refer to Section 4.2 on page 90. An exponential dispersion family is just an indexed set of natural families, indexed by the index or precision parameter, $\lambda > 0$ and, therefore, this family shares fundamental properties with the natural exponential family.

Important members of the family of densities with a discrete support is described in Section B.1 to B.3, followed by members of the families of densities with a continuous support in Section B.4 to B.7. It is described how to formulate the individual densities as a member of the exponential families of densities, and in each case important aspects like the canonical parameter and the unit variance function are described.

Examples of applications of the distributions are briefly mentioned in order to indicate appropriate applications of the distribution models.

Table B.1 on the following page briefly summarizes densities and the first two moments for a number of the common and more fundamental distributions encountered in the book. Some of the derived distributions like the geometric, log-normal, χ^2, F, and t-distributions are not mentioned here but briefly described in specific sections.

The gamma-function is used to describe many of the densities, and hence this function is shown seperately in Section B.8.

Table B.1: *Density, support, mean value, and variance for a number of common and more fundamental distributions.*

Name & page		Support	Density	E[Y]	Var[Y]
Bernoulli Bern(p)	259	$0,1$ $p \in [0,1]$	$p^y(1-p)^{1-y}$	p	$p(1-p)$
Binomial Bin(n,p)	259	$0,1,2,\ldots,n$ $n \in \mathbf{N}, p \in [0,1]$	$\binom{n}{y} p^y(1-p)^{n-y}$	np	$np(1-p)$
Poisson P(λ)	262	$0,1,2,\ldots$ $\lambda \in \mathbb{R}_+$	$\dfrac{\lambda^y}{y!}\exp(-\lambda)$	λ	λ
Neg. Bin. NB(r,p)	264	$0,1,2,\ldots$ $r \in \mathbb{R}_+, p \in]0,1]$	$\binom{r+y-1}{y} p^r(1-p)^y$	$\dfrac{r(1-p)}{p}$	$\dfrac{r(1-p)}{p^2}$
Exponential Ex(β)	266	$\mathbb{R}_+$ $\beta \in \mathbb{R}_+$	$\dfrac{1}{\beta}\exp(-y/\beta)$	β	β^2
Gamma G(α,β)	266	$\mathbb{R}_+$ $\alpha \in \mathbb{R}_+, \beta \in \mathbb{R}_+$	$\dfrac{1}{\Gamma(\alpha)\beta}\left(\dfrac{y}{\beta}\right)^{\alpha-1}\exp(-y/\beta)$	$\alpha\beta$	$\alpha\beta^2$
Inverse Gaussian IG(μ,λ)	275	$\mathbb{R}_+$ $\mu \in \mathbb{R}_+, \lambda \in \mathbb{R}_+$	$\left(\dfrac{\lambda}{2\pi}\right)^{1/2}\exp\left\{-\dfrac{\lambda}{2\mu^2 y}(y-\mu)^2\right\}\dfrac{1}{y^{3/2}}$	μ	μ^3/λ
Normal N(μ,σ^2)	91	$\mathbb{R}$ $\mu \in \mathbb{R}, \sigma^2 \in \mathbb{R}_+$	$\dfrac{1}{\sigma\sqrt{2\pi}}\exp\left(-\dfrac{(y-\mu)^2}{2\sigma^2}\right)$	μ	σ^2

B.1   The binomial distribution model

A discrete random variable Z with support $0, 1, 2, \ldots, n$, is said to be distributed according to the binomial distribution model if the density function is

$$g(z) = \binom{n}{z} p^z (1-p)^{n-z} \tag{B.3}$$

The notation $Z \sim \mathrm{B}(n, p)$ is short for the above expression.

For $n = 1$ the binomial distribution describes a *Bernoulli trial*, and the distribution is called a *Bernoulli distribution* with density function

$$g(y) = p^y (1-p)^{1-y} = \begin{cases} 1-p & \text{for } y = 0 \\ p & \text{for } y = 1 \end{cases} \tag{B.4}$$

In short we write $Y \sim \mathrm{Bern}(p)$ or $Y \sim \mathrm{B}(1, p)$.

The binomial distribution model $Z \sim \mathrm{B}(n, p)$ models the distribution of the total number of "successes" in n independent Bernoulli trials where the probability of "success" in each trial is p.

The characteristic function for the $\mathrm{B}(n, p)$-distribution is

$$\phi(t) = [1 - p + p \exp(it)]^n.$$

If $Z \in \mathrm{B}(n, p)$, then we have that

$$\mathrm{E}[Z] = np \qquad \mathrm{Var}[Z] = np(1-p).$$

The binomial distribution as exponential dispersion model

For any fixed $n \in \mathbf{N}$, the family of binomial distributions, $\{\mathrm{B}(n, p)\}_{p \in]0,1[}$, is a natural exponential family with the canonical parameter $\theta = \log(p/(1-p))$, the canonical parameter space $D = \mathbb{R}$, and the cumulant generating function $\kappa(\theta) = n \log(1 + \exp(\theta))$.

Let Z be binomially distributed, $Z \in \mathrm{B}(n, p)$. The density function (B.3) of Z can be expressed as

$$g(z; p, n) = \binom{n}{z} (1-p)^n [p/(1-p)]^z, \quad \text{for } z \in \{0, 1, \ldots, n\} \tag{B.5}$$

By introducing

$$\theta = \log(p/(1-p)) \tag{B.6}$$

and

$$\kappa(\theta) = \log(1 + \exp(\theta))$$

it is seen that the density (B.5) is of the form (B.2) with θ and $\kappa(\cdot)$ as above, and the index parameter λ equal to the sample size parameter n, so that $\Lambda = \mathbf{N}$.

Table B.2: *The $B(n, p)$-distribution as an exponential dispersion model.*

Canonical parameter θ	Cumulant-generating-function $\kappa(\theta)$	Mean value mapping $\mu = \tau(\theta)$	Unit variance function $V_P(\mu)$	Index parameter λ
$\log(p/(1-p))$	$\log\left(1 + \exp(\theta)\right)$	$\exp(\theta)/[1 + \exp(\theta)]$	$\mu(1-\mu)$	n

Thus, the family of $B(n, p)$-distributions with $0 < p < 1$ and $n \in \mathbf{N}$ is an additive exponential dispersion model with canonical parameter $\theta = \log(p/(1-p))$ and unit cumulant generating function $\kappa(\theta) = \log(1 + \exp(\theta))$.

The family can be generated by adding n independent, identically Bernoulli distributed random variables, i.e., $B(1, p)$ distributed random variables and, thus, the Bernoulli distribution is considered an elementary density.

We have the mean value mapping

$$\mu = \tau(\theta) = \frac{\exp(\theta)}{1 + \exp(\theta)} = p$$

The parameter p is, thus, the mean value of a *unit observation*.

The canonical link, as defined in general in Definition 4.5 on page 95, is the *logit function*

$$\theta = \tau^{-1}(\mu) = \log\left(\frac{\mu}{1-\mu}\right) = \log\left(\frac{p}{1-p}\right)$$

The unit variance function, as defined in Definition 4.3 on page 92, is

$$V_{Bin}(\mu) = \tau'(\tau^{-1}(\mu)) = \mu(1-\mu) ,$$

which is the variance of a $B(1, \mu)$-distribution.

The most important parameters in the interpretation of the family of $B(n, p)$-distributions as an exponential dispersion model are given in Table B.2.

We note that, even though the binomial distribution is well defined and has finite moments for $p = 0$ and $p = 1$, the interpretation as an exponential dispersion model is only valid for $0 < p < 1$, corresponding to $\theta \in \mathbb{R}$. The degenerate distributions corresponding to the values $p = 0$ and $p = 1$ result from a compactification of $\mathbb{R}$ with the points $\theta = -\infty$ and $\theta = \infty$.

The unit deviance corresponding to the binomial distributions is

$$d(y; \mu) = 2\left\{y \log\left(\frac{y}{\mu}\right) + (1-y) \log\left(\frac{1-y}{1-\mu}\right)\right\} , \tag{B.7}$$

where $y = z/n$. See Definition 4.4 on page 93.

We can express the density function for the $B(n, p)$-distribution by the unit deviance in the following way:

$$f(z; \xi, n) = a^*(z; n) \exp\left(-n\left\{y \log\left(\frac{y}{\mu}\right) + (1-y) \log\left(\frac{1-y}{1-\mu}\right)\right\}\right) , \tag{B.8}$$

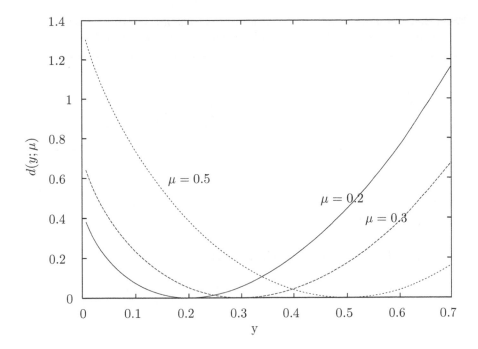

Figure B.1: *Unit deviance of the binomial distribution for some values of μ.*

where
$$a^*(z; n) = \binom{n}{z}$$

for $z = 0, 1, \ldots, n$ and where $y = z/n$ and $\xi = n\mu$. This is in accordance with the general discussion and remark related to (4.11) on page 93.

Figure B.1 shows the unit deviance corresponding to $\mu = 0.2$, $\mu = 0.3$ and $\mu = 0.5$.

Reproductive properties

The addition theorem is expressed as

THEOREM B.1 – SUM OF BINOMIAL DISTRIBUTED RANDOM VARIABLES
Let $Z_1, Z_2, \ldots, Z_k$ be mutually independent with $Z_i \in B(n_i, p)$, and let

$$Z_+ = Z_1 + Z_2 + \cdots + Z_k \, .$$

Then it holds that $Z_+ \in B(n_1 + \cdots + n_k, p)$.

The distribution of a sum of binomially distributed random variables with the same probability parameter, p, is, thus, another binomial distribution

Table B.3: *The $P(\mu)$-distribution as additive exponential dispersion model.*
* The index parameter and the mean value parameter cannot be readily separated.

Canonical parameter θ	Cumulant-generating-function $\kappa(\theta)$	Mean value mapping $\mu = \tau(\theta)$	Unit variance function $V_P(\mu)$	Index para-meter λ
$\log(\mu)$	$\exp(\theta)$	$\exp(\theta)$	μ	*

resulting from addition of the sample-size parameters, n_i. This also illustrates that the Bernoulli density $\mathrm{Bern}(p) = \mathrm{B}(1,p)$ is the elementary density by which all $\mathrm{B}(n,p)$ densities can be constructed.

B.2 The Poisson distribution model

A discrete random variable Y, that can assume the values $0, 1, 2, \ldots$ is said to follow a Poisson distribution with the parameter λ, if the density function for Y is of the form

$$g(y) = \frac{\lambda^y}{y!} e^{-\lambda} \tag{B.9}$$

for $y = 0, 1, 2, \ldots$, where $0 < \lambda$. The notation $Y \sim \mathrm{P}(\lambda)$ is short for the above. The characteristic function for the $\mathrm{P}(\lambda)$-distribution is

$$\phi(t) = \exp\left(\lambda e^{it} - 1\right) .$$

The mean and variance are

$$\mathrm{E}[Y] = \lambda \; ; \mathrm{Var}[Y] = \lambda .$$

Poisson distributions as exponential dispersion model

Let the random variable Z be Poisson distributed, $Z \sim \mathrm{P}(\mu)$.

The family of $\mathrm{P}(\mu)$-distributions for $\mu \in \mathbb{R}_+$ is a natural exponential family with canonical parameter $\theta = \log(\mu)$, canonical parameter space $\Theta = \mathbb{R}$, cumulant generating function $\kappa(\theta) = \exp(\theta)$, and mean value mapping $\tau(\theta) = \exp(\theta)$.

The additive exponential dispersion model generated by a $\mathrm{P}(\mu)$-distribution is precisely this natural exponential family.

The most important parameters in the interpretation of the family of $\mathrm{P}(\mu)$-distributions as an exponential dispersion model are given in Table B.3. The canonical link for the Poisson distribution is the logarithm function

$$\theta = \tau^{-1}(\mu) = \log(\mu)$$

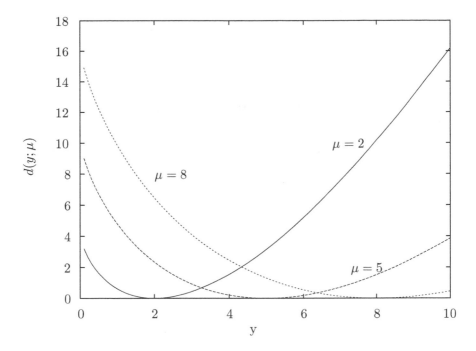

Figure B.2: *Unit deviance of the Poisson distribution for different values of μ.*

The unit deviance corresponding to the Poisson distribution is

$$d(y;\mu) = 2\left\{y\log\left(\frac{y}{\mu}\right) - (y-\mu)\right\}. \qquad (B.10)$$

Figure B.2 shows the unit deviance corresponding to $\mu = \{2,5,8\}$.

We note that no index parameter is included here. The family of Poisson distributions is the only exponential dispersion model with this property. This phenomenon can be explained by the fact that the variance function $V(\mu)$ for the Poisson distribution is the identity.

THEOREM B.2 – SUM OF POISSON DISTRIBUTED RANDOM VARIABLES
Let $Z_1, Z_2, \ldots, Z_k$ be independent with $Z_i \in P(\mu)$. Then the sum $Z_+ = Z_1 + \cdots + Z_k$ follows a $P(k\mu)$-distribution with expectation $\xi = k\mu$ and variance, $\mathrm{Var}[Z_+] = \xi$.

Hence the functional equation

$$\mathrm{Var}[Z] = kV(\xi/k)$$

becomes

$$\xi = k\xi/k \ ,$$

which is trivially satisfied for any factorization, $\xi = k\mu$.

Parameterization of Poisson distributions with mean values proportional to known values; offset of canonical parameter

At times (for example in the analysis of generalized models) it is necessary to parameterize the Poisson distribution so that the mean value is proportional to a known proportionality factor, w. Since the index parameter is undetermined, this cannot be achieved through the usual weighting. However, it can be achieved through an offset of the canonical parameter.

Let $Z \in P(w\mu)$. Then the density function for the distribution of Z can be expressed as

$$f^*(z; \theta, w) = \frac{w^z}{z!} \exp[\theta z - w \exp(\theta)] \ . \tag{B.11}$$

By changing the point of origin from θ to $\theta' = \theta + \log(w)$ the expression for the density becomes

$$f^*(z; \theta', w) = \frac{1}{z!} \exp[\theta' z - \exp(\theta')] \ .$$

This change of point of origin for the canonical parameter is also called an offset of the canonical parameter.

B.3   The negative binomial distribution model

A discrete random variable Z with support $0, 1, 2, \ldots$, is said to be distributed according to the negative binomial distribution model if the probability mass function for Z is

$$g(z) = \binom{z + r - 1}{z} (1 - p)^r p^z \ . \tag{B.12}$$

Short notation for this is $Z \sim \mathrm{NB}(r, p)$. If Z is a discrete random variable with $Z \sim \mathrm{NB}(r, p)$, then

$$\mathrm{E}[Z] = \frac{rp}{1 - p} \ ; \mathrm{Var}[Z] = \frac{rp}{(1 - p)^2}$$

The negative binomial distribution describes the "waiting time", i.e., the number of "successes" before the r'th failure in a sequence of independent Bernoulli trials where the probability of success in each trial is p.

The negative binomial distribution corresponding to $r = 1$ is also called the *geometric distribution*. A short notation for the geometric distribution is $Z \sim \mathrm{Geo}(p)$.

Using the relation

$$\binom{y}{x} = \frac{\Gamma(y+1)}{\Gamma(y+1-x)x!} = \frac{y(y-1)\cdots(y-x+1)}{x!}$$

which is valid for y real, and x an integer, it is seen that we can write

$$\binom{r+x-1}{x} = \frac{\Gamma(r+x)}{\Gamma(r)x!},$$

which illustrates how the definition of the negative binomial distribution may be extended to real, positive values of r. The gamma function, $\Gamma(\cdot)$ is further described in Section B.8 on page 284.

The negative binomial distribution as exponential dispersion family

In connection with the treatment of negative binomial distributions as an exponential dispersion model we will employ a parameterization a little different from what is typically seen. To distinguish between the two parameterizations, the designation NB^* will be used to denote the additive dispersion model generated by the corresponding $Geo^*(p)$-distribution.

A discrete random variable, Z which can take on the values $0, 1, \ldots$, is said to follow a negative binomial distribution with the parameters α and p if the distribution of Z is described by the density function

$$g(z; \alpha, p) = \frac{\Gamma(\alpha+z)}{\Gamma(\alpha)\, z!}\, p^\alpha (1-p)^z \quad \text{for} \quad z \in \{0, 1, \ldots, \}, \qquad (B.13)$$

where $\alpha \in \mathbb{R}_+$ and $0 \le p \le 1$.

We use the notation $Z \in NB^*(\alpha, p)$ when the density function for the distribution of Z has the form (B.13).

For integer values of α we have that

$$Z \in NB_{traditional}(\alpha, 1-p) \Longleftrightarrow Z \in NB^*(\alpha, p)$$

Note that as is the case for the $Geo^*(p)$-distribution, the $NB^*(\alpha, p)$-distribution is parameterized by the probability p of the event for which the number of realizations is being counted, rather than the number of failures.

The family of $NB^*(\alpha, p)$-distributions for $\alpha \in \mathbb{R}_+$ and $0 < p < 1$ is precisely the additive exponential dispersion model generated by a $Geo^*(p)$-distribution. The canonical parameter is $\vartheta = \log(p)$, index set $\Lambda = \mathbb{R}_+$, and canonical parameter space $\Theta =]-\infty, 0[$. Note that the exponential dispersion model does not include the distributions corresponding to $p = 0$ and $p = 1$.

The most important quantities in the interpretation of the family of $NB^*(\alpha, p)$-distributions as an additive exponential dispersion model are provided in Table B.4.

Table B.4: *The NB*$^*(\alpha, p)$*-distribution as additive exponential dispersion model*

Canonical parameter θ	Cumulant-generating function $\kappa(\theta)$	Mean value mapping $\mu = \tau(\theta)$	Unit variance function $V_{NB}(\mu)$	Index parameter λ
$\log(p)$	$-\log\left(1 - \exp(\theta)\right)$	$\exp(\theta)/[1 - \exp(\theta)]$	$\mu(1 + \mu)$	α

Specifically, the mean value and variance for the NB(α, p)-distribution are found to be

$$\mathrm{E}[Z] = \alpha\mu ; \quad \mathrm{Var}[Z] = \alpha V_{NB}(\mu) = \alpha\mu(1 + \mu) ,$$

where $\mu = p/(1 - p)$.

The canonical link for the negative binomial distribution is

$$\theta = \tau^{-1}(\mu) = \log\left(\frac{\mu}{1 + \mu}\right)$$

The unit deviance corresponding to the NB$^*(\alpha, p)$-distribution is

$$d(y; \mu) = 2\left\{y\log\left(\frac{y(1 + \mu)}{(1 + y)\mu}\right) + \log\left(\frac{1 + \mu}{1 + y}\right)\right\} , \qquad (B.14)$$

where $y = z/\alpha$.

The density for the NB(α, p)-distribution can be expressed through the unit deviance, in accordance with (4.11), as

$$f(z; \xi, \alpha) = a^*(z; \alpha) \exp\left[-\alpha\left\{y\log\left(\frac{y(1 + \mu)}{(1 + y)\mu}\right) + \log\left(\frac{1 + \mu}{1 + y}\right)\right\}\right], \quad (B.15)$$

where

$$a^*(z; \alpha) = \frac{\Gamma(\alpha + z)}{\Gamma(\alpha)z!}$$

for $z = 0, 1, \ldots, n$, and where $y = z/\alpha$ and $\xi = \alpha\mu$.

B.4   The exponential distribution model

The continuous counterpart to the geometric distribution is the *exponential distribution*.

A continuous random variable X, which can take on all real non-negative values, is said to follow the exponential distribution with parameter β if the distribution of X is of the form

$$g(x) = \frac{1}{\beta} \cdot \exp(-x/\beta) \text{ for } x \in \mathbb{R}_+ , \qquad (B.16)$$

where $\beta \in \mathbb{R}_+$. In short we write $X \in \text{Ex}(\beta)$.

The exponential distribution is typically used to describe the times between events in a *Poisson process*, i.e., a process in which the events occur continuously and independently at a constant rate $\lambda = \beta^{-1}$. Sometimes the exponential distribtion is parameterized alternatively using the rate parameter, λ.

The characteristic function is

$$\phi(t) = \frac{1}{1 - i\,\beta t}$$

Given $X \in \text{Ex}(\beta)$, and

$$Y = \xi X$$

where $0 < \xi$, then $Y \in \text{Ex}(\xi\beta)$ and, hence, it is seen that β is a scale parameter.

For $X \in \text{Ex}(\beta)$ we have

$$E[X] = \beta; \qquad \text{Var}[X] = \beta^2 \tag{B.17}$$

Exponential distribution as exponential family model

The family of $\text{Ex}(\beta)$ distributions for $0 < \beta < \infty$ is a natural exponential family with the canonical parameter $\vartheta = -1/\beta$, canonical parameter space $D =]-\infty, 0[$ and the cumulant generating function

$$\kappa(\vartheta) = \ln(-\vartheta) \,,$$

and, hence, $\tau(\vartheta) = -1/\vartheta$ and $\text{Var}[X] = 1/\vartheta^2$. The most commonly used parameterization is by the mean value parameter $\mu = E[X] = \beta = -1/\vartheta$.

Since $\vartheta = \tau^{-1}(\mu) = -1/\mu$ it is seen that the canonical link is the reciprocal function.

The variance is $\text{Var}[X] = \beta^2 = \mu^2$, i.e., the variance function is

$$V_G(\mu) = \mu^2$$

Reproductive properties

The exponential distribution is the fundamental distribution for the Gamma distribution (see Section B.5), since for a set of given independent variables $X_1, X_2, \ldots, X_k$, where $X_i \in \text{Ex}(\beta)$, for $i = 1, 2, \ldots, k$, then

$$Y = X_1 + X_2 + \cdots + X_k,$$

will be Gamma distributed; more specifically $Y \in G(k, \beta)$.

Assume that $X_1, X_2, \ldots$ is an infinite sequence of independent variables, where $X_i \in \text{Ex}(\beta)$, for $i = 1, 2, \ldots$. Then, if we consider

$$Y = \max\{n : X_1 + X_2 + \cdots + X_n \le t\},$$

we have that $Y \in P(t/\beta)$, i.e., if the waiting time between successive events are independent, then the number of events in a given time interval will be Poisson distributed with the intensity t/β. The parameter β is the expected waiting time between two successive events, and in some contexts the parameter β is thus called the Mean Time Between Failures (MTBF), and $1/\beta$ is called the failure-rate.

B.5   The gamma distribution model

A continuous random variable Y, which can take all real non-negative values, is said to follow the gamma distribution with parameters α and β if the distribution of Y can be described by a density function of the form

$$g(y) = \frac{1}{\beta\Gamma(\alpha)} \left(\frac{y}{\beta}\right)^{\alpha-1} \exp(-y/\beta) \text{ for } y \in \mathbb{R}_+ \qquad (B.18)$$

with $\alpha \in \mathbb{R}_+$ and $\beta \in \mathbb{R}_+$. Short notation for the above is $Y \sim G(\alpha, \beta)$. The parameter α is called the *shape parameter*, and β is a scale parameter for the distribution.

The characteristic function for the $G(\alpha, \beta)$-distribution is

$$\phi(t) = \frac{1}{(1 - it\beta)^\alpha}$$

If $Y \sim G(\alpha, \beta)$, then

$$E[Y] = \alpha\beta \; ; \operatorname{Var}[Y] = \alpha\beta^2$$

The $G(1, \beta)$-distribution is precisely the $Ex(\beta)$-distribution and, hence, the *exponential distribution*, $Ex(\beta)$, makes up a one-dimensional natural exponential family.

The $G(\nu/2, 2)$-distribution is the χ^2-*distribution with ν degrees of freedom*, which is well known within statistics. The χ^2-distribution is related to the normal distribution and is further described on page 281.

Figure B.3 shows the densities for gamma distributions with the same scale parameter, $\beta = 1$, but with different values of the shape parameter, α.

Cumulative distribution function and incomplete moments

The cumulative distribution function for the $G(k, \beta)$-distribution is expressed as

$$G(c; k, \beta) = \int_0^c \frac{1}{\beta\Gamma(k)} \left(\frac{x}{\beta}\right)^{k-1} \exp(-x/\beta)dx$$

Since β is a scale parameter, we have that

$$G(c; k, \beta) = G(c/k; 1) = \int_0^{c/\beta} \frac{1}{\Gamma(k)} x^{k-1} \exp(-x)dx = \frac{\Gamma_{c/\beta}(k)}{\Gamma(k)}$$

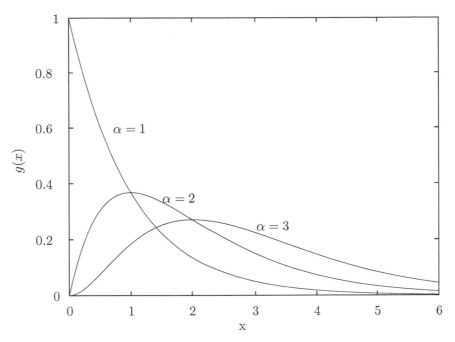

Figure B.3: *The density for the $G(\alpha, 1)$-distribution, corresponding to different values of the shape parameter α. Notice that the exponential distribution is obtained for $\alpha = 1$.*

where the incomplete gamma function $\Gamma_x(k)$ is defined by

$$\Gamma_x(k) = \int_0^x \frac{1}{\Gamma(k)}\, u^{k-1} \exp(-u)\, du$$

The gamma distribution as exponential family

By rewriting the gamma distribution (B.18), it is seen that the density for $G(\alpha, \beta)$-distribution can be written in the form

$$g(x) = \frac{1}{x} \exp\{\alpha \log(x) - \frac{x}{\beta} - \alpha \log(\beta) - \log(\Gamma(\alpha))\} \quad \text{for } x \in \mathbb{R}_+ \qquad \text{(B.19)}$$

with respect to the Lebesgue measure.

The family of gamma distributions is, thus, an exponential family of order 2 with canonical parameters $(\theta_1, \theta_2) = (\alpha, -1/\beta)$, canonical sample statistics $(t_1(x), t_2(x)) = (\log(x), x)$, canonical parameter space $\Omega = \mathbb{R}_+ \times]-\infty, 0[$, and with $\kappa(\theta) = \log \Gamma(\theta_1) - \theta_1 \log(-\theta_2)$, whereby

$$\tau(\theta) = \mathrm{E}\begin{pmatrix} \log(X) \\ X \end{pmatrix} = \begin{pmatrix} \Psi(\theta_1) - \log(\theta_2) \\ -\theta_1/\theta_2 \end{pmatrix}$$

and

$$\mathbf{V}(\boldsymbol{\theta}) = \mathrm{D}\begin{pmatrix} \log(X) \\ X \end{pmatrix} = \begin{pmatrix} \Psi'(\theta_1) & -1/\theta_2 \\ -1/\theta_2 & \theta_1/\theta_2^2 \end{pmatrix}$$

where $\Psi(\cdot)$ denotes the digamma function, $\Psi(x) = \Gamma'(x)/\Gamma(x)$.

The gamma distribution as exponential dispersion model

We remind the reader that the family of $\mathrm{Ex}(\beta)$-distributions makes up a one-dimensional natural exponential family. It follows from the addition properties for the exponential distribution (see Section B.4) that the additive exponential dispersion model generated by the exponential distribution is precisely the family of gamma distributions.

The index set is $\Lambda = \mathbb{R}_+$, the canonical parameter is $\theta = -1/\beta$, the unit cumulant generating function

$$\kappa(\theta) = \log(-\theta),$$

and the unit deviance function is

$$V_G(\mu) = \mu^2.$$

If $Z \sim \mathrm{G}(\alpha, \beta)$ is regarded as an additive dispersion model, we find that the density function for Z is of the form

$$f^*(z; \theta, \alpha) = \frac{1}{\Gamma(\alpha)} z^{\alpha-1} \exp\{\theta z + \alpha \log(-\theta)\} \qquad \text{(B.20)}$$

with $\theta < 0$. The expression is of the form (B.2), where the symbol α is used to denote the index parameter.

The additive model is mapped to a reproductive exponential dispersion model for Y by the usual transformation $Y = Z/\alpha$, which is just a scale transformation of the distribution. In this manner, the distribution of Y becomes a $\mathrm{G}(\alpha, \beta/\alpha)$-distribution.

If $Y \sim \mathrm{G}(\alpha, \beta/\alpha)$, then the density of Y can be expressed using the mean value parameter μ, as

$$f(y; \mu, \sigma^2) = h(y; \sigma^2) \exp\left\{ -\frac{1}{\sigma^2}\left(\frac{y}{\mu} + \log(\mu)\right)\right\} \qquad \text{(B.21)}$$

with

$$h(y; \sigma^2) = \frac{1}{\Gamma(1/\sigma^2)}\left(\frac{y}{\sigma^2}\right)^{1/\sigma^2} \frac{1}{y}$$

and where

$$\mu = \beta, \quad \sigma^2 = 1/\alpha$$

This is precisely of the form (4.11), as introduced on page 93.

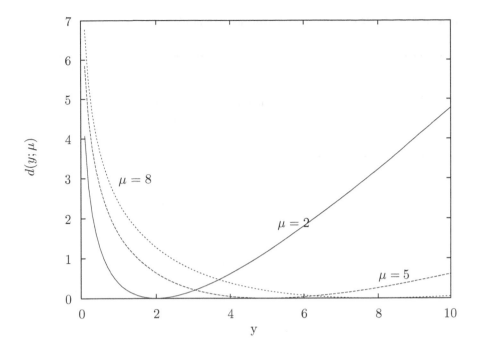

Figure B.4: *Unit deviance of the gamma distribution for different values of μ.*

The family of gamma distributions $G(\alpha, \beta/\alpha)$ with $\alpha \in \mathbb{R}_+$ and $\beta \in \mathbb{R}_+$ is a reproductive exponential dispersion model with canonical parameter $\theta = -1/\beta$ and dispersion parameter $\sigma^2 = 1/\alpha$.

The unit cumulant generating function is $\kappa(\theta) = -\log(-\theta)$, with mean value $\mu = \beta$, and variance

$$\mathrm{Var}[Y] = \sigma^2 V_G(\mu) = \sigma^2 \mu^2 = \frac{\beta^2}{\alpha}.$$

The unit deviance for the $G(\alpha, \beta/\alpha)$-distribution is

$$d(y; \mu) = 2\left(\frac{y}{\mu} - \log\left(\frac{y}{\mu}\right) - 1\right) \tag{B.22}$$

Figure B.4 shows the unit deviance corresponding to different values of the mean value μ. It is seen than the deviance curve is flatter for larger values of μ.

The density of a $G(\alpha, \beta/\alpha)$-distributed random variable in the form (4.11) can be expressed as

$$f(y; \mu, \sigma^2) = a(y; \sigma^2) \, \exp\left\{-\frac{1}{2\sigma^2}\left(\frac{y}{\mu} - \log\left(\frac{y}{\mu}\right) - 1\right)\right\}$$

Table B.5: *The $G(\alpha, \beta/\alpha)$-distribution as a reproductive exponential dispersion model.*

Canonical parameter θ	Cumulant generating function $\kappa(\theta)$	Mean value mapping $\mu = \tau(\theta)$	Unit deviance function- $V_G(\mu)$	Dispersion parameter σ^2
$-1/\beta$	$-\log(-\theta)$	$-1/\theta$	μ^2	$1/\alpha$

with

$$a(y; \sigma^2) = \frac{\exp(-1/\sigma^2)}{(\sigma^2)^{1/\sigma^2}\Gamma(1/\sigma^2)} \frac{1}{y}$$

The most important quantities in the interpretation of the family of $G(\alpha, \beta/\alpha)$-distributions as an exponential dispersion model are given in Table B.5.

The square root, σ, of the dispersion parameter is the coefficient of variation

$$\sigma = \frac{\sqrt{\mathrm{Var}[Y]}}{\mathrm{E}[Y]} = \frac{1}{\sqrt{\alpha}}$$

The shape parameter α in the gamma distribution expresses the precision $1/\sigma^2$ when the family of gamma distributions is interpreted as an exponential dispersion parameter family.

Figure B.5 shows densities of gamma distributions with the same mean value μ and different values of the dispersion parameter σ^2.

▸ **Remark B.1 – Relations between the usual parameterization and parameterization as a reproductive exponential dispersion model**
Since a number of results concerning the gamma distributions have a simple interpretation when represented as a reproductive exponential dispersion model, it is useful to be able to switch between the two parameterizations. If Y follows a reproductive dispersion model with mean value μ, variance function $V_G(\mu) = \mu^2$ and dispersion parameter $1/\sigma^2$, then we have $Y \sim G(1/\sigma^2, \mu\sigma^2)$. If, on the other hand, $Y \sim G(\alpha, \beta/\alpha)$, then Y follows a reproductive exponential dispersion model with mean value $\mu = \mathrm{E}[Y] = \beta$, unit variance function $V(\mu) = \mu^2$ and dispersion parameter $\sigma^2 = 1/\alpha$. If $Z \sim G(\alpha, \beta)$ then the distribution of Z follows an additive dispersion model with index parameter α, mean value $\mathrm{E}[Z] = \alpha\mu$, unit variance function $V(\mu) = \mu^2$ and $\mathrm{Var}[Z] = \alpha V(\mu)$. ◂

▸ **Remark B.2 – Dispersion is not changed by a scale transformation**
If Y follows a gamma distribution with mean value μ and dispersion parameter σ^2, then cY follows a gamma distribution with mean value $c\mu$ and unchanged

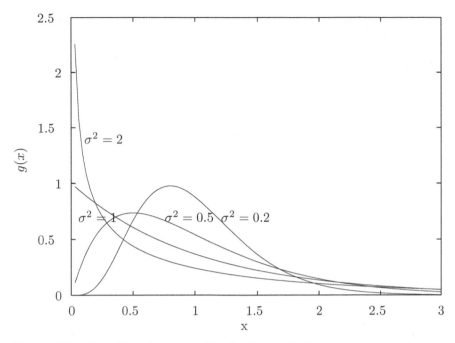

Figure B.5: *Densities of gamma distributions with the same mean value μ and different values of the dispersion parameter σ^2.*

dispersion parameter, σ^2. This result follows from the usual scale transformation property for the gamma distribution by noting that $Y \sim G(1/\sigma^2, \mu\sigma^2)$. Amongst other things, the result expresses that the shape parameter of the gamma distribution does not change by a scale transformation. ◄

Reproductive properties

THEOREM B.3 – ADDITION THEOREM FOR THE GAMMA DISTRIBUTION
Assuming that $Z_1, Z_2, \ldots, Z_k$ are independent random variables, where $Z_i \sim G(\alpha_i, \beta)$ for $i = 1, 2, \ldots, k$, and consider

$$Z_+ = X_1 + X_2 + \cdots + X_k,$$

then $Z_+ \sim G(\alpha_+, \beta)$, where $\alpha_+ = \alpha_1 + \alpha_2 + \cdots + \alpha_k$.

If the addends follow gamma distributions with the same scale parameter, then the sum follows a gamma distribution produced by adding the shape parameters.

In particular, we get the addition theorem for the $Ex(\beta)$-distribution by letting $\alpha_1 = \alpha_2 = \cdots = \alpha_k = 1$. Thus, the $G(\alpha, \beta)$-distribution can be understood as the sum of α independent $Ex(\beta)$-distributed variables for integer values of α. The $Ex(\beta)$-distribution is said to be the elementary distribution for the gamma distribution.

If expressed in terms of the reproductive exponential dispersion model, we get the general reproductive property:

THEOREM B.4 – CONVOLUTION (ADDITION) OF DISTRIBUTIONS WITH EQUAL MEAN VALUES

Let $Y_1, \ldots, Y_k$ be independent, and let Y_i follow a $G(w_i/\sigma^2, \mu\sigma^2/w_i)$-distribution and consider

$$\overline{Y}_w = \frac{1}{w_{tot}} \sum_{i=1}^{k} w_i Y_i$$

with

$$w_{tot} = \sum_{i=1}^{k} w_i$$

then $\overline{Y}_w$ follows a $G(w_{tot}/\sigma^2, \mu\sigma^2/w_{tot})$-distribution.

It is precisely this result that is used when estimating the common variance, s_+^2, of a normal distribution based on a number of independent estimates s_i^2.

Estimation in the gamma distribution

THEOREM B.5 – MAXIMUM LIKELIHOOD ESTIMATORS FOR INDEPENDENT, IDENTICALLY DISTRIBUTED OBSERVATIONS

Let $Y_1, Y_2, \ldots, Y_k$ be independent, and let Y_i follow a $G(1/\sigma^2, \mu\sigma^2)$-distribution, then the maximum likelihood estimator for μ is

$$\widehat{\mu} = \overline{Y}$$

and the maximum likelihood estimator for σ^2 is found as the solution of

$$\Psi\left(\frac{1}{\sigma^2}\right) - \log\left(\frac{1}{\sigma^2}\right) = \log\left(\overline{Y}/\overline{Y}_G\right),$$

where $\Psi(\cdot)$ denotes the digamma function (see page 270), $\overline{Y}$ denotes the arithmetic mean of Y_i, $i = 1, \ldots, k$ and $\overline{Y}_G$ denotes the geometric mean of the Y's,

$$\overline{Y}_G = \left(\prod_{i=1}^{k} Y_i\right)^{1/k}$$

Proof See Cox and Lewis (1966).   ∎

B.6 The inverse Gaussian distribution model

We will briefly introduce the family of inverse Gaussian distributions (also known as the Wald distribution), a family of distributions with support from the non-negative real numbers.

A continuous random variable X which can assume all real non-negative values, is said to follow the inverse Gaussian distribution with parameters μ and λ if the density of X with respect to the Lebesgue measure is of the form

$$g(x; \mu, \lambda) = \left(\frac{\lambda}{2\pi}\right)^{1/2} \exp\left\{-\frac{\lambda}{2\mu^2 x}(x - \mu)^2\right\} \frac{1}{x^{3/2}} \quad \text{for } x \in \mathbb{R}_+ \quad \text{(B.23)}$$

where $\mu \in \mathbb{R}_+$ and $\lambda \in \mathbb{R}_+$. Short notation for the above is $X \sim \text{IG}(\mu, \lambda)$.

The parameter λ can be expressed as

$$\lambda = \text{E}\left[\frac{1}{X}\right] - \frac{1}{\text{E}[X]}$$

The density (B.23) is unimodal with mode

$$x_m = -\frac{3\mu^2}{2\lambda} + \mu\left(1 + \frac{9\mu^2}{4\lambda^2}\right)^{1/2}$$

The term "inverse" originates from the fact that while the Gaussian distribution describes the distance at a fixed time of a Brownian motion, the inverse Gaussian distribution describes the *time* a Brownian motion takes to reach a fixed distance.

Estimation of the parameters μ and λ

If $X_1, X_2, \ldots, X_n$ are independent and $X_i \sim \text{IG}(\mu, \lambda)$, then the maximum likelihood estimates $\widehat{\mu}$ and $\widehat{\lambda}$ for μ and λ are given by

$$\widehat{\mu} = \overline{X}.$$

$$\frac{1}{\widehat{\lambda}} = \frac{1}{n}\sum_{i=1}^{n}\left(\frac{1}{X_i} - \frac{1}{\overline{X}.}\right)$$

with $\overline{X}. = \sum X_i/n$.

The estimators $\widehat{\mu}$ and $\widehat{\lambda}$ are independent, and it can be shown (Tweedie (1957)) that

$$\widehat{\mu} \sim \text{IG}(\mu, n\lambda) \quad \text{(B.24)}$$

$$\frac{n\lambda}{\widehat{\lambda}} \sim \chi^2(n-1) \quad \text{(B.25)}$$

Note the similarity to the distribution of $\overline{X}.$ and S^2 for normally distributed random variables.

Alternative parameterization of the inverse Gaussian distribution

By setting $\omega = \lambda/\mu$ in (B.23), we find the following expression for the density

$$f(x; \mu, \omega) = \left(\frac{\mu\omega}{2\pi x^3}\right)^{1/2} \exp\left\{-\frac{\omega}{2}\left(\frac{x}{\mu} + \frac{\mu}{x}\right) + \omega\right\} \qquad (B.26)$$

which shows that the parameter $\omega = \lambda/\mu$ is a *shape parameter* for the distribution, and that μ is a *scale parameter*. With this parameterization, the mode becomes

$$x_m = -\mu\left[\frac{3}{2\omega} + \sqrt{1 - \left(\frac{3}{2\omega}\right)^2}\right].$$

Since the parameterization in μ and λ is often used in the literature, we have chosen to keep this parameterization instead of the more convenient parameterization in the shape and scale parameters.

Figure B.6 shows densities of some inverse Gaussian distributions with the same mean value $\mu = 1$ and various values of ω.

THEOREM B.6 – MOMENTS FOR THE $IG(\mu, \lambda)$ DISTRIBUTION
Let $X \sim IG(\mu, \lambda)$. Then the characteristic function is

$$\phi(t) = \exp\left[\frac{\lambda}{\mu}\left\{1 - \left(1 + \frac{2\mu^2}{\lambda}it\right)^{1/2}\right\}\right]$$

The mean value and variance are

$$E[X] = \mu; \quad \text{Var}[X] = \mu^3/\lambda \qquad (B.27)$$

Proof Follows trivially.   ∎

Since it is often of interest to consider the reciprocal value of an IG-distributed variable, we state the first moments of the reciprocal variable:

THEOREM B.7 – RECIPROCAL MOMENTS FOR $IG(\mu, \lambda)$ DISTRIBUTION
Let $X \sim IG(\mu, \lambda)$. Then it holds that

$$E[1/X] = \frac{1}{\mu} + \frac{1}{\lambda}; \quad \text{Var}[1/X] = \frac{\lambda + 2\mu}{\mu\lambda^2} \qquad (B.28)$$

Proof Follows trivially.   ∎

Expressed through the shape parameter $\omega = \lambda/\mu$ and the scale parameter μ, we find the expressions for the moments

$$E[X] = \mu \qquad\qquad E[1/X] = \frac{\omega + 1}{\mu\omega}$$

$$\text{Var}[X] = \mu^2/\omega \quad \text{Var}[1/X] = \frac{\omega + 2}{\mu^2\omega^2}$$

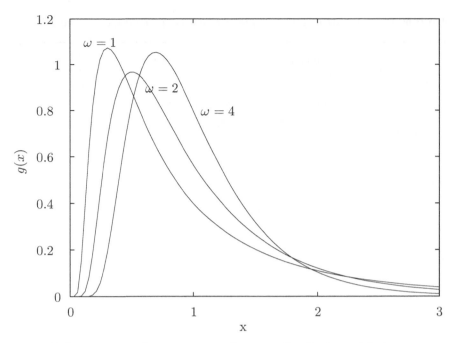

Figure B.6: *Densities of some inverse Gaussian distributions with the same mean value $\mu = 1$ and some values of ω.*

Cumulative distribution function

The cumulative distribution function for the IG-distribution can be determined through the cumulative distribution function for the standard normal distribution.

For $X \sim \mathrm{IG}(\mu, \lambda)$ it holds that

$$\mathrm{P}[X \leq x] = \Phi\{\sqrt{\lambda/x}\,(x/\mu - 1)\} + \exp(2\lambda/\mu)\,\Phi\{-\sqrt{\lambda/x}\,(x/\mu + 1)\}$$

See Chhikara and Folks (1977) regarding proof.

The inverse Gaussian distribution as exponential family

By rewriting the density (B.23) as

$$f(x; \alpha, \lambda) = \left(\frac{\lambda}{2\pi}\right)^{1/2} \times \frac{1}{x^{3/2}} \times \exp\left\{\sqrt{\alpha\,\lambda} - \alpha x/2 - \lambda/(2x)\right\} \qquad \text{(B.29)}$$

with $\alpha = \lambda/\mu^2$, it is seen that the family of $\mathrm{IG}(\mu, \lambda)$-distributions with $(\mu, \lambda) \in \mathbb{R}_+ \times \mathbb{R}_+$ makes up an exponential family of order 2 with canonical parameter

$$\boldsymbol{\theta} = \begin{pmatrix} \theta_1 \\ \theta_2 \end{pmatrix} = \begin{pmatrix} -\alpha/2 \\ -\lambda/2 \end{pmatrix},$$

Table B.6: *$IG(\mu, \lambda)$-distribution as reproductive exponential dispersion model.*

Canonical parameter θ	Cumulant generating function $\kappa(\theta)$	Mean value mapping $\mu = \tau(\theta)$	Unit deviance function $V_{IG}(\mu)$	Dispersion parameter σ^2
$-1/(2\mu^2)$	$-\sqrt{-2\theta}$	$1/\sqrt{-2\theta}$	μ^3	$1/\lambda$

canonical sample statistics

$$\mathbf{t} = \begin{pmatrix} t_1(x) \\ t_2(x) \end{pmatrix} = \begin{pmatrix} x \\ 1/x \end{pmatrix}$$

and cumulant generating function

$$\kappa(\alpha, \lambda) = \frac{1}{2}\,\log(\lambda) + \sqrt{\alpha\lambda}\,.$$

This family is not complete. The complete family appears by allowing the value $\alpha = 0$, corresponding to the stable distribution with index $1/2$ and scale parameter $1/\lambda$.

The family has a large fourth order moment, but the complete family is not regular.

The inverse Gaussian distribution as exponential dispersion model

The family of $IG(\mu, \lambda)$-distribution with $(\mu, \lambda) \in \mathbb{R}_+ \times \mathbb{R}_+$ makes up a reproductive exponential dispersion model with mean value μ, unit deviance function $V_{IG}(\mu) = \mu^3$, and dispersion parameter $\sigma^2 = 1/\lambda$. The canonical parameter is $\theta = -1/(2\mu^2)$ with the canonical parameter space $D =]-\infty, 0]$. The mean value space is $\mathcal{M} = \mathbb{R}_+$ and the model has a large fourth order moment. The density expressed through the canonical parameter $\theta = -1/(2\mu^2)$ and λ is

$$g(y; \theta, \lambda) = \left(\frac{\lambda}{2\pi y^3}\right)^{1/2} \exp\left\{-\frac{\lambda}{2y} + \lambda\left(\theta y + \sqrt{-2\theta}\right)\right\} \quad \text{for } y \in \mathbb{R}_+ \quad (B.30)$$

The most important quantities in the interpretation of the family of $IG(\mu, \lambda)$-distributions as a reproductive exponential dispersion model are provided in Table B.6.

The canonical link is the square of the reciprocal mapping:

$$\theta = \tau^{-1}(\mu) = -\frac{1}{2\mu^2}$$

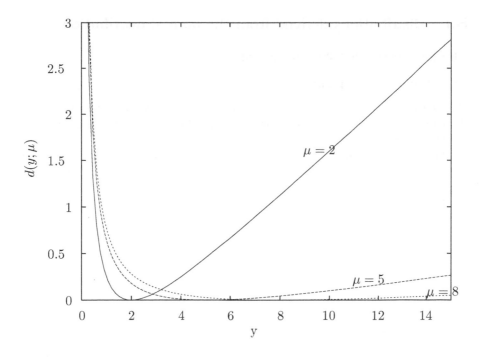

Figure B.7: *Unit deviance of the inverse Gauss distribution for different values of μ.*

The unit deviance for the $IG(\mu, \lambda)$-distribution is

$$d(y; \mu) = \frac{(y - \mu)^2}{y\mu^2} \tag{B.31}$$

We can express the density of a $IG(\mu, \lambda)$-distribution through the unit deviance in the form (4.11) as

$$f(y; \mu, \sigma^2) = a(y; \sigma^2) \exp\left\{ - \frac{(y - \mu)^2}{2y\mu^2\sigma^2} \right\}, \tag{B.32}$$

where

$$a(y; \sigma^2) = \frac{1}{\sqrt{2\pi\sigma^2 y \sqrt{y}}}$$

Figure B.7 shows the unit deviance corresponding to $\mu = 2$, $\mu = 5$ and $\mu = 8$.

▶ **Remark B.3 – Scale transformation and the inverse Gaussian distribution**

If Y follows a $IG(\mu, 1/\sigma^2)$-distribution, then the distribution of cY for $c > 0$ will also be an inverse Gaussian distribution. This distribution will have mean value $c\mu$ and dispersion parameter σ^2/c.   ◀

B.7   Distributions derived from the normal distribution

The log-normal distribution model

Definition and basic properties

The log-normal distribution model is used quite extensively to model the distribution of highly skewed, positive data, like distributions of income, concentrations, growth, etc.

The log-normal distribution model is characterized by the fact that when the distribution model for the raw data is a log-normal distribution, then the logarithmically transformed data follows a normal distribution.

The density function for a log-normal distribution model is given by

$$
p_Y(y) = \frac{1}{y\beta\sqrt{2\pi}} \exp\left[-\frac{1}{2}\left(\frac{\log(y) - \alpha}{\beta}\right)^2\right] \quad \text{for} \quad y > 0
$$

and $p_Y(y) = 0$ otherwise, and with $\beta > 0$, and α denoting *parameters* of the model. Short notation is $Y \sim \text{LN}(\alpha, \beta^2)$.

Mean and variance

Let $Y \sim \text{LN}(\alpha, \beta^2)$, then the mean and variance of Y are given by

$$
\mu_Y = \text{E}[Y] = \exp\left(\alpha + \frac{1}{2}\beta^2\right))
$$

$$
\sigma_Y^2 = \text{Var}[Y] = \mu_Y^2\left(\exp(\beta^2) - 1\right)
$$

Thus, the *coefficient of variation* (relative standard deviation) for the distribution of Y is

$$
\gamma_Y = \frac{\sigma_Y}{\mu_Y} = \sqrt{\exp(\beta^2) - 1}
$$

i.e., a dimensionless quantity that does not depend on the parameter α.

Relation to the normal distribution model

Given $Y \sim \text{LN}(\alpha, \beta^2)$, and let $Z = \log(Y)$, then $Z \sim \text{N}(\alpha, \beta^2)$, i.e., $Z = \log(Y)$ follows a normal distribution model with mean $\mu_Z = \alpha$, and variance, $\sigma_Z^2 = \beta^2$.

Note on transformations

Note from the expressions above that although $Y = \exp(Z)$, then

$$
\mu_Y = \text{E}[Y] = \exp(\mu_Z)\exp\left(\frac{1}{2}\sigma_Z^2\right)
$$

which is *greater than* $\exp(\mu_Z)$. The term $\exp\left(\frac{1}{2}\sigma_Z^2\right)$ is a bias correction.

Cumulative distribution function

When $Y \sim \mathrm{LN}(\alpha, \beta^2)$, the cumulative distribution function is

$$F_Y(y) = \Phi\left(\frac{\log(y) - \alpha}{\beta}\right)$$

with $\Phi(\cdot)$ denoting the cumulative distribution function for the standardized normal distribution. Thus, all probabilities related to log-normal distribution models may be determined from the standardized normal distribution.

The χ^2-distribution model

Definition and basic properties

The family of χ^2-distribution models is a family of distribution models on the positive real line with probability density function

$$f_Y(y) = \frac{1}{2\Gamma(d/2)}\left(\frac{y}{2}\right)^{d/2-1}\exp(-y/2) \qquad \text{for } y \in \mathbb{R}_+$$

and $p_Y(y) = 0$ otherwise, and with $\Gamma(\cdot)$ denoting the *Gamma*-function (see Section B.8).

The symbol ν denotes a positive integer-valued *parameter* that determines which of the many χ^2-distributions is considered. The parameter ν is termed the *degrees of freedom* for the distribution. In applications, the degrees of freedom are usually known.

A random variable that is distributed according to a χ^2-distribution model with ν degrees of freedom is often denoted by the symbol χ^2_ν.

The χ^2-distribution model is a special case of a larger class of distribution models, the so-called *gamma*-distribution models, since for $Y \sim \chi^2_\nu$ we have $Y \sim \mathrm{G}(\alpha, \beta)$ with $\alpha = \nu/2$ and $\beta = 2$.

Mean and variance

Consider $Y \sim \chi^2_d$, then the mean and variance of Y are given by

$$\mathrm{E}[Y] = d; \qquad \mathrm{Var}[Y] = 2d \tag{B.33}$$

Relation to the normal distribution model

In the data distribution in normal distribution models, the χ^2-distribution model is often encountered as a distribution model for the variance, S^2.

Remember that the following assertion holds: Given $Y \sim \mathrm{N}(\mu, \sigma^2)$, then

$$\frac{(Y - \mu)^2}{\sigma^2} \sim \chi^2_1 \tag{B.34}$$

The t-distribution model

Definition and basic properties

The family of t-distribution models is a family of distribution models on the real axis with probability density function

$$f(t) = \frac{\Gamma((d+1)/2)}{\sqrt{d\pi}\Gamma(d/2)} \left(1 + \frac{t^2}{d}\right)^{-(d+1)/2}$$

with $\Gamma(\cdot)$ denoting the gamma-function; see Section B.8 on page 284.

The symbol ν denotes a positive integer-valued *parameter* that determines which of the many t-distributions is being considered. The parameter ν is termed the *degrees of freedom* for the distribution. In applications the degrees of freedom are usually known.

A random variable that is distributed according to a t-distribution model with ν degrees of freedom is often denoted by the symbol t_ν.

Mean and variance

Let $Z \sim t_\nu$. For $\nu > 1$ the mean of the distribution exists, and

$$E[Z] = 0, \quad \text{for } \nu > 1$$

For $\nu > 2$ also the variance exists, and

$$\text{Var}[Z] = \frac{d}{d-2}, \quad \text{for } \nu > 2$$

This result is utilized when making statistical inference in normal distribution models: Let $Y_1, Y_2, \ldots, Y_n$ be independent normally distributed random variables with the same mean and variance, i.e., $Y_i \sim N(\mu, \sigma^2)$, and let, as usual, $\overline{Y}$ denote the mean, and S^2 denote the variance in the data distribution. It has already been observed that

$$U = \frac{\overline{Y} - \mu}{\sigma / \sqrt{n}} \sim N(0,1)$$

$$Z = \nu \cdot S^2 / \sigma^2 \sim \chi_\nu^2$$

with $\nu = n - 1$, and with U and Z being independent.

Now consider

$$T = \frac{\overline{Y} - \mu}{S / \sqrt{n}} \tag{B.35}$$

and note that T may be expressed as

$$T = \frac{\overline{Y} - \mu}{S / \sqrt{n}} = \frac{\left(\overline{Y} - \mu\right) / (\sigma / \sqrt{n})}{\sqrt{S^2 / \sigma^2}} = \frac{U}{\sqrt{Z / \nu}}$$

It then follows from the theorem above that $T \sim t_\nu$.

Small and large sample properties

When comparing U and T, it is noticed that the difference between these two variables is that the denominator in U is the *known* value of the standard deviation $\sigma/\sqrt{n}$ for $\overline{Y}$, whereas the denominator in T has replaced this value by an *estimate* of the standard deviation, $\widehat{\sigma}(\overline{Y}) = S/\sqrt{n}$. Consequently, the distribution of T will have heavier tails than the normal distribution of U, but as the degrees of freedom, ν, increase, the estimate, $\widehat{\sigma}(\overline{Y})$ becomes more precise and, hence, the distribution of T will approach a standard normal distribution. Therefore, the normal distribution may be used to approximate the t-distribution when the degrees of freedom are high, e.g., greater than 30.

Degrees of freedom and sample size for mean of data

Also note from the relation

$$T = \frac{\overline{Y} - \mu}{\widehat{\sigma}(\overline{Y})} = \frac{\overline{Y} - \mu}{S/\sqrt{n}} \sim t_\nu \tag{B.36}$$

that the *degrees of freedom, d,* refers to the *estimate for σ^2*, whereas n refers to the number of observations that are used to determine the mean $\overline{Y}$.

This distinction is useful when the variance estimate, S^2, has been obtained from other sources than the estimate, $\overline{Y}$.

The F-distribution model

Definition and basic properties

The family of F-distribution models is a family of distribution models on the positive real axis with probability density function

$$f_X(x) = \frac{\Gamma((n+m)/2)}{\Gamma(n/2)\Gamma(m/2)} \left(\frac{n}{m}\right)^{n/2} \frac{x^{n/2-1}}{(1+nx/m)^{(n+m)/2}} \qquad \text{for } 0 < x < \infty$$

with $\Gamma(\cdot)$ denoting the *Gamma*-function.

The symbols n and m denote positive integer-valued *parameters* that determine which of the many F-distributions is being considered. The parameters are termed the *degrees of freedom* for the distribution. In applications the degrees of freedom are usually known.

A random variable that is distributed according to a F-distribution model with (n, m) degrees of freedom is often denoted by the symbol $F_{n,m}$.

Sometimes the following relation between percentiles of the F-distribution may be useful

$$F_{n,m;q} = \frac{1}{F_{m,n;1-q}} \tag{B.37}$$

Mean and variance

Let $Y \sim \mathrm{F}_{n,m}$. For $m > 2$ the mean of the distribution exists, and

$$E[Y] = \frac{m}{m-2} \quad \text{for } m > 2$$

For $m > 4$ the variance exists, and

$$\mathrm{Var}[Y] = \frac{m^2(2n + 2m - 4)}{(m-2)^2(m-4)n} \quad \text{for } m > 4$$

Considering linear normal distribution models, the following results are often used.

B.8   The Gamma-function

The gamma-function, $\Gamma(\cdot)$, is a function of a real positive argument, given by

$$\Gamma(x) = \int_0^\infty t^{x-1} \exp(-t) dx$$

The function satisfies the recursion

$$x\Gamma(x) = \Gamma(x+1)$$

In particular, $\Gamma(1) = 1$ and $\Gamma(1/2) = \sqrt{\pi}$.

For integer values, n, we therefore have

$$\Gamma(n) = 1 \cdot 2 \cdot 3 \cdots (n-1) = (n-1)!$$
$$\Gamma(n + 1/2) = \frac{1 \cdot 3 \cdot 5 \cdots (2n-1)}{2^n} \sqrt{\pi}$$

APPENDIX C

List of symbols

The number following the symbol description marks the page where the symbol is first mentioned.

n	Number of observations, 7
$\boldsymbol{Y}$	n-variate random variable, 10
$f_Y(\boldsymbol{y})$	Joint density function of $\boldsymbol{Y}$, 10
$\boldsymbol{y}$	observation set, 11
$\widehat{\boldsymbol{\theta}}(\boldsymbol{Y})$	Estimator, 11
$\widehat{\boldsymbol{\theta}}(\boldsymbol{y})$	Estimate, 11
$\mathrm{E}[\boldsymbol{X}]$	Expectation of $\boldsymbol{X}$, 11
$\mathrm{Var}[X_i]$	Variance of X_i, 12
$\mathrm{Var}[\boldsymbol{X}]$	Covariance matrix (or variance-covariance matrix), 12
$\boldsymbol{i}(\boldsymbol{\theta})$	Expected information or Fisher information matrix, 12
$\boldsymbol{I}$	Identity matrix, 13
$L(\boldsymbol{\theta};\boldsymbol{y})$	Likelihood function, 14
$\ell(\boldsymbol{\theta};\boldsymbol{y})$	Log-likelihood function, 14
$\ell'_\theta(\boldsymbol{\theta};\boldsymbol{y})$	Score function, 17
$\boldsymbol{j}(\boldsymbol{\theta};\boldsymbol{y})$	Observed information, 18
$\mathrm{D}[\cdot]$	Dispersion (variance-covariance) matrix, 19
$\boldsymbol{J}$	Jacobian, 21
$D(Y;\cdot)$	Deviance, 23
Ω	Parameter set, 25
$\mathcal{H}$	Hypothesis, 25
C	Critical region, 26
$\lambda(\boldsymbol{y})$	Likelihood ratio, 26
$L_P(\tau;\boldsymbol{y})$	Profile likelihood, 35
$\ell'_\tau(\tau,\widehat{\zeta}_\tau)$	Profile score function, 35
$L_M(\tau;u)$	Marginal likelihood function, 37
$L_C(\tau;\boldsymbol{y}|u)$	Conditional likelihood function based on conditioning on u, 38
$\delta_\Sigma(\boldsymbol{y}_1,\boldsymbol{y}_2)$	Inner product, 43
$\|\boldsymbol{y}\|_\Sigma$	Norm, 43
$\boldsymbol{H}$	Hat-matrix, 51
r_i	Observed residual, 51
R^2	Coefficient of determination, 57
R^2_{adj}	Adjusted coefficient of determination, 58

r_i^{rs}	Standardized residual, 73	
r_i^{rt}	Studentized residual, 74	
$V(\cdot)$	Variance function, 93	
$d(y; \cdot)$	Unit deviance, 93	
η	Link function, 102	
$D(\boldsymbol{y}; \cdot)$	Residual deviance, 105	
$D^*(\boldsymbol{y}; \cdot)$	Scaled residual deviance, 105	
r_i^R	Response residual, 108	
r_i^D	Deviance residual, 108	
r_i^P	Pearson residual, 108	
r_i^W	Working residual, 109	
γ	Signal to noise ratio, 162	
ρ	Coefficient of correlation, 164	
w	Shrinkage factor, 166	
n_0	Weighted average group size, 167	
$R(\cdot, \boldsymbol{d}(.))$	Risk of an estimator, 187	
$B(n, p)$	Binomial distribution with parameters n and p, 259	
$\text{Bern}(p)$	Bernoulli distribution, 259	
$P(\lambda)$	Poisson distribution with parameter λ, 262	
$\text{NB}(r, p)$	Negative binomial distribution with parameters r and p, 264	
$\text{Geo}(p)$	Geometric distribution with parameter p, 264	
$G(\alpha, \beta)$	Gamma distribution with parameters α and β, 268	
$\text{Ex}(\beta)$	Exponential distribution with parameter β, 268	
ν	degrees of freedom, 268	
$\text{IG}(\mu, \lambda)$	Inverse Gaussian distribution with parameters μ and λ, 275	
$\text{LN}(\alpha, \beta^2)$	Log-normal distribution with parameters α and β^2, 280	
χ_d^2	χ^2 distribution with d degrees of freedom, 281	
t_ν	t-distribution with ν degrees of freedom, 282	
$F_{n,m}$	F-distribution with n and m degrees of freedom, 284	

Bibliography

Anscombe, F. J. (1973) "Graphs in statistical analysis." In: *American Statistician* 27, pp. 17–21.

Arlien-Søborg, P., L. Henriksen, A. Gade, C. Gyldensted, and O. B. Paulson (1982) "Cerebral blood flow in chronic toxic encephalopathy in house painters exposed to organic solvents." In: *Acta Neurol Scand.* 66, pp. 34–41.

Bartlett, M. S. (1935) "Contingency table interaction." In: *Supplement to the Journal of the Royal Statistical Society* 2, pp. 248–252.

——— (1937) "Properties of Sufficiency and Statistical Tests." In: *Proceedings of the Royal Society of London. Series A, Mathematical and Physical Sciences* 160, pp. 268–282.

Bartlett, M. S. and D. G. Kendall (1946) "The statistical analysis of variance-heterogeneity and the logarithmic transformation." In: *Supplement to the Journal of the Royal Statistical Society* 8, pp. 128–138.

Bates, D. M. and D. G. Watts (1988) *Nonlinear Regression Analysis and Its Applications.* Wiley, New York.

Bayarri, M. J., M. H. DeGroot, and J. B. Kadane (1987) "What is the likelihood function?" In: *Statistical Decision Theory and Related Topics IV* 1.

Beal, S. and L. Sheiner (1980) "The NONMEM system." In: *The American Statistician* 34, pp. 118–119.

——— (2004) *NONMEM Users Guide.* University of California San Francisco.

Berger, J. O., B. Liseo, and R. L. Wolpert (1999) "Integrated likelihood methods for eliminating nuisance parameters." In: *Statist. Sci.* 14, pp. 1–28.

Bickel, P. J. and K. A. Doksum (2000) *Mathematical Statistics: Basic Ideas and Selected Topics.* Prentice Hall.

Box, G. E. P. and G. M. Jenkins (1970/1976) *Time Series Analysis, Forecasting and Control.* San Francisco: Holden-Day.

Box, G. E. P. and G. C. Tiao (1975) "Intervention analysis with applications to economic and environmental problems." In: *Journal of the American Statistical Association* 70, pp. 70–79.

Brockwell, P. J. and R. A. Davis (1987) *Time Series; Theory and Methods.* New York: Springer-Verlag.

Brown, S. D., R. Tauler, and B. Walczak, Eds. (2009) *Comprehensive Chemometrics: Chemical and Biochemical Data Analysis*. Elsevier.

Cameron, J. M. (1951) "The use of Components of Variance in preparing schedules for sampling of baled wool." In: *Biometrics* 7, pp. 83–96.

Cappé, O., E. Moulines, and T. Rydén (2005) *Inference in Hidden Markob Models*. New York: Springer.

Casella, G. and R. L. Berger (2002) *Statistical Inference*. Duxbury Press.

Chan, K. S. and L. Ledolter (1995) "Monte Carlo FM estimation for time series models involving counts." In: *Journal of the American Statistical Association* 90, pp. 242–252.

Chhikara, R. S. and J. L. Folks (1977) "The inverse Gaussian distribution as a lifetime model." In: *Technometrics* 19, pp. 461–468.

Consonni, G. and P. Veronese (1992) "Conjugate priors for exponential families having quadratic variance functions." In: *Journal of the American Statistical Association* 87, pp. 1123–1127.

Cox, D. R. and P. A. W. Lewis (1966) *The Statistical Analysis of Series of Events*. London: Methuen & Company.

Cox, D. and D. Hinkley (2000) *Theoretical Statistics*. Chapmann & Hall.

Diggle, P., P. Heagerty, K. Y. Liang, and S. Zeger (2002) *Analysis of Longitudinal Data*. 2nd ed. Oxford University Press.

Draper, N. R. and H Smith (1981) *Applied Regression Analysis*. Wiley, New York.

Efron and Morris (1972) In:

Eisenhart, C. and P. Wilson (1943) "Statistical methods and control in bacteriology." In: *Bacterial Rev.* 7, pp. 57–137.

Elston, R. C. and J. E. Grizzle (1962) "Estimation of time-response curves and their confidence bands." In: *Biometrics* 18, pp. 148–159.

Ephrain, Y. and N. Merhav (2002) "Hidden Markov processes." In: *IEEE Trans. Inform. Theory* 48, pp. 1518–1569.

Fisher, R. A. (1922) "On the Mathematical Foundations of Theoretical Statistics." In: *Philosopical Transactions of the Royal Society of London, Series A*, pp. 309–368.

Geisser, S. (1993) *Predictive Inference*. Chapman & Hall.

Gilbert, P. (2009) *numDeriv: Accurate Numerical Derivatives*. R package version 2009.2-1 URL: http://www.bank-banque-canada.ca/pgilbert.

Griewank, A. (2000) *Evaluating Derivatives: Principles and Techniques of Algorithmic Differentiation*. Philadelphia: SIAM.

Gutierez-Pena, E. and A. F. M. Smith (1995) "Conjugate parametrizations for natural exponential families." In: *Journal American Statistical Association* 90, pp. 1347–1356.

Harvey, A. C. (1996) *Forecasting, Structural Time Series Models and the Kalman Filter*. Cambridge: Cambridge University Press.

Harville, D. (1977) "Maximum likelihood approaches to variance component estimation." In: *Journal of the Americal Statistical Association* 72, pp. 320–340.

Hastie, T., R. Tibshirani, and J. Friedman (2001) *The Elements of Statistical Learning; Data Mining, Inference and Prediction.* New York: Springer.

Heyde, C. C. (1997) *Quase-likelihood and Its Application: A General Approach to Optimal Parameter Estimation.* New York: Springer.

Hoblyn, T. N. and R. C. Palmer (1934) "A complex experiment in the propagation of plum rootstocks from root cuttings, season 1931-32." In: *Journ. Pom. and Hort. Sci.* 12, pp. 36–56.

Hoerl, A. E. (1962) "Application of ridge analysis to regression problems." In: *Chemical Engineering Progress* 58, pp. 54–59.

Jazwinski, A. H. (1970) *Stochastic Processes and Filtering Theory.* New York: Academic Press.

Jørgensen, B. (1993) *The Theory of Linear Models.* Chapman & Hall.

Klim, S., S. B. Mortensen, N. R. Kristensen, R. V. Overgaard, and H. Madsen (2009) "Population stochastic modelling (PSM) – An R package for mixed-effects models based on stochastic differential equations." In: *Computer Methods and Programs in Biomedicine* 94, pp. 279–289.

Koenker, R. (2005) *Quantile regression.* New York: Cambridge University Press.

Kramer, R. (1998) *Chemometric Techniques for Quantitative Analysis.* CRC Press.

Kristensen, N. R., H. Madsen, and S. B. Jørgensen (2004) "Parameter estimation in stochastic grey-box models." In: *Automatica* 40, pp. 225–237.

Kuk, A. Y. C. and Y. W. Cheng (1999) "Pointwise and functional approximations in Monte Carlo maximum likelihood estimation." In: *Statistics and Computing* 9, pp. 91–99.

Lee, Y. and J. Nelder (1996) "Hierarchical generalized linear models." In: *Journal of the Royal Statistical Society* B,58, pp. 619–678.

——— (2000) "Two ways of modelling overdispersion in non-normal data." In: *Applied Statistics* 49, pp. 591–598.

——— (2001) "Hierarchical generalized linear models: a synthesis of generalised linear models, random-effect models and structured dispersions." In: *Biometrika* 88, pp. 987–1006.

Lehmann, E. L. (1986) *Testing Statistical Hypotheses.* 2nd ed. Wiley.

Lehmann, E. L. and G. Casella (1998) *Theory of Point Estimators.* Springer.

Likert, R. (1932) "A technique for the Measurement of Attitudes." PhD thesis Department of Psychology, Columbia University.

Lindsey, J. K. (1997) *Applying Generalized Linear Models.* New York: Springer.

Lindstrom, M. J. and D. M. Bates (1990) "Nonlinear Mixed Effects Models for Repeated Measures Data." In: *Biometrics* 46, pp. 673–687.

Ljung, L. (1976) *System Identification; Advances and Case Studies.* New York: Academic Press Chap. On the consistency of prediction error identification methods.

——— (1987) *System Identification: Theory for the User.* New York: Prentice Hall.

Madsen, H. (1992) *Projection and Separation in Hilbert Spaces*. Lyngby, Denmark: IMM, DTU.

––––––– (2008) *Time Series Analysis*. Chapman & Hall.

Madsen, H., P. Pinson, G. Kariniotaktis, H. A. Nielsen, and T. S. Nielsen (2005) "Standardizing the performance evaluation of short-term wind prediction models." In: *Wind Engineering* 29, pp. 475–489.

Martens, H. and T. Naes (1989) *Multivariate Calibration*. Wiley.

Maybeck, P. S. (1982) *Stochastic Models, Estimation and Control; Vol 1, 2, 3*. New York: Academic Press.

McCrady, M. H. (1915) "The numerical interpretation of fermentation tube results." In: *Journ. Infectious Diseases* 17, pp. 183–212.

McCullagh, P. and J. A. Nelder (1989) *Generalized linear models*. 2nd ed. London: Chapman & Hall.

McCulloch, C. E. and S. R. Searle (2001) *Generalized, Linear, and Mixed Models*. John Wiley & Sons, Inc.

Millar, R. B. (2004) "Simulated Maximum Likelihood Applied to Non-Gaussian and Nonlinear Mixed Effects and State-space Models." In: *Australian & New Zealand Journal of Statistics* 46, pp. 543–554.

Montgomery, D. C. (2005) *Design and Analysis of Experiments*. 6th ed. Wiley.

Morris, C. N. (1982) "Natural exponential families with quadratic variance functions." In: *Ann. Statist.* 10, pp. 65–80.

Mortensen, S. B. (2009) "Markov and mixed models with applications." PhD thesis Lyngby, Denmark: Technical University of Denmark (DTU).

Mortensen, S. B., S. Klim, B. Dammann, N. R. Kristensen, H. Madsen, and R. V. Overgaard (2007) "A Matlab framework for estimation of NLME models using stochastic differential equations." In: *Pharmakinetics and Pharmadynamics* 34, pp. 623–642.

Müller-Funk, U. and F. Pukelsheim (1987) "How Regular are Conjugate Exponential Families?" In: *Statistics and Probability Letters* 7, pp. 327–333.

Nelder, J. A. and R. W. M. Wedderburn (1972) "Generalized linear models." In: *Journal of the Royal Statistical Society* A 135, pp. 370–384.

Öjelund, H., H. Madsen, and P. Thyregod (2001) "Calibration with absolute shrinkage." In: *Journal of Chemometrics* 15, pp. 497–509.

Öjelund, H., P. J. Brown, H. Madsen, and P. Thyregod (2002) "Prediction based on mean subset." In: *Technometrics* 44, pp. 369–374.

Patterson, H. D. and R. Thompson (1971) "Recovery of inter-block information when block sizes are unequal." In: *Biometrika* 58, pp. 545–554.

Pawitan, Y. (2001) *In All Likelihood, Statistical Modelling and Inference Using Likelihood*. New York: Oxford University Press.

Pinheiro J. C. Bates, D. M. (2000) *Mixed-Effects Models in S and S-PLUS*. New York: Springer.

Pinheiro, J. C. and D. M. Bates (1995) "Approximations to the Nonlinear Mixed-Effects Model." In: *Jounal of Computational and Graphical Statistics* 4, pp. 12–35.

Pinheiro, J. C., D. M. Bates, S. DebRoy, D. Sarkar, and the R Core team (2008) *nlme: Linear and Nonlinear Mixed Effects Models*. R package version 3.1-90.

Price, C. J., C. Kimmel, J. D. George, and M. C. Marr (1987) "The developmental toxity of diethylene glycol dimethyl ether in mice." In: *Fundamental and Applied Toxicology* 8, pp. 115–126.

Proschan, F (1963) "Theoretical explanation of observed decreasing failure rate." In: *Techometrics* 5, pp. 375–383.

Ripley, B. D. (1987) *Stochastic Simulation*. Wiley.

SAS Institute Inc. (2004) *SAS/Stat 9.1 User's Guide*. Cary, NC: SAS Institute Inc.

Searle, S. R., G. Casella, and C. E. McMulloch (1992) *Variance Components*. New York: Wiley.

Shao, J. (1999) *Mathematical Statistics*. Springer.

Stein (1955) In:

Tikhonov, A. N. and V. A. Arsemin (1977) *Solution of Ill-Posed Problems*. Winston & Sons.

Tornøe, C. W., J. Jacobsen, and H. Madsen (2004) "Grey-box pharmacokinetic/pharmacodynamic modelling of euglycamemic clamp study." In: *Mathematical Biology* 48, pp. 591–604.

Tornøe, C. W., H. Agersø, H. Madsen, E. N. Jonsson, and H. A. Nielsen (2004) "Non-linear mixed effect pharmacokinetic/pharmacodynamic modelling in NLME using differential equations." In: *Computer Methods and Programs in Biomedicine* 76, pp. 31–40.

Tweedie, M. C. K. (1957) "Statistical properties of inverse Gaussian distribution." In: *Ann. Math. Stat.* 28, pp. 362–377.

Verbeke, G. and G. Molenberghs (2000) *Linear Mixed Models for Longitudinal Data*. Springer.

Vonesh, E. F. (1996) "A note on the use of Laplace's approximation for nonlinear mixed-effects models." In: *Biometrika* 83, pp. 447–52.

Wasserman, L. A. (2004) *All of Statistics – A Concise Course in Statistical Inference*. New York: Springer.

Wilkinson, G. N. and C. E. Rogers (1973) "Symbolic description of factorial models for analysis of variance." In: *Applied Statistics* 22, pp. 392–399.

Williford, W. O., M. C. Carter, and P. Hsieh (1974) "A Bayesian analysis of two probability models describing thunderstorm activity at Cape Kennedy, Florida." In: *Journal of Applied Meteorology* 13, pp. 718–725.

Wolfinger, R. D. and X. Lin (1997) "Two Taylor-series approximation methods for nonlinear mixed models." In: *Computational Statistics & Data Analysis* 25, pp. 465–490.

Zeger, S. L. (1988) "A regression-model for time series of counts." In: *Biometrika* 75, pp. 621–629.

Index

A

AD Model Builder, 211, 213
additive exponential dispersion
  model, 91
adjusted coefficient of determination,
  58
ADMB, 211
alternative hypothesis, 25
analysis of covariance, 41
analysis of deviance table, 113
analysis of variance, 41
ANCOVA, 41
ANOVA, 41
  one-way, 48
anova table, 55, 57
arithmetic mean, 274
autocorrelation, 77
automatic differentiation, 211
autoregressive process, 216

B

back-fitting
  iterative, 184
backward selection, 30
balanced design, 166
balanced experiment, 157
Bartlett correction, 140
Bayes estimator
  in the one-way random effects
    model, 188
Bayes risk, 188
Bayes' Theorem, 185
Bernoulli data, 2
Bernoulli distribution, 92, 259

Bernoulli trial, 259
best linear unbiased estimator, 12
best linear unbiased predictor, 172,
  183
best subset selection, 32
bias correction, 280
binary data, 2
binary response, 118
binomial beta model, 237
binomial distribution, 87, 89, 92,
  118, 259
  as exponential dispersion model,
    259
  canonical parameter, 92
  canonical parameter space, 92
  cumulant generator, 92
  precision parameter, 92
  reproductive properties, 261
  two-factor model, 140
block design, 175
blocked experiments, 175
BLUP, 172, 183
Box-Cox transformation, 40
Brownian motion, 275

C

canonical link, 6, 102
canonical link function, 95
canonical parameter, 90
  offset, 264
canonical parameterization, 90
CAR process, 204
categorical varible, 157
chain of hypotheses, 58

characteristic function
    binomial distribution, 259
    Poisson, 262
χ^2-distribution, 268, 281
classical GLM, 41
classification, 157
Cochran's theorem, 51
coefficient of correlation, 164
coefficient of determination, 57
collinearity, 64
complementary log-log transforma-
    tion, 129
conditional distribution, 185
conditional likelihood, 38
conditional test, 56
confidence interval, 73
    profile likelihood, 34
    asymmetric, 36
    for individual parameters, 117
    likelihood based, 117
    profile likelihood based, 36
    Wald-type, 117
confidence region, 36
confidence regions, 72
conjugate, 230
conjugate distribution, 242
    for an exponential dispersion
        family, 233
conjugate prior distributions, 233
consistency, 11
consistent estimator, 11
continuous auto-regressive process,
    204
contrasts, 78
    estimable, 79
    Helmert-transformation, 79
    in R, 83
    sum-coding, 80
    treatment-coding, 80
corrected test, 29
correction for effects, 64
correlation
    intraclass, 164, 235
correlation structure, 163
covariance matrix

    sparce, 217
Cramer-Rao inequality, 12
critical region, 26
cross-validation, 32
cumulant generator, 90
curvature
    of the log-likelihood function, 18

D

data
    binary, 2
    continuous, 2
    counts, 2
    Gaussian, 2
    nominal, 2
    positive, 2
    quantal, 2
data mining, 32
Delta method, 255
design matrix, 7, 47, 79, 99
    not full rank, 47
deviance, 23, 27, 44, 89
    for normal distribution, 43
    for the normal distribution, 94
    incremental, 59
    partitioning, 59
    quasi, 110
deviance residual, 108
deviance table, 55, 57
digamma function, 270
dimension
    of general linear model, 46
    of generalized linear model, 99
dispersion matrix, 13, 19, 43
    decomposition of, 23
    of the score function, 19
dispersion parameter, 89, 100, 114,
    225
distribution
    Bernoulli, 92, 259
    binomial, 87, 89, 92, 259
    χ^2, 281
    conditional, 185
    exponential, 266, 268
    F-distribution, 283

gamma, 89, 268
geometric, 228, 264
inverse Gaussian, 89, 275
log-normal, 280
marginal, 185
multinomial, 89
negative binomial, 89, 227, 264
normal, 89, 91
overview of, 257
Poisson, 88, 89, 91, 262
Polya, 239
simultaneous, 185
t-distribution, 282
Wald, 275

E

efficient estimator, 13
empirical Bayes estimator, 190
entropy, 23
equicorrelation-matrix, 164
error propagation law, 255
estimable contrasts, 79
estimable function, 47
estimate, 7, 11
    maximum likelihood, 9, 21
estimating equation, 111
estimation equations
    for ML-estimator, 21
estimator, 11
    BLUE, 12
    efficient, 13
    empirical Bayes, 190
    maximum likelihood, 10, 21
    minimum mean square error, 12
    ML, 21
    MVU, 39
    MVUE, 12
    unbiased, 11
    uniformly minimum mean
        square error, 12
Euclidean distance, 75
events
    independent, 267
    time between, 267
Ex(β)-distribution, 267

expected information, 18
explanatory variables, 46
exponential dispersion family, 89, 90,
    225
    expected information, 98
    joint density, 95
    log-likelihood function, 97
    mean and variance, 92
    observed information, 98
    precision parameter, table of, 95
    score function, 97
    unit deviance, table of, 95
    variance function, table of, 95
exponential distribution, 266, 268
exponential family, 88
    dispersion, 89, 90
    natural, 89, 90
exponential family of densities, 1

F

F-distribution, 283
factor level, 157
factorial experiment, 157
failure-rate, 268
first stage model, 199
Fisher information, 12, 18
Fisher scoring algorithm, 110
fitted values, 107
fitted values in the classical GLM,
    50
fixed effects model, 7, 48, 157
FOCE, 203
forward selection, 30
full model, 28, 32, 44–47, 100, 112

G

gamma distribution, 89, 268
    addition theorem, 273
    as exponential dispersion model,
        270
    estimation, 274
    reparameterization of, 230
gamma function, 281, 284
    incomplete, 269
gamma regression, 125

Gauss-Newton approximation, 203
Gaussian data, 2
Gaussian mixed model, 159
general linear model, 6, 41, 46
    confidence intervals, 71
    confidence region, 71
    estimation of the residual vari-
        ance, 53
    fitted values, 50
    in R, 81
    inference on individual parame-
        ters, 70
    prediction, 72
    residuals, 51
    test for model sufficiency, 56
    tests for model reduction, 58
general mixed effects models, 199
generalization error, 32
generalized linear mixed models, 225
generalized linear model, 6, 88, 99
    analysis of deviance table, 113
    canonical link, 102
    confidence intervals, 117
    design matrix, 99
    dimension of, 99
    estimation, 104
    fitted values, 105
    in R, 152
    initial test for goodness of fit,
        112
    likelihood ratio tests, 111
    link function, 100
    link functions, table of, 102
    local design matrix, 101
    overdispersion, 113
    Poisson regression, 123
    predicted values, 105
    residual deviance, 105
    residuals, 108
    specification of, 100
    test for model reduction, 115
    test of individual parameters,
        116
    types of response variables, 89
generalized loss-function, 23

generalized one-way random effects
    model, 234
generalized performance, 32
geometric distribution, 228, 264
geometric mean, 274
GLMM, 225
GLS, 181
Godambe Information, 182
Gompit regression, 129
goodness of fit, 57, 112
    residual deviance, 113
goodness of fit statistic
    Pearson, 115
grouped data, 157
grouping structure, 200
Gumbel regression, 130

H

hat-matrix, 51, 75
Helmert-transformation, 79
Hessian matrix, 18
heteroscedasticity, 40
hidden Markov model, 185
hierarchical binomial-beta distribu-
    tion model, 239
hierarchical generalized linear mod-
    els, 233, 242
hierarchical likelihood, 171, 182
hierarchical model, 7, 8, 48, 158, 159,
        180, 199, 225
    Bayesian interpretation, 185
    normally distributed observa-
        tions, 163
hierarchical Poisson gamma model,
        226
HMM, 185
hypothesis
    chain of, 27, 58, 144
hypothesis chain
    variable selection, 30
hypothesis testing, 25

I

idempotent, 50
importance sampling, 213

inclusion diagram, 29, 66, 144
independent variables, 46
index parameter, 90
    undetermined, 264
influence, 73
influential observations, 75
Information
    Godambe, 182
information
    expected, 18, 44
    Fisher, 18
    observed, 18, 44
information matrix, 18, 44
inner product, 43
inner product spaces, 45
interactions in tables of counts, 147
intraclass correlation, 164, 235
invariance property, 21
inverse Gaussian distribution, 89,
    275
    alternative parameterization,
        276
    as exponential dispersion model,
        278
iterative back-fitting algorithm, 184
iteratively reweighted least squares,
    104, 109

J
Jacobian, 21
joint density, 10
joint likelihood, 200, 202

K
K-fold cross validation, 32
Kalman filter, 185

L
lack-of-fit, 77
Lagrange Multiplier test, 27
Laplace approximation, 201
Laplace likelihood, 203
latent variables, 158
least squares
    ordinary, 41

weighted, 181
leave-one-out, 33
leverage, 75
likelihood, 14
    alternative parameterizations,
        20
    conditional, 38
    hierarchical, 182
    Laplace, 203
    marginal, 36
    profile, 34
likelihood function, 9, 14, 43
    quadratic approximation, 23
    mean of normal distribution, 15
    regular, 10, 24
likelihood ratio, 26
likelihood ratio test, 25, 29, 53, 63,
        111
    partial, 28
linear component, 6
linear mixed effects model, 160, 179
linear models, 46
linear prediction, 107
linear predictor, 95
link function, 88, 95, 100
    canonical, 95
    complementary log-log, 129
    for binary response regression,
        126
    logit, 118, 127
    probit, 128
link functions, table of, 102
local design matrix, 101
log-likelihood, 14
log-likelihood function, 9
log-normal distribution, 280
logistic function, 119
logistic regression, 5, 118, 127
    in AD model builder, 214
    with ordered categorical response,
        133
logit function, 260
logit transformation, 119
loss function, 32
    generalized, 23

squared error, 187

M

Mahalanobis distance, 76
marginal distribution, 185
    in the generalized one-way random effects model, 234
    normal-normal sampling, 163
    under exponential family sampling, 234
marginal likelihood, 36, 199
    for the variance of a normal distribution, 37
    importance sampling, 213
    one level of grouping, 200
marginal sum of squares, 64
maximum likelihood, 9
    residual, 182
    restricted, 182
maximum likelihood estimate, 21
mean
    arithmetic, 274
    geometric, 274
mean Time Between Failures, 268
mean value equation, 105
mean value function, 160
mean value space, 92
minimum variance unbiased estimator, 12
mixed effects model, 158
    Gaussian, 159
    grouping structures, 200
    likelihood estimation, 181
    linear, 8, 160
    nonlinear, 160
mixed effects models
    in R, 218
mixed model equations, 184
ML equations, 21
ML estimates
    of mean value parameters, 48
ML estimator, 21, 106
    distribution of, 22
    properties of, 49, 106
MLE, 9, 21

model
    fixed effects, 48, 157
    full, 28, 44–47, 100, 112
    hierarchical, 48, 158, 159
    mixed effects, 158
    nested, 33
    non-nested, 33
    null, 28, 46, 47, 56
    predictive performance, 33
    random effects, 48, 158
    saturated, 44, 100, 112
    sufficient, 25, 56, 58, 66
    three-factor, 30
    two-factor, 29
    validation of, 77
    variance components, 48
model diagnostics, 73
model reduction
    test for, 55, 58
model selection, 30
model vector, 7, 47, 100
modified profile log-likelihood, 181
moment estimates
    for the multivariate random effects model, 194
    in normal-normal distribution, 167
    random effects model, 167
moments
    conditional, 185
    marginal, 185
most probable number, 133
MTBF, 268
multicollinearity, see collinearity, 64
multinomial distribution, 89
multiple linear regression, 47
multivariate normal distribution, 42
multivariate random effects model
    MLE, 195
    moment estimates, 194

N

natural exponential family, 89, 90
negative binomial distribution, 89, 227, 264

as exponential dispersion family, 265
nested effects, 200
Newton Raphson algorithm, 110
noise structure, 45
nominal data, 2
nonlinear mixed effects model, 160
NONMEM, 158
norm, 43
normal distribution, 89, 91
    canonical parameter, 91
    canonical parameter space, 91
    cumulant generator, 91
    multivariate, 42
    precision parameter, 91
    unit deviance, 93
normal equation, 49, 105
nuisance parameter, 33
null hypothesis, 25
null model, 28, 30, 46, 47, 56

O

observation set, 11
observed information, 18
ODE, 158
offset, 46, 47, 99, 123, 264
OLS, 41
one-way random effects model, 160
    estimation of parameters, 166
    maximum likelihood estimation, 169
ordinary differential equations, 158
orthogonal parameterization, 65
orthogonality, 43, 45
outliers, 74
over-dispersion, 214
overdispersion, 113, 151, 225, 235

P

p-value, 25
parameter
    nuisance, 33
parameterization
    orthogonal, 65
Partial Least Squares, 65

partial likelihood ratio test, 28
partial test, 29, 63
Pearson X^2, 115
Pearson residual, 108
performance evaluation, 58
PK/PD studies, 158
PLS, 65
point estimate, 9
point estimation theory, 10
poisson data, 2
Poisson distribution, 88, 89, 91, 262
    as exponential dispersion model, 262
    canonical parameter, 91
    canonical parameter space, 91
    cumulant generator, 91
    overdispersion, 151
    two-factor model, 148
Poisson process, 267
Poisson regression, 123
Poisson-gamma hierarchical generalized linear model, 242
Polya distribution, 239
posterior distribution, 185, 186, 230
    for multivariate normal distributions, 191
    in regression model, 192
posterior variance, 188
precision, 197
precision parameter, 89, 90, 95, 125
precision weighted average, 57
predicted values, 72, 105
predicted values in the classical GLM, 50
prediction error, 72
prediction interval, 73
predictor
    BLUP, 172, 183
prior distribution, 185
probit transformation, 128
profile likelihood, 34, 181, 210
    for the variance of a normal distribution, 35
profile likelihood based confidence intervals, 36

profile score function, 35
projection matrix, 50
projections
    for normally distributed vari-
        ables, 45

Q

quantal data, 2
quantile regression, 6
quasi-deviance, 110
quasi-likelihood, 110

R

R
    anova, 82, 119, 144
    aov, 82
    coef, 82, 146
    confint, 146
    confint.**default**, 146
    contr.helmert, 83
    contr.poly, 83
    contr.sum, 83
    contr.treatment, 83
    dnorm, 207
    factor, 81
    fitted, 82, 121
    fixef, 219
    general linear models, 81
    generalized linear models, 152
    glm, 119, 123, 127–129, 132, 138,
        142, 149, 151, 152, 172
    glmer, 218
    hatvalues, 83
    hessian, 207
    jacobian, 207
    Laplace approximation, 203
    library, 218
    lm, 82, 152, 172
    lme, 172, 178, 218
    lme4, 218
    lmer, 218, 226
    mixed effects model, 218
    model.matrix, 82, 143
    nlme, 172, 218
    nlmer, 218

    nlminb, 207
    plot, 82
    predict, 82, 121
    predict.glm, 153
    quasibinomial, 153
    quasipoisson, 153
    ranef, 219
    residuals, 83, 121, 153
    rstandard, 83
    rstudent, 83
    str, 81
    summary, 82, 120, 145, 153
    update, 82
random coefficient regression lines,
    175
random effects, 160
    regression line, 175
random effects model, 8, 158
    for multivariate measurements,
        192
random effects models, 48
rate parameter, 267
regression
    Gamma, 125
    linear, 41
regression analysis, 41
regression model
    gamma distribution, 125
    inverse Gaussian distribution,
        125
    normal distribution, 125
    Poisson distribution, 125
regular, 10, 24
regularization, 64, 191
    mean subset, 65
REML, 170, 182
repeatability, 197
repeated measurements, 175
repetitions, 157
reproducibility, 197
reproductive exponential dispersion
    model, 91
residual analysis, 77
residual deviance, 105
    partitioning of, 112

residual maximum likelihood, 170, 182
residual sum of squares, 43, 89
residual variance, 59
residuals, 51, 73, 108
    autocorrelated, 204
    deviance, 108
    Pearson, 108
    response, 108
    scatter plot, 77
    standardized, 73
    studentized, 74
    types of, 121
    working, 109
response residual, 108
response variable, 46
restricted maximum likelihood, 182
Ridge regression, 64
Risk of the ML-estimator, 187
RSS, 43, 44, 89

S

saturated model, 44, 100, 112
scale parameter, 89
scaled deviance, 115, 152
scaled residual deviance, 105
score function, 17, 21, 44
    for the canonical link, 104
    profile, 35
SDE, 158
sequential sum of squares, 64
shrinkage factor, 166, 172, 189
signal to noise ratio, 162, 228
    confidence interval, 166
    generalized, 191, 194
    normal distribution, 162
simultaneous distribution, 185
spline, 211
split plot, 175
squared error loss function, 187
standard conjugate distribution, 226
    for an exponential dispersion family, 233
standardized residuals, 73
state space model, 185, 216

statistic, 16
statistical learning, 32
Stein's paradox, 191
stepwise selection, 32
stochastic differential equations, 158
studentized residuals, 74
sufficiency
    of the sample mean, 17
sufficient model, 25, 56, 58, 66
sufficient statistic, 16
sum of squares
    marginal, 64
    sequential, 64
sum-coding, 80

T

t-distribution, 282
test
    conditional, 56
    in-sample, 32
    Lagrange Multiplier, 27
    likelihood ratio, 25, 29, 111
    out-of-sample, 32
    partial, 29
    Wilk's likelihood ratio, 26
test for model sufficiency, 112
test of homogeneity
    normal distribution, 166
test set, 32
testing
    type I error, 26
    type II error, 26
Tikhonov matrix, 64
Tikhonov regularization, 64
time series
    integer valued, 216
time series data, 215
training data, 32
transformation, 40
treatment, 157
treatment-coding, 80
two-factor model
    binomial distribution, 140
    Poisson distribution, 148
type I error, 26

Type I partitioning, 58
type I partitioning, 59, 66, 69, 144
type II error, 26
type III partitioning, 63, 66, 144

U

unbiased estimator, 11
uniroot, 211
unit deviance, 93, 95
    for normal distribution, 93
unit observation, 260
unit variance function, 92
unweighted problem, 50

V

validation, 77
validation set, 32
variable
    categorical, 41
    continuous, 41
    discrete, 41
    explanatory, 46
    independent, 46
    nominal, 41
    ordinal, 41

response, 46
variance
    pooling of, 274
variance components model, 48
variance covariance matrix, 13
variance function, 92, 95, 100
variance separation theorem, 185

W

Wald distribution, 275
Wald statistic, 23
Wald test, 71, 116
Wald-confidence interval
    example, 146
weight matrix, 42
weighted average group size, 167
weighted average sample size, 236
weighted least squares, 181
weighted problem, 50
white noise, 216
Wilk's test, 26
WLS, 181
working residual, 109
working response, 110